VOLUME FIFTY TWO

ADVANCES IN
ECOLOGICAL RESEARCH
Trait-Based Ecology - From Structure
to Function

ADVANCES IN ECOLOGICAL RESEARCH

Series Editor

GUY WOODWARD

*Department of Life Sciences,
Imperial College London, Silwood Park,
Ascot, Berkshire, United Kingdom*

VOLUME FIFTY TWO

ADVANCES IN
ECOLOGICAL RESEARCH
Trait-Based Ecology - From Structure to Function

Edited by

SAMRAAT PAWAR
Grand Challenges in Ecosystems and the Environment, Silwood Park, Department of Life Sciences, Imperial College London, Ascot, Berkshire, United Kingdom

GUY WOODWARD
Department of Life Sciences, Imperial College London, Silwood Park, Ascot, Berkshire, United Kingdom

ANTHONY I. DELL
National Great Rivers Research and Education Center (NGRREC), East Alton, Illinois, USA

AMSTERDAM • BOSTON • HEIDELBERG • LONDON
NEW YORK • OXFORD • PARIS • SAN DIEGO
SAN FRANCISCO • SINGAPORE • SYDNEY • TOKYO
Academic Press is an imprint of Elsevier

ELSEVIER

Academic Press is an imprint of Elsevier
225 Wyman Street, Waltham, MA 02451, USA
525 B Street, Suite 1800, San Diego, CA 92101-4495, USA
125 London Wall, London, EC2Y 5AS, UK
The Boulevard, Langford Lane, Kidlington, Oxford OX5 1GB, UK

First edition 2015

Notices
Knowledge and best practice in this field are constantly changing. As new research and experience broaden our understanding, changes in research methods, professional practices, or medical treatment may become necessary.

Practitioners and researchers must always rely on their own experience and knowledge in evaluating and using any information, methods, compounds, or experiments described herein. In using such information or methods they should be mindful of their own safety and the safety of others, including parties for whom they have a professional responsibility.

To the fullest extent of the law, neither the Publisher nor the authors, contributors, or editors, assume any liability for any injury and/or damage to persons or property as a matter of products liability, negligence or otherwise, or from any use or operation of any methods, products, instructions, or ideas contained in the material herein.

ISBN: 978-0-12-802445-4
ISSN: 0065-2504

For information on all Academic Press publications
visit our website at store.elsevier.com

Printed in the United States of America

Working together
to grow libraries in
developing countries

www.elsevier.com • www.bookaid.org

CONTENTS

CONTRIBUTORS

Ross A. Alford
College of Marine and Environmental Sciences and Centre for Tropical Biodiversity and Conservation, James Cook University, Townsville, Queensland, Australia

Alison M. Bell
School of Integrative Biology, University of Illinois, Champaign, Illinois, USA

Miles Bensky
School of Integrative Biology, University of Illinois, Champaign, Illinois, USA

Stephen P. Bonser
Evolution and Ecology Research Centre and School of Biological, Earth and Environmental Sciences, University of New South Wales, Sydney, New South Wales, Australia

Ulrich Brose
Systemic Conservation Biology, Department of Biology, Georg-August University Göttingen, Göttingen, Germany

Robert W. Buchkowski
School of Forestry and Environmental Studies, Yale University, New Haven, Connecticut, USA

Karin T. Burghardt
Department of Ecology and Evolutionary Biology, Yale University, New Haven, Connecticut, USA

Anthony I. Dell
National Great Rivers Research and Education Center (NGRREC), East Alton, Illinois, USA

John P. DeLong
School of Biological Sciences, University of Nebraska-Lincoln, Lincoln, Nebraska, USA

Colin M. Donihue
School of Forestry and Environmental Studies, Yale University, New Haven, Connecticut, USA

Clare Duncan
Institute of Zoology, Zoological Society of London, and UCL Department of Geography, University College London, London, United Kingdom

Sarah M. Durant
Institute of Zoology, Zoological Society of London, London, United Kingdom

Brian J. Enquist
Department of Ecology and Evolutionary Biology, University of Arizona, Bioscience West, Tucson, Arizona, and Santa Fe Institute, Santa Fe, New Mexico, USA

André Frainer
Department of Ecology and Environmental Science, Umeå University, Umeå, and Department of Aquatic Sciences and Assessment, Swedish University of Agricultural Sciences, Uppsala, Sweden

Ewen Georgelin
Institute of Ecology and Environmental Sciences—Paris (UPMC-CNRS-IRD-INRA-UPEC-Paris Diderot), Université Pierre et Marie Curie, UMR 7618, Paris, and Ecology, Systematic and Evolution (UPSUD, CNRS, AgroParisTech), Université Paris Sud, UMR 8079, Orsay, France

Jean P. Gibert
School of Biological Sciences, University of Nebraska-Lincoln, Lincoln, Nebraska, USA

Amanda Henderson
Department of Ecology and Evolutionary Biology, University of Arizona, Bioscience West, Tucson, Arizona, USA

Anne Hilborn
Institute of Zoology, Zoological Society of London, London, United Kingdom, and Department of Fish and Wildlife Conservation, Virginia Tech, Blacksburg, Virginia, USA

Grigoris Kylafis
Institute of Ecology and Environmental Sciences—Paris (UPMC-CNRS-IRD-INRA-UPEC-Paris Diderot), Université Pierre et Marie Curie, UMR 7618, Paris, France

Kate L. Laskowski★
School of Integrative Biology, University of Illinois, Champaign, Illinois, USA

Nicolas Loeuille
Institute of Ecology and Environmental Sciences—Paris (UPMC-CNRS-IRD-INRA-UPEC-Paris Diderot), Université Pierre et Marie Curie, UMR 7618, Paris, France

Brendan G. McKie
Department of Aquatic Sciences and Assessment, Swedish University of Agricultural Sciences, Uppsala, Sweden

Jon Norberg
Department of Systems Ecology, and Stockholm Resilience Centre, Stockholm University, Stockholm, Sweden

Samraat Pawar
Grand Challenges in Ecosystems and the Environment, Silwood Park, Department of Life Sciences, Imperial College London, Ascot, Berkshire, United Kingdom

Simon Pearish
School of Integrative Biology, University of Illinois, Champaign, Illinois, USA

★Current address: Department of Biology & Ecology of Fishes, Leibniz Institute of Freshwater Ecology and Inland Fisheries, Berlin, Germany.

Richard G. Pearson
College of Marine and Environmental Sciences and Centre for Tropical Biodiversity and Conservation, James Cook University, Townsville, Queensland, Australia

Nathalie Pettorelli
Institute of Zoology, Zoological Society of London, London, United Kingdom

Van M. Savage
Santa Fe Institute, Santa Fe, New Mexico; Department of Biomathematics, David Geffen School of Medicine at UCLA, and Department of Ecology and Evolutionary Biology, UCLA, Los Angeles, California, USA

Oswald J. Schmitz
School of Forestry and Environmental Studies, and Department of Ecology and Evolutionary Biology, Yale University, New Haven, Connecticut, USA

Lindsey L. Sloat
Department of Ecology and Evolutionary Biology, University of Arizona, Bioscience West, Tucson, Arizona, USA

Cyrille Violle
Department of Ecology and Evolutionary Biology, University of Arizona, Bioscience West, Tucson, Arizona, USA, and CNRS, Centre d'Ecologie Fonctionnelle et Evolutive, UMR 5175, Montpellier, France

Colleen T. Webb
Department of Biology, Colorado State University, Fort Collins, Colorado, USA

Lei Zhao
Research Centre for Engineering Ecology and Nonlinear Science, North China Electric Power University, Beijing, China, and Department of Life Sciences, Imperial College London, London, United Kingdom

CHAPTER ONE

Scaling-up Trait Variation from Individuals to Ecosystems

Jean P. Gibert*,1, Anthony I. Dell†, John P. DeLong*, Samraat Pawar‡

*School of Biological Sciences, University of Nebraska-Lincoln, Lincoln, Nebraska, USA
†National Great Rivers Research and Education Center (NGRREC), East Alton, Illinois, USA
‡Grand Challenges in Ecosystems and the Environment, Silwood Park, Department of Life Sciences, Imperial College London, Ascot, Berkshire, United Kingdom
1Corresponding author: e-mail address: jeanphisth@gmail.com

Contents

Abstract

Ecology has traditionally focused on species diversity as a way of characterizing the health of an ecosystem. In recent years, however, the focus has increasingly shifted towards trait diversity both within and across species. As we increasingly recognize that ecological and evolutionary timescales may not be all that different, understanding the ecological effects of trait variation becomes paramount. Trait variation is thus the keystone to our understanding of how evolutionary processes may affect ecological dynamics as they unfold, and how these may in turn alter evolutionary trajectories. However, a multi-level understanding of how trait variation scales up from individuals to whole communities or ecosystems is still a work in progress. The chapters in this volume

Advances in Ecological Research, Volume 52
ISSN 0065-2504
http://dx.doi.org/10.1016/bs.aecr.2015.03.001

1

explore how functional trait diversity affects ecological processes across levels of biological organization. This chapter aims at binding the messages of the different contributions and considers how they advance our understanding of how trait variation can be scaled up to understand the interplay between ecological and evolutionary dynamics from individuals to ecosystems.

1. WHY IS IT IMPORTANT TO UNDERSTAND TRAITS AND TRAIT VARIATION?

Evolutionary theory has long recognized the importance of heritable individual (or intraspecific) variation in phenotypic traits (Fordyce, 2006; Lande, 2013). At the same time, ecology has historically focused on mean traits as both a characterization of populations and a response variable (Araújo et al., 2011; Bolnick et al., 2011; Sherratt and Macdougal, 1995; Violle et al., 2012). This difference in focus, to a large extent, stems from the viewpoint that ecology and evolution operate at vastly different time scales. And the origins of this viewpoint, in turn, are probably to be found in the fact that including Darwin's work, early evolutionary ideas were based on the fossil record and tended to be 'gradualist' (Stanley, 1989). In Darwin's (1859) words '*I do believe that natural selection will generally act very slowly, only at long intervals of time, and only on a few of the inhabitants of the same region. I further believe that these slow, intermittent results accord well with what geology tells us of the rate and manner at which the inhabitants of the world have changed*'. This view was reinforced by the fact that examples of evolutionary change mostly came from observations of gradual change over millions of years in the fossil record.

We are now becoming increasingly aware that evolutionary and ecological processes do not occur in isolation and that the time scales of ecological (changes in population sizes) and evolutionary (changes in allele frequencies or trait distributions) rates of change often overlap (Hairston et al., 2005; Schoener, 2011; Schoener et al., 2014). Indeed, feedback loops between ecological and evolutionary processes, or 'eco-evolutionary feedbacks', may be common in natural systems (Fussmann et al., 2003; Jones et al., 2009; Yoshida et al., 2003). Evolutionary biologists have long recognized that the variation in (heritable) individual traits can change during the course of evolution and can affect the strength of selection (Dobzhansky, 1937)—a process central to the evolutionary component of the eco-evolutionary feedbacks. But the potential effects of this individual variation on ecological processes *per se* are less well understood (Araújo et al., 2011; Bolnick et al.,

2011; Lomnicki, 1988; McGill et al., 2006). To fully grasp how individual variation in functional traits can affect ecological dynamics and processes (and thus potentially eco-evolutionary feedbacks), we need to develop a mechanistic understanding of how trait variation 'scales up' from individuals, through species interactions, to ecosystem dynamics (Pawar et al., 2014). The goal of this volume is to advance these ideas by proposing ways to assess how variation in functional traits may alter the outcome of ecological interactions. In this introductory chapter, we present a brief description of each of the contributions to the volume, including a discussion about how they fit into a broader perspective of ecological processes, and how they contribute to a better understanding of how trait variation effects scale up from individuals to ecosystems.

2. TRAITS AND INDIVIDUAL-LEVEL VARIATION

To understand community structure and ecosystem processes, ecologists have long focused on species diversity as an important explanatory mechanism where, for example, decomposition rates, primary production or food web topology results from the number and types of species present (Chapin et al., 1997, 2000; Naeem et al., 2012). This approach has been the basis for some of the most successful ecological theories, such as Tilman's R^* competition theory (Tilman, 1982, 1986). Species-centric approaches like these build upon the idea that groups of organisms differing in species composition will differentially impact higher levels of biological organization such as communities or ecosystems. Focusing on groups of species with similar trophic positions or feeding types (functional groups) has also yielded important and powerful insights (Hooper et al., 2005; Loreau et al., 2001), but can mask information regarding the effect of particular species (Naeem and Wright, 2003; Reich et al., 2004), and its predictive capacity has been difficult to assess (McGill et al., 2006; Schmitz, 2010).

As a consequence, alternative approaches for understanding the emergence of complex properties of communities and ecosystems have been sought, and many ecologists now consider that it is the specific traits that species have that are largely responsible for determining the properties and dynamics of ecological systems (Eviner and Chapin, 2003; Lavorel and Garnier, 2002; McGill et al., 2006; Mlambo, 2014; Naeem and Wright, 2003; Violle et al., 2007). This perspective suggests that in order to understand and predict community and ecosystem organization, ecologists should also focus on the mechanistic basis of 'functional traits' of species in the focal system, instead of simply categorizing their broad functional role (Eviner and Chapin, 2003; Mlambo, 2014). Such a mechanistic trait-based approach should be generalizable across taxa and habitats and may yield general predictions about how ecosystems respond to environmental effects, such as climate change or overharvesting of animals or plants.

The definition of exactly what are functional traits remains controversial, and a historical perspective of this issue is provided by Schmitz et al. (2015) in their chapter "Functional Traits and Trait-Mediated Interactions: Connecting Community-Level Interactions with Ecosystem Functioning". Adopting Schmitz et al.'s definition, a functional trait represents any given trait (whether physiological, behavioural or morphological) that, in the course of maximizing fitness, impacts or regulates higher-level ecological processes and patterns (Mlambo, 2014). At the same time, a functional trait also affects the absolute fitness of individuals, and thus the mean fitness of the population. This is no different from the traditional, evolutionary, quantitative-genetic definition of a trait, but in an ecological framework, heritability of traits is no longer a pre-requisite as purely plastic change can have important ecological implications as well (Gibert and Brassil, 2014). Nor is it necessary to restrict focus of trait variation to phenotypic distributions within populations—as shown by Norberg et al. (Norberg et al., 2001; Savage et al., 2007), it is possible to meaningfully study the effects of across-species trait distributions, especially when it is necessary to tractably link trait variation to ecosystem function.

3. POPULATION-LEVEL EFFECTS OF TRAIT VARIATION

3.1 Functional Response and Prey Selection

In the chapter "Individual Variability: The Missing Component to our Understanding of Predator -Prey Interactions", Pettorelli et al. (2015) explore how individual variation in traits controlling 'predation risk' in prey and 'prey selection' in predators can alter population dynamics. They argue that trait variation can have important yet poorly understood consequences for the shape of predator functional responses, which can in turn affect population dynamics. The authors discuss examples where individual variation in body size, age, sex, condition, behavioural type and territory location can increase or decrease predation risk in a wide range or organisms, including ungulates, cetaceans, lagomorphs and birds. For example, individuals that are older, less healthy or larger are more prone to predation, potentially due to their re-duced ability to perform defensive manoeuvres (see Laskowski et al. (2015)

in this volume for an example of how variation in body size can alter mobility and dispersal). Thus, ignoring individual variation in functional traits can deeply affect our ability to understand predator–prey interactions and dynamics.

Pettorelli et al. also point out that patterns of prey selection among conspecific predators are highly variable and can change throughout ontogeny. Individual specialization, therefore, can directly impact interaction strengths between predators and prey (Pettorelli et al., 2011). For example, individual southern sea otters (*Enhydra lutris nereis*) can learn to selectively hunt different prey items, resulting in strong within individual specialization and decreased intraspecific competition (Tinker et al., 2008). Different populations of sea otters with different degrees of specialization result in different individual-level interaction networks, which in turn affect the structure of the overall community in which the species reside (Tinker et al., 2012).

Finally, Pettorelli et al. discuss how individual variation in functional traits can alter the strength and shape of predator functional responses. Such variation can alter the magnitude of the attack rate in a predator, impacting the strength of the trophic interaction, or lower prey handling time, which can alter the form of the response, e.g., from type II to type I. Some aspects of these effects were revealed by Gibert and Brassil (2014) in a simple pairwise consumer–resource model that accounts for individual variation in a trait simultaneously controlling attack rate and handling time such as body size. The authors find that increased individual variation decreased interaction strengths under some assumptions by lowering attack rates and increasing handling times. Consequently, increased intraspecific trait variation can lead to more stable population dynamics with a higher probability of persistence. Together, these results stress the need to study the effect of variation in multiple functional traits simultaneously as different traits can have potentially antagonistic effects (see Gibert and DeLong, 2015 in the paper entitled "Individual Variation Decreases Interference Competition but Increases Species Persistence").

3.2 Functional Response, Interference Competition and Species Interactions

In this paper Gibert and DeLong (2015) assess how variation in a trait simultaneously controlling attack rate, handling time and interference competition can affect population persistence, and subsequently the competitive ability of the population and community structure. The authors extend a predator–prey model to incorporate individual phenotypic variation in attack rate and handling time (Gibert and Brassil, 2014) and use an empirically quantified relationship between attack rate and interference competition (DeLong and Vasseur, 2013) to incorporate individual variation in interference. They then assess the effect of variation in consumer–resource dynamics and competitive ability via its joint effects on attack rate, handling time and interference competition.

Interference competition is thought to be mostly stabilizing in natural systems (Arditi et al., 2004; DeLong and Vasseur, 2013), while attack rate is mostly destabilizing (Rosenzweig and MacArthur, 1963). However, an increase in attack rate can be accompanied by an increase in interference competition (DeLong and Vasseur, 2013). Thus, individual trait variation could potentially have opposing effects on predator–prey dynamics. Their results suggest that while trait variation mostly decreases interference competition, it often decreases attack rate to a larger extent, thus being largely stabilizing. Moreover, increased variation reduces the chance of species extinction due to demographic stochasticity. These results make an interesting comparison with those of Pawar (2015) in this volume, who shows that the scaling (or lack thereof) of interference competition with body size has a strong influence on interaction-driven community assembly dynamics and outcomes.

The authors also show how trait variation can have important effects upon competitive ability and community structure. When predators compete for a common resource, those that can reduce resource levels the most will competitively exclude all other predators (Tilman, 1982, 1986). Because individual variation hinders their ability to reduce resource levels, greater trait variation leads to predators with a larger chance of persisting, but with a lower competitive ability, thus leading to a fundamental trade-off between persistence and competitive ability. This chapter shows that intermediate levels of individual variation optimize that trade-off, which further deserves experimental investigation.

4. META-POPULATION EFFECTS OF TRAIT VARIATION

4.1 Dispersal Ability

In their chapter "Predictors of Individual Variation in Movement in a Natural Population of Threespine Stickleback", Laskowski et al. (2015) consider variation in the dispersal behaviour of individuals within a population of threespine sticklebacks (*Gasterosteus aculeatus*). In a first experiment, the authors use a controlled, open field arena and then test whether those results could be generalized to movements within a natural stream. In both instances, they find consistent levels of individual variation, with individual fish tending to repeat their dispersal behaviour across contexts. Similar results have been observed in other systems (Bell et al., 2009; Sih et al., 2004). Also, dispersal in the stream was strongly influenced by variation in body condition and habitat type, which further suggests that behaviour depends upon key features of both the organism and the environment. As other studies in this volume suggest (Georgelin et al., 2015; Gibert and DeLong, 2015; Schmitz et al., 2015), dispersal behaviour is also affected by other ecological factors, such as time of the day, individual size, sex and year. Thus, the key to understanding the ecological effects of behavioural variation between individuals is likely to be an approach that integrates both the variation itself and other ecological factors (Gibert and DeLong, 2015; Pettorelli et al., 2015; Schmitz et al., 2015 of this same issue). Interestingly, Laskowski et al. find that dispersal distance did not depend on time to recapture, which strongly suggests that the observed variation in dispersal behaviour is a core component of the ecology of this species.

Their results suggest that variation in dispersal behaviour can have important yet poorly understood effects upon meta-population dynamics. Indeed, it is possible that meta-populations with larger behavioural variation in dispersal abilities may have increased chances of persisting than meta-populations with less variation, only because larger individual variation in dispersal may result in larger colonization rates. Systems such as the threespine stickleback are well suited to empirically test some these predictions. These kinds of studies are also crucial for parameterizing more general models that incorporate dispersal kernels (e.g. Pawar, 2015 in this volume).

5. COMMUNITY-LEVEL EFFECTS OF TRAIT VARIATION

5.1 Eco-Evolutionary Dynamics of Traits in Tri-Trophic Systems

In their chapter "Eco-Evolutionary Dynamics of Plant—Insect Communities Facing Disturbances: Implications for Community Maintenance and Agricultural Management", Georgelin et al. (2015) study how functional trait evolution and eco-evolutionary dynamics can impact species persistence and community structure in a tri-trophic system involving plants, pollinators and herbivores. They also assess how this might occur when environmental disturbances are frequent and severe, such as in agricultural landscapes where populations of the interacting species might be affected by frequent exposure to pesticides.

The authors use an adaptive dynamics eco-evolutionary model that follows the abundance of a plant species, a pollinator and an herbivore, as well as the pollinator and herbivore sensitivities to pesticides. To understand the effect of the evolution of these traits on the tri-trophic interaction, they devise three separate models for comparison. A first model where the sensitivity of the herbivore only is allowed to evolve, a second model where the sensitivity of the pollinator only is allowed to evolve and a third model where both sensitivities are allowed to evolve. Their results suggest that when herbivore sensitivities evolve, pollinators can go extinct due to low plant densities. When pollinators evolve but not herbivores, all species co-exist. When the two species evolve, however, the model predicts more diverse systems than can be accounted for based on eco-evolutionary dynamics, and they suggest multiple mechanisms that could explain that pattern based on density-dependent effects and individual phenotypic variation.

These results have important implications for the maintenance of pollinators in frequently disturbed habitats, a major contemporary problem. Ecosystems across the world are facing pollinator losses, and while the causes of this depletion are largely unknown, many authors linked this pattern to the abuse of pesticides in agricultural landscapes (Barnett et al., 2007; Porrini et al., 2003, and also Frainer and McKie, 2015 in this volume). The authors also show that to understand the joint effect of frequent disturbances and eco-evolutionary dynamics on population persistence and species richness, it is paramount to account for the broader network of interacting species in which each pairwise interaction is embedded. More importantly, they show that while antagonistic interactions are important (e.g. predation), positive interactions (e.g. mutualism) can have large yet poorly understood effects for the persistence of the overall community, as other studies have also argued (Guimarães et al., 2011; Saavedra et al., 2011; Staniczenko et al., 2013).

5.2 Patterns of Functional Trait Distributions in Real Systems

In their chapter "Population and Community Body Size Structure Across a Complex Environmental Gradient", Dell et al. (2015) quantify functional trait distributions for each species in an experimental intermittent pool bed to understand how processes at underlying levels of biological organization (i.e. individuals) can affect patterns at higher levels of organization (i.e. populations and communities). Their experimental design allows them to do so at different successional stages across both aquatic and terrestrial habitat.

They find that species abundance and total biomass strongly depend on the type of community considered, with both factors increasing with the amount of moisture/water present. For example, for the same given area, terrestrial communities were less rich in species and had a smaller total biomass than moist (ecotonal) communities, but these were in turn less rich than aquatic communities. On the other hand, terrestrial communities tended to be more even, which suggests that as moisture increased, communities tended to be increasingly dominated by a few superabundant species. This chapter also presents novel results with respect to community behaviour in the ecotone habitat along the transition from aquatic to terrestrial. These ecotonal communities tended to have more species with more individuals because they were pooling species from both the aquatic and terrestrial ecosystems. More importantly, their results show how total community biomass peaks at the ecotone, suggesting interesting interactions between ecological constraints, ecosystem type and functional traits of the ecotonal species.

Their results are not only consistent with previous studies (e.g. Heliölä et al., 2001), but also the comprehensiveness of the data collected allowed the authors to show patterns that could not have been noticed without a size-explicit description of individuals in populations spanning multiple complex ecosystems. For instance, the authors show that both the mean and the variance of the population size distributions within a given community change with moisture; mean, range and variance of size distributions were larger in moist than in terrestrial communities. This further suggests the existence of a feedback between environmental conditions and functional trait distributions (also see Georgelin et al., 2015; Gibert and DeLong, 2015; Schmitz et al., 2015 of this volume).

5.3 Functional Traits and Community Assembly

In the paper entitled "The Role of Body Size Variation in Community Assembly", Pawar (2015) takes into account the fact that body sizes can span as much as 20 orders in magnitude in natural communities (Brose et al., 2006; Cohen et al., 2003; Jonsson et al., 2005, and also Dell et al. in this same volume) to study what role the distribution of body sizes in the immigrating species pool plays in the dynamics of non-neutral (interaction driven) community food web assembly. Pawar uses a size-constrained mathematical model of food web assembly and shows that assembled food webs at quasi-equilibrium (i.e. where species numbers remains relatively constant; Bastolla et al., 2005; Pawar, 2009) are expected to exhibit 'signatures' of non-neutral assembly in a number of aspects: (i) the distribution of body sizes, (ii) the distribution of size-ratios between consumers and resources and (iii) the distribution of size and size-ratios across trophic levels. Interestingly, the results remain robust, and the signatures emerge consistently across a wide range of size distribution types—ranging from distributions that impose immigration bias towards small species to those that result in a bias towards large species. That is, species interactions impose a very strong 'filter' on functional trait (body size) distributions during assembly. The author also evaluates the predictions of the model using food web data from nine terrestrial and aquatic communities and finds that the predicted signatures are indeed observed in most of them.

It is worth noting that Pawar considers both body sizes and consumer–resource size-ratios. Size-ratio can arguably be considered a trait in itself because it strongly determines trophic level (Yvon-Durocher et al., 2011) and interaction strength (Berlow et al., 2009; Pawar et al., 2012), both key factors driving individual invasion fitness as well as community stability (Brose et al., 2006; Otto et al., 2007; Tang et al., 2014). Size-ratios are also known to change with temperature (Gibert and DeLong, 2014), which suggests that body-size distributions may have an important role to play in a context of global warming.

Overall, Pawar's results suggest that body-size variation has important implications for community food web assembly and recovery (also see Dell et al., 2015 of this same volume). This adds a much-needed assembly-oriented perspective to current knowledge of the effects of trait distribution and variation on community and ecosystem dynamics.

6. ECOSYSTEM-LEVEL EFFECTS OF TRAIT VARIATION

6.1 Functional Traits and Their Effect on Ecosystem Functioning

In their chapter "Shifts in the Diversity and Composition of Consumer Traits Constrain the Effects of Land Use on Stream Ecosystem Functioning", Frainer and McKie (2015) study the effect of agricultural land use on the distribution and diversity of functional traits across 10 boreal streams covering a gradient of agricultural use and subsequent effects on ecosystem processes such as litter decomposition. The study was conducted in two different seasons (Fall and Spring) to account for the fact that decomposition rates change dramatically across seasons.

The authors show that disturbed sites, such as agricultural lands, tend to have diverse but less even species composition (Stirling and Wilsey, 2001). Less even communities may result in lower ecosystem functioning (e.g. decomposition rates) because these systems tend to be dominated by generalist leaf-consumers rather than obligate leaf-consumers (McKie and Malmqvist, 2009). The authors showed that leaf decomposition was positively related to diversity in functional traits and that this relationship declined along an agricultural land use gradient. However, agricultural use did not favour dominance of generalist traits associated with non–obligate leaf-litter consumers. Actually, obligate consumers were more abundant in agricultural streams. In a sense, these results suggest the existence of buffering mechanisms that mitigate the effect of human–related disturbances on agricultural streams.

Finally, this contribution provides empirical evidence that variation in species traits are linked with ecosystem process rates, suggesting context-dependent effects of functional diversity in ecosystem processes (also see Schmitz et al., 2015 in this same volume). It also shows how the effects of human disturbance on ecosystem functioning were buffered by concurrent shifts in the functional diversity and composition of a key consumer guild, which highlights the value of a trait-based framework for understanding ecosystem-level responses to environmental change.

6.2 Functional Traits, Variation and Metabolic Theory: A Trait-Driver Approach to Ecosystems

In chapter "Scaling from Traits to Ecosystems: Developing a General Trait Driver Theory via Integrating Trait-Based and Metabolic Scaling Theories", Enquist et al. (2015) argue that there is a need to move beyond species richness and into an integrative and predictive framework that takes into account the mechanisms generating species diversity via trait composition, distribution and diversity (also see Dell et al., 2015; Gibert and DeLong, 2015; Pawar, 2015; Pettorelli et al., 2015 in this same volume). To do so, building upon previous work (Norberg et al., 2001; Savage et al., 2007), they formulate and propose a framework, which they name 'Traits Drivers Theory' (TDT), that is applicable across different geographic and temporal scales and gradients. Their TDT unifies trait-based approaches and the Metabolic Theory of Ecology to understand and predict ecosystem processes and patterns.

TDT makes several predictions with respect to the feedbacks between traits, trait distributions and ecosystem patterns. First, it predicts that shifts in environments will cause shifts in trait distributions. Second, it predicts that the difference between the optimal trait, the observed mean trait value and the individual variation in the trait provides a measure of the capacity of a community to respond to environmental change (also see Gibert and Brassil, 2014; Schreiber et al., 2011; Vasseur et al., 2011, and Gibert and DeLong, 2015 of this same volume). Third, the skewness of the trait distribution can be an indicator of recent immigration or environmental change (also see Pawar, 2015 in this volume). Fourth, the rate of change of ecosystem productivity in response to an environmental change is a function of the community biomass-trait distribution. And last, an increase in individual variation in a trait controlling primary production will lead to a decrease in primary production (see Gibert and Brassil, 2014 and Gibert and DeLong in this volume).

The authors test these predictions with an extensive dataset of shifts in trait distribution and ecosystem productivity across an elevational gradient and a 140-year long ecological experiment spanning local and global gradients. They argue that their framework (i) provides predictions of ecological patterns based on the shape of trait distributions, (ii) integrates how specific traits and functional diversity influence the dynamics of species assemblages across gradients and (iii) provides predictions as to how shifts in functional composition can influence ecosystem functioning.

6.3 A Relational Approach to Trait Ecology

In their chapter "Functional Traits and Trait-Mediated Interactions: Connecting Community-Level Interactions with Ecosystem Functioning", Schmitz et al. (2015) argue that there is a need to move beyond functional groups to understand ecosystem processes (also see Enquist et al., 2015 of this volume). They thus propose a framework that links functional traits and food web structure to understand and predict ecosystem-level processes and patterns. Their approach builds upon the notion that plastic traits of intermediate trophic levels can have ecosystem-wide effects (Schmitz, 2010). Biologists thus need to take these into account to explain most of the residual variation that ends up largely unexplained in classic approaches. To do so, they propose the use of a 'relational' approach that involves focusing on the plastic response of functional traits in intermediate trophic levels in different contexts to understand how and why ecosystem function is context dependent. They illustrate their approach by applying it to four different empirical examples of plant-based and detritus-based food chains: a carnivore–herbivore–plant pathway, an herbivore–plant–detritus pathway, a carnivore–detritus/microbivore–detritus pathway and a detritivore/microbivore–soil microbe–detritus pathway.

This contribution sets the stage for understanding how organismal-level processes, community-level processes and ecosystem-level processes may interplay to yield the patterns we observe in nature. More importantly, their approach suggests new potential ways in which ecologists could approach questions and devise experiments when trying to understand upper-level patterns. This experimental take on trait ecology can provide important clues as to how context-dependency may come about in ecosystems in a changing world.

Together, the contributions by Enquist et al. (2015) and Schmitz et al. (2015) provide a new integrative approach that merges quantification of temporal changes in trait distributions and plastic trait responses of intermediate trophic levels.

7. CONCLUDING REMARKS

The contributions in this volume not only show the importance of taking into account functional traits to understand ecological patterns and processes, but also how individual variation in these functional traits may have paramount effects upon population dynamics (Pettorelli et al., 2015) and community structure (Gibert and DeLong, 2015). These contributions also suggest that individual variation in functional traits can have pervasive effects upon meta-population persistence through variation in dispersal rates (Laskowski et al., 2015). One contribution shows how variation in functional traits can affect eco-evolutionary dynamics in a tri-trophic system (Georgelin et al., 2015), while another contribution quantifies body-size distributions across ecotones (Dell et al., 2015). Variation in body size is later shown to determine food web assembly dynamics and outcomes (Pawar, 2015) and to be linked to ecosystem process rates (Frainer and McKie, 2015). Finally, this volume provides two integrative frameworks: one that aims at making testable quantitative predictions of how trait distributions can affect upper-level patterns (Enquist et al., 2015), and another that aims at providing with a common experimental approach to assess the effect of phenotypically plastic traits on ecosystem context-dependency (Schmitz et al., 2015; Fig. 1).

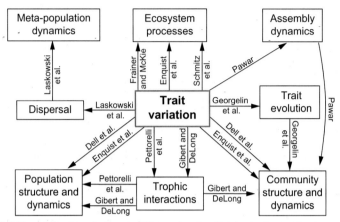

Figure 1 Effects of trait variation at different levels of biological organization, as shown by the chapters in this volume.

REFERENCES

Araújo, M.S., Bolnick, D.I., Layman, C.A., 2011. The ecological causes of individual specialisation. Ecol. Lett. 14, 948–958.

Arditi, R., Callois, J., Tyutyunov, Y., Jost, C., 2004. Does mutual interference always stabilize predator-prey dynamics? A comparison of models. C. R. Biol. 327, 1037–1057.

Barnett, E.A., Charlton, A.J., Fletcher, M.R., 2007. Incidents of bee poisoning with pesticides in the United Kingdom, 1994–2003. Pest Manag. Sci. 1057, 1051–1057.

Bastolla, U., Lassig, M., Manrubia, S.C., Valleriani, A., 2005. Biodiversity in model ecosystems, II: species assembly and food web structure. J. Theor. Biol. 235, 531–539.

Bell, A.M., Hankison, S.J., Laskowski, K.L., 2009. The repeatability of behaviour: a meta-analysis. Anim. Behav. 77, 771–783.

Berlow, E.L., Dunne, J.A., Martinez, N.D., Stark, P.B., Williams, R.J., Brose, U., 2009. Simple prediction of interaction strengths in complex food webs. Proc. Natl. Acad. Sci. U.S.A. 106, 187–191.

Bolnick, D.I., Amarasekare, P., Araújo, M.S., Bürger, R., Levine, J.M., Novak, M., Rudolf, V.H.W., Schreiber, S.J., Urban, M.C., Vasseur, D.A., 2011. Why intraspecific trait variation matters in community ecology. Trends Ecol. Evol. 26, 183–192.

Brose, U., Jonsson, T., Berlow, E.L., Warren, P.H., Banasek-Richter, C., Bersier, L.-F., Blanchard, J.L., Brey, T., Carpenter, S.R., Cattin, M.-F., Cushing, L., Hassan, A.D., Dell, A.I., Edwards, F., Harper-Smith, S., Jacob, U., Ledger, M.E., Martinez, N.D., Memmott, J., Mintenbeck, K., Pinnegar, J.K., Rall, B.C., Rayner, T.S., Reuman, D.C., Ruess, L., Ulrich, W., Williams, R.J., Woodward, G., Cohen, J.E., 2006. Consumer–resource body-size relationships in natural food webs. Ecology 87, 2411–2417.

Chapin, F.S., Walker, B.H., Hobbs, R.J., Hooper, D.U., Lawton, J.H., Sala, O.E., Tilman, D., 1997. Biotic control over the functioning of ecosystems. Science 277, 500–504.

Chapin, F.S., Zavaleta, E.S., Eviner, V.T., Naylor, R.L., Vitousek, P.M., Reynolds, H.L., Hooper, D.U., Lavorel, S., Sala, O.E., Hobbie, S.E., Mack, M.C., Díaz, S., 2000. Consequences of changing biodiversity. Nature 405, 234–242.

Cohen, J.E., Jonsson, T., Carpenter, S.R., 2003. Ecological community description using the food web, species abundance, and body size. Proc. Natl. Acad. Sci. U.S.A. 100, 1781–1786.

Darwin, C., 1859. On the Origin of Species by Means of Natural Selection, or the Preservation of Favoured Races in the Struggle for Life, sixth ed. P. F. Collier & Son co., New York.

DeLong, J.P., Vasseur, D.A., 2013. Linked exploitation and interference competition drives the variable behavior of a classic predator-prey system. Oikos 122, 1393–1400.

Dobzhansky, T., 1937. Genetics and the Origin of Species. Columbia University Press, New York, p. 364.

Eviner, V.T., Chapin, F.S., 2003. Functional matrix: a conceptual framework for predicting multiple plant effects on ecosystem processes. Annu. Rev. Ecol. Evol. Syst. 34, 455–485.

Fordyce, J.A., 2006. The evolutionary consequences of ecological interactions mediated through phenotypic plasticity. J. Exp. Biol. 209, 2377–2383.

Fussmann, G.F., Ellner, S.P., Hairston Jr., N.G., 2003. Evolution as a critical component of plankton dynamics. Proc. R. Soc. B Biol. Sci. 270, 1015–1022.

Gibert, J.P., Brassil, C.E., 2014. Individual phenotypic variation reduces interaction strengths in a consumer-resource system. Ecol. Evol. 4, 3703–3713.

Gibert, J.P., DeLong, J.P., 2014. Temperature alters food web body-size structure. Biol. Lett. 10, 20140573.

Guimarães Jr., P.R., Jordano, P., Thompson, J.N., 2011. Evolution and coevolution in mutualistic networks. Ecol. Lett. 14, 877–885.

Hairston, N.G., Ellner, S.P., Geber, M.A., Yoshida, T., Fox, J.A., 2005. Rapid evolution and the convergence of ecological and evolutionary time. Ecol. Lett. 8, 1114–1127.

Heliölä, J., Koivula, M., Niemelä, J., 2001. Distribution of carabid beetles (Coleoptera, Carabidae) across a boreal forest—clearcut ecotone. Conserv. Biol. 15, 370–377.

Hooper, D.U., Chapin, F.S., Ewel, J.J., Hector, A., Inchausti, P., Lavorel, S., Lawton, J.H., Lodge, D.M., Loreau, M., Naeem, S., Schmid, B., Setälä, H., Symstad, A.J., Vandermeer, J., Wardle, D.A., 2005. Effects of biodiversity on ecosystem functioning: a concensus of current knowledge. Ecol. Monogr. 75, 3–35.

Jones, L.E., Becks, L., Ellner, S.P., Hairston, N.G., Yoshida, T., Fussmann, G.F., 2009. Rapid contemporary evolution and clonal food web dynamics. Philos. Trans. R. Soc. B-Biol. Sci. 364, 1579–1591.

Jonsson, T., Cohen, J., Carpenter, S., 2005. Food webs, body size, and species abundance in ecological community description. Adv. Ecol. Res. 36, 1–84.

Lande, R., 2013. Neutral theory of quantitative genetic variance in an Island model with local extinction and colonization. Evolution 46, 381–389.

Lavorel, S., Garnier, E., 2002. Predicting changes in community composition and ecosystem functioning from plant traits: revisiting the Holy Grail. Funct. Ecol. 16, 545–556.

Lomnicki, A., 1988. Population Ecology of Individuals. Princeton University Press, Princeton.

Loreau, M., Naeem, S., Inchausti, P., Bengtsson, J.P., Grime, A., Hector, D.U., Hooper, M.A., Huston, D., Raffaelli, B., Schmid, D. Tilman, Wardle, D.A., 2001. Biodiversity and ecosystem functioning: current knowledge and future challenges. Science 294, 804–808.

McGill, B.J., Enquist, B.J., Weiher, E., Westoby, M., 2006. Rebuilding community ecology from functional traits. Trends Ecol. Evol. 21, 178–185.

McKie, B.G., Malmqvist, B., 2009. Assessing ecosystem functioning in streams affected by forest management: increased leaf decomposition occurs without changes to the composition of benthic assemblages. Freshw. Biol. 54, 2086–2100.

Mlambo, M.C., 2014. Not all traits are "functional": insights from taxonomy and biodiversity-ecosystem functioning research. Biodivers. Conserv. 23, 781–790.

Naeem, S., Wright, S., 2003. Disentangling biodiversity effects on ecosystem functioning: deriving solutions to a seemingly insurmountable problem. Ecol. Lett. 6, 567–579.

Naeem, S., Duffy, J.E., Zavaleta, E., 2012. The functions of biological diversity in an age of extinction. Science 336, 1401–1407.

Norberg, J., Swaney, D.P., Dushoff, J., Lin, J., Casagrandi, R., Levin, S.A., 2001. Phenotypic diversity and ecosystem functioning in changing environments: a theoretical framework. Proc. Natl. Acad. Sci. U.S.A. 98, 11376–11381.

Otto, S.B., Rall, B.C.B.C., Brose, U., 2007. Allometric degree distributions facilitate food-web stability. Nature 450, 1226–1229.

Pawar, S., 2009. Community assembly, stability and signatures of dynamical constraints on food web structure. J. Theor. Biol. 259, 601–612.

Pawar, S., Dell, A.I.A., Savage, V.M.V., 2012. Dimensionality of consumer search space drives trophic interaction strengths. Nature 486, 485–489.

Pawar, S., Dell, A.I., Savage, V.M., 2014. From metabolic constraints on individuals to the eco-evolutionary dynamics of ecosystems. In: Belgrano, A., Woodward, G., Jacob, U. (Eds.), Aquatic Functional Biodiversity: An Eco-Evolutionary Approach. Elsevier, Waltham, Massachusetts.

Pettorelli, N., Coulson, T., Durant, S.M., Gaillard, J.-M., 2011. Predation, individual variability and vertebrate population dynamics. Oecologia 167, 305–314.

Porrini, C., Sabatini, A.G.S., Girotti, S., Fini, F., Monaco, L., Celli, G., Bortolotti, L., Ghini, S., 2003. The death of honey bees and environmental pollution by pesticides: the honey bees as biological indicators. Bull. Insectol. 56, 147–152.

Reich, P.B., Tilman, D., Naeem, S., Ellsworth, D.S., Knops, J., Craine, J., Wedin, D., Trost, J., 2004. Species and functional group diversity independently influence biomass accumulation and its response to CO2 and N. Proc. Natl. Acad. Sci. U.S.A. 101, 10101–10106.

Rosenzweig, M.L., MacArthur, R.H., 1963. Graphical representation and stability conditions of predator-prey interactions. Am. Nat. 97, 209–223.

Saavedra, S., Stouffer, D.B., Uzzi, B., Bascompte, J., 2011. Strong contributors to network persistence are the most vulnerable to extinction. Nature 478, 233–235.

Savage, V.M., Webb, C.T., Norberg, J., 2007. A general multi-trait-based framework for studying the effects of biodiversity on ecosystem functioning. J. Theor. Biol. 247, 213–229.

Schmitz, O.J., 2010. Resolving Ecosystem Complexity. Princeton University Press, Princeton, p. 192.

Schoener, T.W., 2011. The newest synthesis: understanding the interplay of evolutionary and ecological dynamics. Science 331, 426–429.

Schoener, T.W., Moya-Laraño, J., Rowntree, J., Woodward, G., 2014. Preface. Adv. Ecol. Res. 50, 13–22.

Schreiber, S.J., Bürger, R., Bolnick, D.I., 2011. The community effects of phenotypic and genetic variation within a predator population. Ecology 92, 1582–1593.

Sherratt, T., Macdougal, A.D., 1995. Some population consequences of variation in preference among individual predators. Biol. J. Linn. Soc. 55, 93–107.

Sih, A., Bell, A., Johnson, J.C., 2004. Behavioral syndromes: an ecological and evolutionary overview. Trends Ecol. Evol. 19, 372–378.

Staniczenko, P.P., Kopp, J.C., Allesina, S., 2013. The ghost of nestedness in ecological networks. Nat. Commun. 4, 1391.

Stanley, S.M., 1989. Fossils, macroevolution, and theoretical ecology. In: Roughgarden, J., May, R.M., Levin, S.A. (Eds.), Perspectives in Ecological Theory. Princeton University Press, Princeton, NJ, pp. 125–134.

Stirling, G., Wilsey, B., 2001. Empirical relationships between species richness, evenness, and proportional diversity. Am. Nat. 158, 286–299.

Tang, S., Pawar, S., Allesina, S., 2014. Correlation between interaction strengths drives stability in large ecological networks. Ecol. Lett. 17, 1094–1100.

Tilman, D., 1982. Resource competition and community structure. In: Simon, A.L., Henry, S.H. (Eds.), Monographs in Population Biology, vol. 17. Princeton University Press, Princeton, NJ, p. xi, 296 p.

Tilman, D., 1986. Resources, competition and the dynamics of plant communities. In: Plant Ecology. Blackwell Scientific Publications, Oxford, pp. 51–75.

Tinker, M.T., Bentall, G., Estes, J.A., 2008. Food limitation leads to behavioral diversification and dietary specialization in sea otters. Proc. Natl. Acad. Sci. U.S.A. 105, 560–565.

Tinker, M.T., Guimarães Jr., P.R., Novak, M., Marquitti, F.M.D., Bodkin, J.L., Staedler, M., Bentall, G., Estes, J.A., 2012. Structure and mechanism of diet specialisation: testing models of individual variation in resource use with sea otters. Ecol. Lett. 15, 475–483.

Vasseur, D.A., Amarasekare, P., Rudolf, V.H.W., Levine, J.M., 2011. Eco-Evolutionary dynamics enable coexistence via neighbor-dependent selection. Am. Nat. 178, E96–E109.

Violle, C., Navas, M.-L., Vile, D., Kazakou, E., Fortunel, C., Hummel, I., Garnier, E., 2007. Let the concept of trait be functional! Oikos 116, 882–892.

Violle, C., Enquist, B.J., McGill, B.J., Jiang, L., Albert, C.H., Hulshof, C., Jung, V., Messier, J., 2012. The return of the variance: intraspecific variability in community ecology. Trends Ecol. Evol. 27, 244–252.

Yoshida, T., Jones, L.E., Ellner, S.P., Fussmann, G.F., Hairston, N.G., 2003. Rapid evolution drives ecological dynamics in a predator-prey system. Nature 424, 303–306.

Yvon-Durocher, G., Reiss, J., Blanchard, J., Ebenman, B., Perkins, D.M., Reuman, D.C., Thierry, A., Woodward, G., Petchey, O.L., 2011. Across ecosystem comparisons of size structure: methods, approaches and prospects. Oikos 120, 550–563.

CHAPTER TWO

Individual Variability: The Missing Component to Our Understanding of Predator–Prey Interactions

Nathalie Pettorelli*[,1], Anne Hilborn*[,†], Clare Duncan*[,‡],
Sarah M. Durant*
*Institute of Zoology, Zoological Society of London, London, United Kingdom
†Department of Fish and Wildlife Conservation, Virginia Tech, Blacksburg, Virginia, USA
‡UCL Department of Geography, University College London, London, United Kingdom
[1]Corresponding author: e-mail address: nathalie.pettorelli@ioz.ac.uk

Contents

Abstract

Predator–prey interactions are central to our understanding of adaptive evolution and community ecology. A growing body of research indicates that predation risk and prey selection can be highly variable from one individual to another; nonetheless, individual variability both within predators and within prey is still classically ignored when attempting to model predator–prey dynamics. This chapter explores how our current knowledge of the factors shaping prey selection and predation risk relate to current modelling approaches of predator–prey dynamics. It also discusses how dismissal of inherent individual heterogeneity in predator–prey interactions may be impacting our ability to advance food web theory as well as our understanding of evolutionary trajectories in predator and prey populations. It finally reviews possible methodological frameworks that could help integrate individual variability into the modelling of predator–prey interactions.

Advances in Ecological Research, Volume 52
ISSN 0065-2504
http://dx.doi.org/10.1016/bs.aecr.2015.01.001
19

1. INTRODUCTION

Interactions, should they be among organisms or between abiotic environmental conditions and organisms, define the processes that shape the diversity of life on Earth at both the ecological and the evolutionary scales. Predator–prey interactions, in particular, are powerful forces that underpin the behaviour and ecology of all organisms (Arditi and Ginzburg, 2012; Drossel et al., 2004), being at the heart of our understanding of adaptive evolution (Mousseau et al., 2000). Predation can indeed change the distribution of life history traits over generations and influence prey evolution, through prey selectivity, the direct induction of traits and the indirect induction of traits via reduced or increased competition (Relyea, 2002; Reznick et al., 1990; Wittmer et al., 2005). Moreover, predation may increase stability in trophic interactions, and the exertion of top-down control from secondary consumers is a crucial link in structuring ecological communities (Hairston et al., 1960; May, 1973; Ripple et al., 2014; Yeakel et al., 2014). A large part of our ability to explain how biodiversity is distributed, how communities are structured and how ecosystems function is thus directly linked to our ability to decipher predator–prey interactions (Litvaitis, 2000).

Central to theories underpinning our current understanding of predator–prey interactions are the concepts of predation risk, prey selection, numerical response and functional response. Predation risk refers to an individual's propensity to be predated upon by another organism. This risk can compromise an organism's ability to acquire and maintain body reserves by hindering foraging time and efficiency and increasing physiological stress (Creel and Christianson, 2008), making it a key determinant of energy intake per unit time and individual fitness. To avoid being predated, organisms can indeed modify their habitat selection patterns (Gilliam and Fraser, 1987) and change trophic flows by altering the selection of their diet (Schmitz, 1998); such anti-predator tactics can lead to prey reducing their energy intake, and/or prey increasing their energetic allocation to predator avoidance strategies. These shifts in energy acquisition and allocation can directly impact individuals' reproductive abilities and survival (Magnhagen, 1991; Preisser et al., 2005). Trade-offs between reproductive investment and predation risk and between starvation and predation risk are thus central to many decisions individual prey make regarding habitat choice, foraging and mating (Krams et al., 2013), ultimately leading to predation risk being key to our understanding of how predators shape the life histories of prey.

Prey selection focuses on the relationship between predators and prey from the predator's perspective. Because each prey item has particular associated hunting costs and energetic benefits to the predator (Werner and Hall, 1974), prey selection can occur at different levels (e.g. predators selecting for different species, size, age or sex classes; predators selecting physically substandard individuals) and be influenced by various factors such as the presence of competitors, the habitat structure or the season (see e.g. Creel and Creel, 2002; Fitzgibbon and Fanshawe, 1989; Kruger et al., 1999; Mills and Gorman, 1997; Pole et al., 2004; Radloff and du Toit, 2004). Ultimately, prey selection drives hunting success and the ability of predators to acquire and maintain body reserves, making it a key determinant of predators' individual fitness.

The combination of the numerical response and functional response represents the primary framework for studying how predator and prey populations influence the dynamics of each other, being, to some extent, one of the mathematical outcomes of our understanding of predation risk and prey selection. More precisely, the numerical response aims to capture the effect of prey on predator populations, formulating the increase of predator density with prey density. Functional responses, on the other hand, serve to capture the effect of predators on prey populations, designed to formalize the relationship between prey density and the number of prey eaten by a predator (Holling, 1959a,b). Three types of possible functional responses are generally recognized: (1) Type I assumes predators consume a constant proportion of prey, leading to a linear relationship between the number of prey consumed and prey density; (2) Type II is characterized by a decelerating intake rate of prey consumed as prey density increases, up to a maximum beyond which the number of prey consumed reaches a plateau (i.e. the proportion of prey consumed is assumed to decline with increasing prey density) and (3) Type III is similar to Type II in that saturation occurs at high levels of prey density, but differs from Type II in the sense that at low prey density, predator response to prey is depressed (i.e. the proportion of prey consumed is assumed to increase up to a maximum and then decrease with increasing prey density; see Fig. 1 and Appendix for more information on Holling's functional responses).

Functional and numerical responses are currently integral to our ability to predict predator and prey population dynamics, being also fundamental to our quest to bridge current knowledge gaps found at the interface between

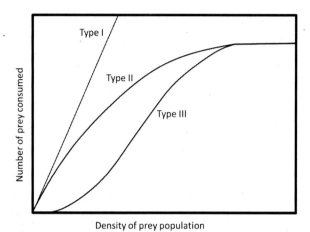

Figure 1 Holling's functional response curves, labelled with type of response.

population ecology and community ecology. Functional responses are, for example, instrumental in research on the long-term evolution of food webs (Drossel et al., 2004), the existence of facilitation or competition in species assemblages (McCoy et al., 2012; van Langevelde et al., 2008), the persistence of some species at low densities despite heavy predation (Lipcius and Hines, 1986), the potential applicability of biocontrol methods (Putra and Yasuda, 2006; Schenk and Bacher, 2002), the effects of climate change on community stability (Vucic-Pestic et al., 2011), the effects of predator reintroduction on prey populations (Varley and Boyce, 2006) and the mortality rates of predators on game or commercially valuable species (Hayes et al., 2000; Van Ballenberghe and Ballard, 1994). Inaccurate or overly simplified modelling of predator–prey interactions, therefore, can seriously hamper efforts to deepen our understanding of basic community ecology and food web theory, while reducing the adequacy of population dynamics models in supporting wildlife management decisions.

This chapter aims to explore how our current knowledge of the factors shaping prey selection and predation risk relate to current modelling approaches of predator–prey dynamics, focusing in particular on the functional response. We will here concentrate our efforts on identifying the overall importance of the interactions between predation and individual heterogeneity, and discuss their possible role in shaping the dynamics of predator and prey populations. We will consider primarily observational studies of free-ranging vertebrate populations, but, where appropriate, will also draw on work from observational and experimental laboratory studies on invertebrates.

2. WHAT SHAPES PREDATION RISK AND PREY SELECTION?

2.1 Predation risk

An overwhelming amount of evidence gathered on free-ranging vertebrates now allows us to conclude that individual prey from a given population (1) do not always face the same average lifetime predation risk and (2) do not always face the same predation risk as they age or as they grow. Such a level of individual variability in predation risk has been linked to a myriad of phenotypic, behavioural and environmental variables (see Table 1 for examples). MacLeod and colleagues (2006), for example, reported that predation risk increases with body mass in individual house sparrows *Passer domesticus*. Individuals allocating less time to vigilance (e.g. individuals of a given sex or juveniles of many species) as well as individuals with less defensive abilities (e.g. old individuals or those with poor body condition) have then been found to experience increased predation risk. For example, old elk *Cervus elaphus canadensis* (Wright et al., 2006) and old female bighorn sheep *Ovis canadensis* (Festa-Bianchet et al., 2006) have been shown to endure higher predation risk than prime-age adults. Behavioural differences according to the sex of the individual have moreover been found to heighten risk of predation. In the Serengeti National Park, Tanzania, male Thomson's gazelles *Gazella thomsonii* have a greater risk of predation from cheetahs *Acinonyx jubatus*, due to their tendency to be positioned alone on the periphery of groups of gazelles as well as their tendency to be less vigilant than female individuals (Fitzgibbon, 1990).

Behavioural differences linked to variation in the phenotypic attributes or personalities of individual prey may moreover alter their risk of predation. For example, boldness in bighorn ewes reduces their susceptibility to predation (Réale and Festa-Bianchet, 2003). Male roe deer *Capreolus capreolus* fawns are more active than females and suffer twice the predation rate from red foxes *Vulpes vulpes* than females (Aanes and Andersen, 1996). Morphologically similar prey can also vary their behaviour as a response to exposure to predators. Squirrel tree frog tadpoles *Hyla squirella* that had been exposed to chemical cues from their odonate predators had lower activity levels than naïve conspecifics and therefore suffered lower attack rates (McCoy and Bolker, 2008).

Furthermore, spatiotemporal variation in environmental conditions, habitat type and resource availability may influence predation risk of

Table 1 Examples of phenotypic and behavioural attributes structuring individual variability in predation risk and prey selection in free-ranging populations

Parameter	Prey species	Individual attribute	Reference
Predation risk	Thomson's gazelle *Gazella thomsonii*	Sex	Fitzgibbon (1990)
Predation risk	Moose *Alces alces*	Year of birth	Thompson and Peterson (1988)
Predation risk	White-tailed deer *Odocoileus virginianus*	Mother's experience	Ozoga and Verme (1986)
Predation risk	Caribou *Rangifer tarandus*	Birth date	Adams et al. (1995)
Predation risk	Bighorn sheep *Ovis canadensis* Elk *Cervus elaphus canadensis*	Age	Réale and Festa-Bianchet (2003) Festa-Bianchet et al. (2006) Wright et al. (2006)
Predation risk	Bighorn sheep	Personality	Réale and Festa-Bianchet (2003)
Predation risk	Moose Elk	Habitat	Berger (2007) Hebblewhite et al. (2005)
Predation risk	House sparrow *Passer domesticus*	Body mass	MacLeod et al. (2006)
Predation risk	Snowshoe hare *Lepus americanus*	Body condition	Murray (2002)
Predation risk	Feral horse *Equus caballus*	Coat colour	Turner and Morrison (2001)
Prey selection	Cougar *Puma concolor*	Age	Ross et al. (1997)
Prey selection	Cheetah *Acinonyx jubatus*	Sex	Cooper et al. (2007)
Prey selection	Cougar Lynx *Lynx lynx*	Reproductive status	Pierce et al. (2000) Nilsen et al. (2009)
Prey selection	Blood python *Python brongersmai*	Colour	Shine et al. (1998)

Table 1 Examples of phenotypic and behavioural attributes structuring individual variability in predation risk and prey selection in free-ranging populations—cont'd

Parameter	Prey species	Individual attribute	Reference
Prey selection	Oystercatcher *Haematopus ostralegus* Killer whale *Orcinus orca*	Dominance/ social status	Sutherland et al. (1996) Sautilis et al. (2000)
Prey selection	American robin *Turdus migratorius*	Body size	Jung (1992)
Prey selection	American pine marten *Martes americana*	Territory location	Ben-David et al. (1997)

individual prey. Habitat characteristics are indeed of major importance to predator–prey interactions (Gorini et al., 2012). Refuges provide areas where prey can escape from predation pressure, reducing prey availability. Habitat characteristics can also influence detection of prey by predators and their hunting success, altogether influencing predation risk. Grey wolves *Canis lupus* were, for example, more likely to encounter elks in lower elevations than in higher ones in Alberta, Canada, while elks had a different chance of being killed depending on whether they were encountered in grasslands, pine stands or open conifer stands (Hebblewhite et al., 2005). In the winter months in Yellowstone National Park, USA, elk also experienced heightened rates of predation from wolves, due to the comparatively limited mobility of elk compared with wolves in deep snow cover (Wilmers and Getz, 2005). Importantly, individuals can use different habitats depending on their age or size (Englund and Krupa, 2000; Heithaus and Dill, 2002; Montgomery et al., 2013; Sweitzer and Berger, 1992), nutritional need (Barten et al., 2001), life history strategy (Daverat et al., 2006) or reproductive status (Berger, 1991), and such co-variation can underpin a certain amount of individual variability in predation risk. Bighorn ewes with offspring are, for example, less likely to utilize dangerous foraging areas than lone females (Berger, 1991), while predation threat by toadfish *Opsanus tau* on mud crabs *Panopeus herbstii* led smaller crabs to consistently use refuges more than larger ones (Toscano et al., 2014).

2.2 Prey selection

Dynamic interactions exist between prey behavioural response to predation and predators' behaviour, resource specialization and distributions, and thus prey selection by individual predators (Abrams, 2000; Lima, 2002). Decades of research on predator–prey interactions has indeed revealed that individual predators (1) do not all display similar patterns in prey selection and (2) can change patterns in prey selection as they age or as they grow (Pettorelli et al., 2011; see Table 1 for examples). Because natural selection acts on individuals, variance in diet among individuals can have several ecological, evolutionary and management implications through its contribution to differences in individual fitness (Bolnick et al., 2003).

Prey selection is indeed shaped by the trade-off between an individual's energetic requirements and the associated costs of hunting, capturing and consuming prey (MacArthur and Pianka, 1966). The major metabolic life history constraint to prey selection as a result of this fundamental trade-off is a predator's body mass, with increasing mass and locomotive capability resulting in a greater benefit to larger predators from selecting larger prey items (Carbone et al., 1999). Large predatory fishes *Crenicichla alta* have, for example, been reported to prey predominantly on large sexually mature size classes of guppies *Poecilia reticulata* (Reznick et al., 1990), meaning that vulnerability to predation increases with body size in guppies. However, such a relationship between prey body size and prey selection by predators does not always necessitate their specialization on larger prey species. Instead, it has been found that larger predators exploit a wider range of prey species and sizes (Radloff and du Toit, 2004), increasing the potential for both individual specialization and opportunistic offtake in prey selection for larger predators (Sinclair et al., 2003). Differences in prey selection have also been reported according to the sex of predators, some linked to differences emerging from the existence of sexual dimorphism in the predator species considered. In vertebrates, the usually larger-bodied male predators require higher protein diets than females of the same species (Nagy, 1987). This can result in greater prey specialization for female individuals while males can have a propensity to opportunistically take higher-quality (higher protein content) prey; male polar bears *Ursus maritimus* were, for example, found to exhibit much greater dietary variability than females, whose diets were more strongly constrained to ringed seal *Pusa hispida* individuals (Thiemann et al., 2011). Likewise, female Adélie penguins *Pygoscelis adeliae* showed much greater specialization on krill *Euphausia* sp. than males (Clarke et al., 1998). Similarly, a predator's age may influence its prey selection, with younger individuals being smaller in size and having less hunting experience to capture larger prey; for example, juvenile bobcats *Felis rufus* have been found to exhibit greater selection for small- and medium-sized prey than adult individuals (Litvaitis et al., 1986). The reproductive status of individual predators can also matter when it comes to prey selection: female cougars *Puma concolor* with offspring preferentially select female deer, while males and females without offspring do not (Pierce et al., 2000). Individual variability in prey selection has yet sometimes been reported without any obvious correlation with phenotypic attributes: neighbouring kestrels *Falco tinnunculus* were, for example, reported to show consistent differences in prey selection, even though the birds were sharing the same hunting grounds (Constantini et al., 2005). This may, however, be a result of the individuals' increasing niche exploitation through reducing intraspecific competition; it is proposed that through this process, populations of generalist predators may be in reality clusters of highly specialized individuals (the niche variation hypothesis: Bolnick et al., 2003, 2007).

Behavioural differences between species, populations and individual predators can then represent an important source of variation in prey selection. The type of prey that it is possible for predators to consume can be influenced by the sociality of the predator in question, with group-living predators more able to capture and kill larger prey. Furthermore, the group size of individual predator groups can influence their prey selection: in the Serengeti National Park, small groups of lions *Panthera leo* seem unable to kill buffalos *Syncerus caffer*, yet the species is an important prey item for larger lion prides (Packer et al., 1990; Scheel and Packer, 1991). Behavioural differences in individuals' prey selection can moreover be driven by learning. For example, distinct specialization on particular prey types in sea otters *Enhydra lutris* has been found to be passed through matriarchal lineages (Estes et al., 2003), and hunting techniques and prey selection can vary markedly between killer whale *Orcinus orca* populations (Sautilis et al., 2000). The presence of competition from other predators of the same species or of other species can also alter individuals' prey selection: for example, Bolnick and colleagues (2007) found that release from inter-specific competition can result in increased dietary generalism.

Environmental factors can also influence prey selection. Indeed, Andruskiw and colleagues (2008) showed that the frequency of prey encounter, prey attack and prey kill by martens *Martes americana* were higher in old uncut forests, despite the fact that red-backed vole *Clethrionomys gapperi* density was similar to that in younger logged forests. Heterogeneous distribution of prey was also reported to result in prey switching and temporary dietary specialism in juvenile silver perch *Bidyanus bidyanus* (Warburton et al., 1998). Work in Alaska then shows that yearling coho salmon *Oncorhynchus kisutch* grow faster in warm streams than in nearby cold streams, making them more likely to grow enough in spring and early summer to be able to exploit the glut of eggs from spawning sockeye salmon in August (Armstrong et al., 2010). Coho that do not grow enough are too small to eat eggs and are limited to a diet of insects. Phenotypic attributes coupled with information on environmental conditions can also be key to understanding patterns of prey selection: in Hwange National Park, Zimbabwe, lions are able to kill juvenile elephants *Loxodonta africana* only during very dry periods, and though sometimes they are killed by solitary male lions, in general it is mostly the larger prides that are able to bring them down (Loveridge et al., 2006).

3. INDIVIDUAL VARIABILITY AND FUNCTIONAL RESPONSES

In Section 2, we established that individual variability in predation risk and prey selection is common in free-ranging populations, and that such variability tends to be shaped by differences in phenotypic and behavioural attributes, as well as environmental factors. How can such variability be expected to influence functional responses? The importance of individual variability in functional responses was actually recognized early on by Holling (1961), who postulated that characteristics of the prey and predator species, as well as the environment, would affect the functional response. There are two main situations that can be expected to arise: in the first case, individual variability in predation risk and/or prey selection leads to different levels of predation rates among predators and thus differences in the strength of the same functional response. In the second case, individual variability in predation risk and/or prey selection leads to individual predators displaying different types of functional responses (Fig. 2).

3.1 Impact on the strength of the response

Has the existence of groups of predators displaying different strengths of the same functional response been reported in experimental or observational studies so far? The answer is yes. Work on larval ladybeetles *Coccinella septempunctata* (Xia et al., 2003) and *Chaoborus americanus* larvae (Spitze, 1985), for example, shows that larger predators or those hunting larger prey tend to have higher predation rates than smaller conspecifics or those hunting smaller prey. Similarly, an increased number of hunting individuals lowers kill rates of lions (Fryxell et al., 2007) and wolves (Hayes et al., 2000). When

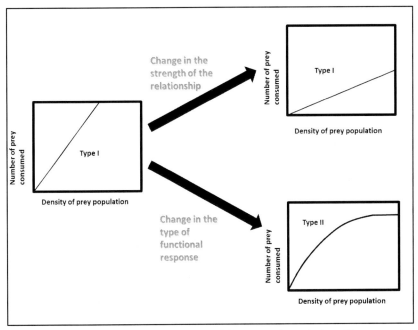

Figure 2 Possible impacts of individual variability in predation risk and/or prey selection on the functional response. In the first case, individual variability in predation risk and/or prey selection leads to different levels of predation rates among predators and thus differences in the strength of the same functional response (in this case, Type I). In the second case, individual variability in predation risk and/or prey selection leads to individual predators displaying different types of functional responses.

hunting individually, male weasels *Mustela nivalis nivalis* had higher predation rates than females (Sundell, 2003), while female wolf spiders *Pardosa vancouveri* had higher rates than males (Hardman and Turnbull, 1974). Both age and hunger level can then impact predation rates: starved damselflies *Ischnura elegans* had higher predation rates than satiated conspecifics (Akre and Johnson, 1979; see also Bressenford and Toft, 2011); sub-adult pumas have lower kill rates than adults, possibly because experience makes pumas more effective predators (Knopff et al., 2010). As with predation risk and prey selection, predation rate thus seems to vary between individuals and such variability can be structured according to phenotypic attributes.

3.2 Impact on the type of response

Few studies have explicitly examined how individuals vary in the type of functional response they exhibit, but there are suggestions that this kind of variability may not be uncommon, especially among species that exhibit prey switching with size or age. For example, owing to changing nutritional needs during different life stages, adult wolf spiders *Pardosa amentata* have a Type II functional response for protein-rich fruit flies *Drosophila melanogaster* and a dome-shaped response for lipid-rich fruit flies, while sub-adult spiders display the opposite pattern (Bressenford and Toft, 2011). Putra and Yasuda (2006) examined larval hoverflies *Eupeodes corolla* preying on aphids *Acyrthosiphon pisum* and found that first instars had a unimodal response to prey density, second instars had a Type II curve, and third instars had a Type I (linear) response. Observations so far also suggest that individual heterogeneity may produce shapes of functional response beyond those commonly recognized. Rudolf (2008) studied dragonfly *Anax junius* and damselfly *Enallagma aspersum* larvae and found that when the prey were considered as an homogeneous whole, the functional response for the dragonflies and damselflies was a Type III, but when the prey were split by size classes, the functional response pattern did not fit any of the standard forms. This, however, may be due to violations of model assumptions, as was suggested when the response for sub-adult bald eagles *Haliaeetus leucocephalus* scavenging sockeye salmon *Oncorhynchus nerka* could not be characterized (Restani et al., 2000). Blue crabs *Callinectes sapidus* are also particularly variable in their responses according to prey type, prey spatiotemporal distribution, sex of crab and ambient environmental conditions. Not only do they have a Type II response preying on clams *Mya arenaria* buried in mud and a Type III when they were buried in sand (Lipcius and Hines, 1986), but at low oyster *Crassostrea virginica* density, male blue crabs exhibited a Type II response and females a Type III (Eggleston, 1990a); they then exhibited a Type II at low and high temperatures and Type III at intermediate temperature (Eggleston, 1990b).

4. ACCOUNTING FOR INDIVIDUAL VARIABILITY WHEN MODELLING PREDATOR–PREY INTERACTIONS

Does the existence of individual variation in prey selection, predation risk and functional responses really matter when it comes to modelling predator and prey population dynamics and understanding the impact of predation on evolutionary trajectories of both predators and prey? Studies so far have emphasized that accounting for individual heterogeneity can have important theoretical and practical implications in terms of sharpening our understanding of evolution, and population and community ecology. From a management perspective, inter-individual differences can affect predator impacts on prey (MacNulty et al., 2009), leading researchers and managers who ignore this type of variability to potentially under- or overestimate the expected impacts of predators on prey populations (Okuyama, 2008). A good example of such a situation is provided by individual cougars specializing on bighorn sheep in Canada (Festa-Bianchet et al., 2006). In this situation, predator control programmes aimed at decreasing cougar densities at large scales are unlikely to reduce predation pressure on bighorn sheep, if the specialized individuals are missed by such programmes; however, such a management action could have drastic consequences on the overall predators' population structure and functioning (Robinson et al., 2008). From a theoretical perspective, Bolnick and colleagues (2011) elegantly showed how manipulating the level of individual variability in predators' attack rate or handling time can actually increase or decrease predation pressure, affecting the severity of predator–prey oscillations. Large population oscillations are frequent in many models of predator–prey systems, yet occur relatively rarely in the wild (but see Arditi and Ginzburg, 2012; Holt, 2011). This may be due to the stabilizing effects of complex individual interactions between predators and prey. Accounting for this individual variability in predator–prey models may thus minimize the disconnect between patterns seen in theoretical/experimental versus wild populations. Individual variability in both predation risk and prey selection could also represent a general mechanism maintaining the diversity of the phenotypic traits associated with this variability (e.g. personality type) within populations, something that is beginning to be explored (Pruitt et al., 2012). Importantly, if individual variability strongly structures variability in handling time and attack rate, current model assumptions are likely to be violated (McPhee et al., 2011). For example, if both attack rate and handling time are highly variable from one individual predator to the next, then population average predation rates are unlikely to be helpful in making mechanistic interpretations about predator–prey interactions, because multiple combinations of attack rate and handling time will result in the same functional response curve (Okuyama, 2012; see also Table A.2 in the Appendix). Violations of assumptions are, however, rarely examined, and functional response models are often used regardless of whether the assumed behavioural mechanisms behind them are actually operating or not (Okuyama, 2009). This can lead to inaccurate or implausible parameter estimates, thus making it impossible to scale up predictions from the individual to the community level (Okuyama, 2008, 2012).

Despite the recognition that accounting for individual variability in predator–prey interactions may be important, only a quick review of the literature demonstrates that most population dynamics models aimed at predicting dynamics of prey in the presence of predators consider predators as being identical individuals. Individual variability in predator choice, for example, is generally overlooked, and conservation strategies are traditionally based on identifying average resource requirements for a population (Bolnick et al., 2003; Durell, 2000). Both individual variability within predators and within prey are classically ignored (see e.g. Fryxell et al., 2007; Nilsen et al., 2005; Post et al., 1999; Stenseth et al., 1997; Vucetich et al., 2005, 2011; but see Chesson, 1978 for an exception); instead a population mean is generally used for parameter values (Messier, 1994). Some models do account for certain types of variability such as age class or sex (see e.g. Nilsen et al., 2007; Post et al., 1999), but models accounting for individual variability within predefined groups are extremely rare, even though such levels of variability can be demonstrably large (Akre and Johnson, 1979; Eggleston, 1990a; Restani et al., 2000).

Do modelling frameworks that enable us to take into account such a level of complexity in predator–prey interactions exist? As discussed above, models have been developed that take into account individual variability in susceptibility to predation: these approaches are generally based on (1) attributing different predation susceptibility to only a limited number of phenotypic categories, or (2) at each time step, randomly attributing to each individual prey a different probability of being predated (see e.g. Abrams, 2007). By randomly attributing probabilities at each time step, the second approach might dismiss potential temporal autocorrelation in each individual's ability to escape predators, but such an approach might be well adapted if a high degree of individual variability in prey selection is expected. Another possible framework enabling explicit consideration of individual variability in predation risk could be based on defining a relative measure of susceptibility to predation for individual prey analogous to frailty, *sensu* Vaupel and colleagues (1979), and consider it as an age-invariant trait (i.e. each individual in a population has a fixed value of relative susceptibility throughout its lifespan). This approach first requires that the level of individual variability in susceptibility to predation is assessed, by collating information on age-specific mortality causes in predated populations and by identifying individuals that died through predation from individuals that died from other causes, such as disease or accidents. Such data can be difficult to access, however, particularly for free-ranging populations. One possible exception is provided by radio-telemetry or GPS-based studies, assuming that radio-collaring does not influence the vulnerability of prey to predation (see e.g. Table 2).

Table 2 Examples of radio-telemetry or GPS-based studies that quantify the relative importance of predation (indexed as the % of deaths due to predation) as the cause of mortality

Species	Parameter	N	Nd	% of deaths due to predation	Reference
Moose *Alces alces*	Calf mortality	62	39	92	Bertram and Vivion (2002)
Moose	Adult female mortality	30	7	86	Bertram and Vivion (2002)
Pronghorn *Antilocapra americana*	Calf mortality	104	87	86	Gregg et al. (2001)
Mule deer *Odocoileus hemionus*	Adult mortality	43	21	62	Robinson et al. (2002)
White-tailed deer *Odocoileus virginianus*	Adult mortality	27	13	46	Robinson et al. (2002)
White-tailed deer	≥0.6-year-old female mortality	153	97	48	DelGuidice et al. (2002)
White-tailed deer	Fawn mortality	29	14	59	Long et al. (1998)
White-tailed deer	Adult mortality	112	46	37	Patterson et al. (2002)
Elk *Cervus elaphus canadensis*	Calf mortality	127	65	45	Singer et al. (1997)
Roe deer *Capreolus capreolus*	Fawn mortality	151	45	73	Panzacchi et al. (2008)
Caribou *Rangifer tarandus*	Calf mortality	46	15	55	Mahoney and Virgl (2003)
Reindeer	Calf mortality	621	43	53	Norberg et al. (2006)
Snowshoe hare *Lepus americanus*	Mortality	177	115	97	Griffin et al. (2005)

In this table, N represents the number of individuals fitted with radio-collars while Nd represents the number of individuals fitted with radio-collars that died.

Recent developments of multi-event capture–recapture models (Pradel, 2005) might also provide a basis for estimating the relative susceptibility of individuals to predation when detection probability is less than 1. Within a defined class (e.g. sex, cohort), the probability of being predated could be modelled as a function of age or size (which are commonly reported to influence absolute susceptibility to predation). Since each individual only contributes to one data point in the data set, the distribution of the residuals (between the observed and predicted probabilities of being predated) provides the distribution of the individual susceptibilities to predation. Vaupel and colleagues (1979) originally assumed these susceptibilities to be gamma distributed. Considering various values for k (k and λ being the parameters of the gamma distribution, with k measuring the degree of individual heterogeneity in susceptibility to predation), it is then possible to compare (1) the distribution of the observed mortality rates according to age within a category and (2) the expected distribution of mortality rates according to age, with the best fit then enabling the parameterization of the degree of individual heterogeneity in susceptibility to predation within an age-invariant phenotypic category.

Incorporating a large number of variables (e.g. size, sex and habitat use) into the functional response itself remains, however, mostly unexplored territory. Several modelling techniques, such as structural equation modelling (Grace et al., 2010), Individual-Based Models (IBMs; Grimm & Railsback, 2013) and linear mixed models, have not been used traditionally in functional response modelling and may hold the potential to begin to address this problem. For situations when there are data on individual animals or groups of animals, mixed-effects models may indeed be used to determine how much of the variance in a parameter, such as attack rate or handling time, is due to differences among individuals. Mixed-effects models can also be used to estimate the importance of factors such as size, sex or habitat use on attack rate or handling time, while controlling for unexplained differences between individuals. IBMs have been used to examine how differences in morphology and behaviour among individuals can impact predator–prey interactions and population dynamics (DeAngelis and Mooij, 2005; McCauley et al., 1993). While there has been limited use of IBMs in functional response research, Petersen and De Angelis (1992) used them to investigate squawfish *Ptychocheilus oregonensis* predation on juvenile salmon in the Columbia River. They incorporated individual differences in predator feeding rate and timing, and predator and prey size to attempt to distinguish between Type II and III responses. IBMs can be used for theoretical and practical explorations of how behaviours that differ among individual predators affect their encounter and attack rates, handling time and predation rate, and, implemented with care, could provide mechanistic insight into the impact of heterogeneity across individuals on population processes. With the addition of parameter estimates from field or experimental data, they have potential to make predictions about how demographic or environmental changes could impact functional response and dynamics of specific populations.

5. CONCLUSIONS

Our message is simple: predator–prey interactions are complex, much more complex than previously thought. Future research efforts should be focused on exploring how this complexity can best be incorporated into rigorous modelling approaches to improve our ability to predict fluctuations in predators and prey numbers. Predation data from current long-term studies that keep track of individuals in the wild, coupled with innovative modelling techniques, are of paramount importance to improving our understanding of the impacts of individual heterogeneity on population responses (Table 3). Most research so far has focused on the differences between individuals of one species of predator and prey. However, a deeper understanding of predator–prey interactions will require careful study of how the differences in morphological traits and/or behavioural plasticity of both individual predators and prey affect their relationships (see McGhee et al., 2013 for an example). Scaling up from the population level to communities, where empirical data on the functional responses of multiple interacting species are rare, will however likely remain a major challenge for some time. Yet the variation seen in predator–prey interactions suggests that each predator–prey relationship can contribute in unexpected ways to increasing the complexity of biological systems, potentially leading to outcomes we currently fail to predict.

Interestingly, the re-establishment of large predators in many ecosystems (e.g. Breitenmoser, 1998; Valière et al., 2003) and the observed increase in population size of a range of predator species following large-scale cessation of predator control (e.g. Wright, 1999) may provide ecologists and wildlife managers with some fantastic opportunities to tackle the current lack of information regarding the importance of predation in shaping life history traits of free-ranging vertebrates (Pettorelli et al., 2011); to test theories developed using experimental settings and simulation work; to examine how individual variability in both predators and prey impact the interactions between them and to develop tools and frameworks allowing for better integration of the complexity of predator–prey interactions in applied situations (see e.g. MacLeod et al., 2014). Of particular interest is the exploration of the

Table 3 Non-exhaustive list of key research questions that still require adequate collection of relevant data in the wild in order to be addressed

Questions	Potential issues
Does the presence of predators lead to differences in the means and variation in survival and recruitment rates among phenotypic categories within a prey population? Does it affect the trade-off between survival and reproduction?	Potential changes to the effects of density dependence on prey populations that the presence of predators may generate should be accounted for
Are there predator species characteristics that influence the level of individual variability in predator diet?	Meta-analyses based on information on the level of individual variability in predators' populations should account for differences in levels of inter-specific competition
How does the level of individual variability in prey selection correlate with the distribution of indirect costs of predation in the prey population and the heritability of anti-predator-induced traits?	Most long-term projects on large vertebrate populations have been carried out in predator-free environments, reducing opportunities to empirically address this question
What is the relative importance of indirect and direct costs of predation in determining prey population dynamics?	The answer might be a function of predator and prey densities
Do changes in the average composition of the predator population lead to differences in the means and variation in survival and recruitment rates among phenotypic categories within a prey population?	This question is likely to be best answered within a single prey-single predator system where all the required information is accessible over a sufficient time frame
How does individual vulnerability to predation correlate with individual predation costs and prey selection?	Prey selection might not reflect prey preference, leading to potential inter-population differences in the relationships between predation susceptibility and predation costs within populations of selected and preferred prey
Is susceptibility to predation mainly determined by phenotypic attributes, should it be considered as an individual age-invariant characteristic, or is it mainly randomly variable over time for each individual without temporal autocorrelation?	Predator characteristics might influence how susceptibility to predation is structured within prey populations

Table 3 Non-exhaustive list of key research questions that still require adequate collection of relevant data in the wild in order to be addressed—cont'd

Questions	Potential issues
How do individual differences in both predators and prey impact their interactions? On what temporal/spatial scales are those differences important and what are the implications for community dynamics?	Difficult to get detailed information on both predator and prey in natural systems; species characteristics and level of habitat heterogeneity might be important when tackling this issue

Here, the direct costs of predation refer to the changes in prey mortality due to predation. The indirect costs of predation, on the other hand, refer to the costs of anti-predator behavioural responses of prey, which can be manifest by reduced prey survival, growth or reproduction (Creel and Christianson, 2008).

relative role of the phenotype and the ecological context (notably, whether or not predation is present) in determining life history traits. A recent study by Cote and colleagues (2013) illustrates the importance of understanding such interactions to better predict population dynamics of both predator and prey. Examining the link between phenotypic specialization and dispersal inclination in mosquitofish *Gambusia affinis*, the authors show that dispersing mosquitofish are less social than non-dispersing fish when predation risk is null. However, they also demonstrate how personality-dependent dispersal is negated under predation risk, with dispersers displaying similar personality types to residents in such conditions (Cote et al., 2013). Such results support previous calls, as well as that made here, for more research into assessing and taking into account the potentially individually variable indirect impacts of predation when modelling predator–prey interactions (Creel and Christianson, 2008; Pettorelli et al., 2011).

 APPENDIX

Table A.1 Holling's three functional responses

Model	Formula	Description
Type I	Kill rate = attack rate*prey density	Linear
Type II	$\text{Kill rate} = \dfrac{\text{attack rate*prey density}}{1 + \text{attack rate*prey density*handling time}}$	Asymptotic
Type III	$\text{Kill rate} = \dfrac{b\text{*prey density}^2}{1 + c\text{*prey density} + b\text{*handling time*prey density}^2}$	Sigmoid

Note that in these equations, b and c are constants.

Table A.2 Definitions of the parameters used in the functional response equations

Parameter	Definition
Encounter rate	Search rate (area searched/unit time)*prey density (individuals/area)
Attack rate	Encounter rate*proportion of encounters that turn into attacks*proportion of attacks that are successful
Handling time	The time required to handle prey so that other prey cannot be concurrently attacked. May include pursuing, subduing, eating, digesting. Constant in Types II and III
Kill/predation rate	Number of prey killed/predator (individual or group)/unit time. Y axis of functional response curves

REFERENCES

Aanes, R., Andersen, R., 1996. The effects of sex, time of birth, and habitat on the vulnerability of roe deer fawns to red fox predation. Can. J. Zool. 74, 1857–1865.

Abrams, P.A., 2000. The evolution of predator-prey interactions: theory and evidence. Annu. Rev. Ecol. Syst. 31, 79–105.

Abrams, P.A., 2007. Habitat choice in predator-prey systems: spatial instability due to interacting adaptive movements. Am. Nat. 169, 581–594.

Adams, L.G., Singer, F.J., Dale, B.W., 1995. Caribou calf mortality in Denali National Park, Alaska. J. Wildl. Manage. 59, 584–594.

Akre, B.G., Johnson, D.M., 1979. Switching and sigmoid functional response curves by damselfly naiads with alternative prey available. J. Anim. Ecol. 48, 703–720.

Andruskiw, M., Fryxell, J.M., Thompson, I.D., Baker, J.A., 2008. Habitat-mediated variation in predation risk by the American marten. Ecology 89, 2273–2280.

Arditi, R., Ginzburg, L.R., 2012. How Species Interact: Altering the Standard View of Trophic Ecology. Oxford University Press, New York.

Armstrong, J.B., Schindler, D.E., Omori, K.L., Ruff, C.P., Quinn, T.P., 2010. Thermal heterogeneity mediates the effects of pulsed subsidies across a landscape. Ecology 91, 1445–1454.

Barten, N.L., Bowyer, R.T., Jenkins, K.J., 2001. Habitat use by female caribou: trade-offs associated with parturition. J. Wildl. Manage. 65, 77–92.

Ben-David, M., Flynn, R.W., Schell, D.M., 1997. Annual and seasonal changes in diets of martens: evidence from stable isotope analysis. Oecologia 111, 280–291.

Berger, J., 1991. Pregnancy incentives, predation constraints and habitat shifts: experimental and field evidence for wild bighorn sheep. Anim. Behav. 41, 61–64.

Berger, J., 2007. Fear, human shields and the redistribution of prey and predators in protected areas. Biol. Lett. 3, 620–623.

Bertram, M.R., Vivion, M.T., 2002. Moose mortality in eastern interior Alaska. J. Wildl. Manage. 66, 747–756.

Bolnick, D.I., Svanbäck, R., Fordyce, J.A., Yang, L.H., Davis, J.M., Hulsey, C.D., Forister, M.L., 2003. The ecology of individuals: incidence and implications of individual specialization. Am. Nat. 161, 1–28.

Bolnick, D.I., Svänback, M., Araujo, J., Persson, J., 2007. More generalized populations are also more heterogeneous: comparative support for the niche variation hypothesis. Proc. Natl. Acad. Sci. U.S.A. 104, 10075–10079.

Bolnick, D.I., Amarasekare, P., Araujo, M.S., Buerger, R., Levine, J.M., Novak, M., Rudolf, V.H.W., Schreiber, S.J., Urban, M.C., Vasseur, D.A., 2011. Why intraspecific trait variation matters in community ecology. Trends Ecol. Evol. 26, 183–192.

Breitenmoser, U., 1998. Large predators in the Alps: the fall and rise of man's competitors. Biol. Conserv. 3, 279–289.

Bressenford, B.B., Toft, S., 2011. Dome-shaped functional response induced by nutrient balance of the prey. Biol. Lett. 23, 517–520.

Carbone, C., Mace, G.M., Roberts, S.C., MacDonald, D.W., 1999. Energetic constraints on the diet of terrestrial carnivores. Nature 402, 286–288.

Chesson, P.L., 1978. Predator-prey theory and variability. Annu. Rev. Ecol. Syst. 9, 323–347.

Clarke, J., Manly, B., Kerry, K., Gardner, H., Franchi, E., Corsolini, S., Focardi, S., 1998. Sex differences in Adélie penguin foraging strategies. Polar Biol. 20, 248–258.

Constantini, D., Casagrande, S., Di Lieto, G., Fanfani, A., Dell'Omo, G., 2005. Consistent differences in feeding habits between neighbouring breeding kestrels. Behaviour 142, 1409–1421.

Cooper, A.B., Pettorelli, N., Durant, S.M., 2007. Large carnivore menus: factors affecting hunting decisions by cheetahs in the Serengeti. Anim. Behav. 73, 651–659.

Cote, J., Fogarty, S., Tymen, B., Sih, A., Brodin, T., 2013. Personality-dependent dispersal cancelled under predation risk. Proc. R. Soc. B 280, 20132349.

Creel, S., Christianson, D., 2008. Relationships between direct predation and risk effects. Trends Ecol. Evol. 23, 194–201.

Creel, S., Creel, N.M., 2002. The African Wild Dog: Behaviour, Ecology and Conservation. Princeton University Press, Princeton.

Daverat, F., Limburg, K.E., Thibault, I., Shiao, J.-C., Dodson, J.J., Caron, F., Tzeng, W.-N., Iizuka, Y., Wickstrom, H., 2006. Phenotypic plasticity of habitat use by three temperate eel species, Anguilla anguilla, A. japonica and A. rostrata. Mar. Ecol. Prog. Ser. 308, 231–241.

DeAngelis, D.L., Mooij, W.M., 2005. Individual-based modeling of ecological and evolutionary processes. Annu. Rev. Ecol. Evol. Syst. 36, 147–168.

DelGuidice, G.D., Riggs, M.R., Jolly, P., Pan, W., 2002. Winter severity, survival, and cause-specific mortality of female white-tailed deer in North-central Minnesota. J. Wildl. Manage. 66, 698–717.

Drossel, B., McKane, A.J., Quince, C., 2004. The impact of nonlinear functional responses on the long-term evolution of food web structure. J. Theor. Biol. 229, 539–548.

Durell, S.E.A., 2000. Individual feeding specialization in shorebirds: population consequences and conservation implications. Biol. Rev. Camb. Philos. Soc. 75, 503–518.

Eggleston, D.B., 1990a. Functional responses of blue crabs Callinectes sapidus Rathbun feeding on juvenile oysters Crassostrea (Gmelin): effects of predator sex and size, and prey size. J. Exp. Mar. Biol. Ecol. 143, 73–90.

Eggleston, D.B., 1990b. Behavioural mechanisms underlying variable functional responses of blue crabs, Callinectes sapidus feeding on juvenile oysters, Crassostrea virginica. J. Anim. Ecol. 59, 615–630.

Englund, G., Krupa, J.J., 2000. Habitat use by crayfish in stream pools: influence of predators, depth and body size. Freshw. Biol. 43, 75–83.

Estes, J.A., Riedman, M.L., Staefler, M.M., Tinker, M.T., Lyon, B.E., 2003. Individual variation in prey selection by sea otters: patterns, causes and implications. J. Anim. Ecol. 72, 144–155.

Festa-Bianchet, M., Coulson, T., Gaillard, J.-M., Hogg, J.T., Pelletier, F., 2006. Stochastic predation events and population persistence in bighorn sheep. Proc. R. Soc. B 273, 1537–1543.

Fitzgibbon, C.D., 1990. Why do hunting cheetahs prefer male gazelles? Anim. Behav. 40, 837–845.

Fitzgibbon, C.D., Fanshawe, J.H., 1989. The condition and age of Thomson's gazelles killed by cheetahs and wild dogs. J. Zool. Soc. Lond. 218, 99–107.

Fryxell, J.M., Mosser, A., Sinclair, A.R.E., Packer, C., 2007. Group formation stabilizes predator-prey dynamics. Nature 449, 1041–1043.

Gilliam, J.F., Fraser, D.F., 1987. Habitat selection under predation hazard: test of a model with foraging minnows. Ecology 68, 1856–1862.

Gorini, L., Linnell, J.D.C., May, R., Panzacchi, M., Boitani, L., Odden, M., Nilsen, E.B., 2012. Habitat heterogeneity and mammalian predator-prey interactions. Mammal Rev. 42, 55–77.

Grace, J.B., Anderson, T.M., Olff, H., Scheiner, S.M., 2010. On the specification of structural equation models for ecological systems. Ecol. Monogr. 80, 67–87.

Gregg, M.A., Bray, M., Kilbride, K.M., Dunbar, M.R., 2001. Birth synchrony and survival of pronghorn fawns. J. Wildl. Manage. 65, 19–24.

Griffin, P.C., Griffin, S.C., Waroquiers, C., Mills, L.S., 2005. Mortality by moonlight: predation risk and the snowshoe hare. Behav. Ecol. 16, 938–944.

Grimm, V., Railsback, S.F., 2013. Individual-Based Modeling and Ecology. Princeton University Press, Princeton, 448 pp.

Hairston, N.G., Smith, F.E., Slobodkin, L.B., 1960. Community structure, population control, and competition. Am. Nat. 94, 421–425.

Hardman, J.M., Turnbull, A.L., 1974. The interaction of spatial heterogeneity, predator competition and the functional response to prey density in a laboratory system of wolf spiders (Araneae: Lycosidae) and fruit flies (Diptera: Drosophilidae). J. Anim. Ecol. 43, 155–171.

Hayes, R.D., Baer, A.M., Wotschikowsky, U., Harestad, A.S., 2000. Kill rate by wolves on moose in the Yukon. Can. J. Zool. 78, 49–59.

Hebblewhite, M., Merrill, E.H., McDonald, T.L., 2005. Spatial decomposition of predation risk using resource selection functions: an example in a wolf/elk predator/prey system. Oikos 111, 101–111.

Heithaus, M.R., Dill, L.M., 2002. Food availability and tiger shark predation risk influence bottlenose dolphin habitat use. Ecology 83, 480–491.

Holling, C.S., 1959a. The components of predation as revealed by a study of small mammal predation of the European Pine Sawfly. Can. Entomol. 91, 293–320.

Holling, C.S., 1959b. Some characteristic of simple types of predation and parasitism. Can. Entomol. 91, 385–398.

Holling, C.S., 1961. Principles of insect predation. Annu. Rev. Entomol. 6, 163–182.

Holt, R.D., 2011. Natural enemy-victim interactions: do we have a unified theory yet? In: Scheiner, S.M., Willig, M.R. (Eds.), The Theory of Ecology. The University of Chicago Press, Chicago, pp. 125–161.

Jung, R.E., 1992. Individual variation in fruit choice by American robins (*Turdus migratorius*). Auk 109, 98–111.

Knopff, K.H., Knopff, A.A., Kortello, A., Boyce, M.S., 2010. Cougar kill rate and prey composition in a multiprey system. J. Wildl. Manage. 74, 1435–1447.

Krams, I., Kivleniece, I., Kuusik, A., Krama, T., Freeberg, T.M., Mand, R., Vrublevska, J., Rantala, M.J., Mand, M., 2013. Predation selects for low resting metabolic rate and consistent individual differences in anti-predator behaviour in a beetle. Acta Ethol. 16, 163–172.

Kruger, S.C., Lawes, M.J., Maddock, A.H., 1999. Diet choice and capture success of wild dog in Hluhluwe-Umfolozi Park, South Africa. J. Zool. Soc. Lond. 248, 543–551.

Lima, S.L., 2002. Putting predators back into behavioural predator-prey interactions. Trends Ecol. Evol. 17, 70–75.

Lipcius, R.N., Hines, A.H., 1986. Variable functional responses of a marine predator in dissimilar homogeneous microhabitats. Ecology 67, 1361–1371.

Litvaitis, J.A., 2000. Investigating food habits of terrestrial vertebrates. In: Boitani, L., Fuller, T.K. (Eds.), Research Techniques in Animal Ecology: Controversies and Consequences. Columbia University Press, New York, pp. 165–190.

Litvaitis, J.A., Clark, A.G., Hunt, J.H., 1986. Prey selection and fat deposits of bobcats Felis rufus during autumn and winter in Maine. J. Mammal. 67, 389–392.

Long, R.A., O'Connell, A.F., Harrison, D.J., 1998. Mortality and survival of white-tailed deer fawns on a north Atlantic coastal island. Wildl. Biol. 4, 237–247.

Loveridge, A.J., Hunt, J.E., Murindagomo, F., MacDonald, D.W., 2006. Influence of drought on predation of elephant (Loxodonta africana) calves by lions (Panthera leo) in an African wooded savannah. J. Zool. 270, 523–530.

MacArthur, R.H., Pianka, E.R., 1966. On the optimal use of a patchy environment. Am. Nat. 100, 603–609.

MacLeod, R., Barnett, P., Clark, J., Cresswell, W., 2006. Mass-dependent predation risk as a mechanism for house sparrow declines? Biol. Lett. 2, 43–46.

MacLeod, C.D., MacLeod, R., Learmonth, J.A., Cresswell, W., Pierce, G.J., 2014. Predicting population-level risk effects of predation from the responses of individuals. Ecology 95, 2006–2015.

MacNulty, D.R., Smith, D.W., Vucetich, J.A., Mech, L.D., Stahler, D.R., Packer, C., 2009. Predatory senescence in ageing wolves. Ecol. Lett. 12, 1347–1356.

Magnhagen, C., 1991. Predation risk as a cost of reproduction. Trends Ecol. Evol. 6, 183–186.

Mahoney, S.P., Virgl, J.A., 2003. Habitat selection and demography of a non-migratory woodland caribou population in Newfoundland. Can. J. Zool. 81, 321–334.

May, R.M., 1973. Stability and Complexity in Model Ecosystems. Princeton University Press, Princeton.

McCauley, E., Wilson, W.G., de Roos, A.M., 1993. Dynamics of age-structured and spatially structured predator-prey interactions: individual-based models and population-level formulations. Am. Nat. 142, 412–442.

McCoy, M.W., Bolker, B.M., 2008. Trait-mediated interactions: influence of prey size, density and experience. J. Anim. Ecol. 77, 478–486.

McCoy, M.W., Stier, A.C., Osenberg, C.W., 2012. Emergent effects of multiple predators on prey survival: the importance of depletion and the functional response. Ecol. Lett. 15, 1449–1456.

McGhee, K.E., Pintor, L.M., Bell, A.M., 2013. Reciprocal behavioral plasticity and behavioral types during predator-prey interactions. Am. Nat. 182, 704–717.

McPhee, H.M., Webb, N.F., Merrill, E.H., 2011. Time-to-kill: measuring attack rates in a heterogenous landscape with multiple prey types. Oikos 121, 711–720.

Messier, F., 1994. Ungulate population models with predation: a case study with the North American moose. Ecology 75, 478–488.

Mills, M.G.L., Gorman, M.L., 1997. Factors affecting the density and distribution of wild dogs in the Kruger National Park. Conserv. Biol. 11, 1397–1406.

Montgomery, R.A., Vucetich, J.A., Peterson, R.O., Roloff, G.J., Millenbah, K.F., 2013. The influence of winter severity, predation and senescence on moose habitat use. J. Anim. Ecol. 82, 301–309.

Mousseau, T.A., Endler, J., Sinervo, B., 2000. Adaptive Genetic Variation in the Wild. Oxford University Press, New York, 276 pp.

Murray, D.L., 2002. Differential body condition and vulnerability to predation in snowshoe hares. J. Anim. Ecol. 71, 614–625.

Nagy, K.A., 1987. Field metabolic rate and food requirement scaling in mammals and birds. Ecol. Monogr. 57, 111–128.

Nilsen, E.B., Pettersen, T., Gundersen, H., Milner, J.M., Mysterud, A., Solberg, E.J., Andreassen, H.P., Stenseth, N.C., 2005. Moose harvesting strategies in the presence of wolves. J. Anim. Ecol. 42, 389–399.

Nilsen, E.B., Milner-Gulland, E.J., Schofield, L., Mysterud, A., Stenseth, N.C., Coulson, T.N., 2007. Wolf introduction to Scotland: public attitudes and consequences for red deer management. Proc. R. Soc. Lond. B 274, 995–1002.

Nilsen, E.B., Gaillard, J.-M., Andersen, R., Odden, J., Delorme, D., van Laere, G., Linnell, J.D.C., 2009. A slow life in hell or a fast life in heaven: demographic analyses of contrasting roe deer populations. J. Anim. Ecol. 78, 585–594.

Norberg, H., Kojola, I., Aikio, P., Nylund, M., 2006. Predation by golden eagle on semi-domesticated reindeer calves in northeastern Finnish Lapland. Wildl. Biol. 12, 393–402.

Okuyama, T., 2008. Individual behavioral variation in predator-prey models. Ecol. Res. 23, 665–671.

Okuyama, T., 2009. Local interactions between predators and prey call into question commonly used functional responses. Ecol. Model. 220, 1182–1188.

Okuyama, T., 2012. Flexible components of functional responses. J. Anim. Ecol. 81, 185–189.

Ozoga, J.J., Verme, L.J., 1986. Relation of maternal age to fawn-rearing success in white-tailed deer. J. Wildl. Manage. 50, 480–486.

Packer, C., Scheel, D., Pusey, A.E., 1990. Why lions form groups: food is not enough. Am. Nat. 136, 1–19.

Panzacchi, M., Linnell, J.D.C., Odden, J., Andersen, R., 2008. When a generalist becomes a specialist: patterns of red fox predation on roe deer fawns under contrasting conditions. Can. J. Zool. 86, 116–126.

Patterson, B.R., MacDonald, B.A., Lock, B.A., Anderson, D.G., Benjamin, L.K., 2002. Proximate factors limiting population growth of white-tailed deer in Nova Scotia. J. Wildl. Manage. 66, 511–521.

Petersen, J.H., De Angelis, D.L., 1992. Functional response and capture timing in an individual-based model: predation by Northern Squawfish (Ptychocheilus oregonensis) on juvenile salmonids in the Columbia River. Can. J. Fish. Aquat. Sci. 49, 2551–2565.

Pettorelli, N., Coulson, T., Durant, S.M., Gaillard, J.-M., 2011. Predation, individual variability and population dynamics. Oecologia 167, 305–314.

Pierce, B.M., Bleich, V.C., Bowyer, R.T., 2000. Selection of mule deer by mountain lions and coyotes: effects of hunting style, body size, and reproductive status. J. Mammal. 81, 462–472.

Pole, A., Gordon, I.J., Gorman, M.L., MacAskill, M., 2004. Prey selection by African wild dogs (Lycaon pictus) in southern Zimbabwe. J. Zool. Soc. Lond. 262, 207–215.

Post, E., Peterson, R.O., Stenseth, N.C., McLaren, B.E., 1999. Ecosystem consequences of wolf behavioural response to climate. Nature 401, 905–907.

Pradel, R., 2005. Multievent: an extension of multistate capture-recapture models to uncertain states. Biometrics 61, 442–447.

Preisser, E.L., Bolnick, D.I., Benard, M.F., 2005. Scared to death? The effects of intimidation and consumption in predator–prey interactions. Ecology 86, 501–509.

Pruitt, J.N., Stachowicz, J.J., Sih, A., 2012. Behavioral types of predator and prey jointly determine prey survival: potential implications for the maintenance of within-species behavioural variation. Am. Nat. 179, 217–227.

Putra, N.S., Yasuda, H., 2006. Effects of prey species and its density on larval performance of two species of hoverfly larvae, Episyrphus balteatus de Geer and Eupeodes corollae Fabricius (Diptera: Syrphidae). Appl. Entomol. Zool. 41, 389–397.

Radloff, F.G.T., du Toit, J.T., 2004. Large predators and their prey in a southern African savanna: a predator's size determines its prey size range. J. Anim. Ecol. 73, 410–423.

Réale, D., Festa-Bianchet, M., 2003. Predator-induced natural selection on temperament in bighorn ewes. Anim. Behav. 65, 463–470.

Relyea, R.A., 2002. The many faces of predation: how induction, selection, and thinning combine to alter prey phenotypes. Ecology 83, 1953–1964.

Restani, M., Harmata, A.R., Madden, E.M., 2000. Numerical and functional responses of migrant bald eagles exploiting a seasonally congregated food source. Condor 102, 561–568.

Reznick, D.A., Bryga, H., Endler, J.A., 1990. Experimentally induced life-history evolution in a natural population. Nature 346, 357–359.

Ripple, W.J., Estes, J.A., Beschta, R.L., Wilmers, C.C., Ritchie, E.G., Hebblewhite, M., Berger, J., Letnic, M., Nelson, M.P., Schmitz, O.J., Smith, D.W., Wallach, A.D., Wirsing, A.J., 2014. Status and ecological effects of the world's largest carnivores. Science 343, 1241484.

Robinson, H.S., Wielgus, R.B., Gwilliam, J.C., 2002. Cougar predation and population growth of sympatric mule deer and white-tailed deer. Can. J. Zool. 80, 556–568.

Robinson, H.S., Wielgus, R.B., Cooley, H.S., Cooley, S.W., 2008. Sink populations in carnivore management: cougar demography and immigration in a hunted population. Ecol. Appl. 18, 1028–1037.

Ross, P.I., Jalkotzy, M.G., Festa-Bianchet, M., 1997. Cougar predation on bighorn sheep in southwestern Alberta during winter. Can. J. Zool. 74, 771–775.

Rudolf, V.H., 2008. Consequences of size structure in the prey for predator-prey dynamics: the composite functional response. J. Anim. Ecol. 77, 520–528.

Sautilis, E., Matkin, C., Barrett-Lennard, L., Heise, K., Ellis, G., 2000. Foraging strategies of sympatric killer whale populations in Prince William Sound, Alaska. Mar. Mamm. Sci. 16, 94–109.

Scheel, D., Packer, C., 1991. Group hunting behaviour of lions: a search for cooperation. Anim. Behav. 41, 697–709.

Schenk, D., Bacher, S., 2002. Functional response of a generalist insect predator to one of its prey species in the field. J. Anim. Ecol. 71, 524–531.

Schmitz, O.J., 1998. Direct and indirect effects of predation and predation risk in old-field interaction webs. Am. Nat. 151, 327–342.

Shine, R., Ambariyanto, Harlow, P.S., Mumpuni, 1998. Ecological divergence among sympatric colour morphs in blood pythons Python brongersmai. Oecologia 116, 113–119.

Sinclair, A.R.E., Mduma, S.A.R., Brashares, J.S., 2003. Patterns of predation in a diverse predator-prey system. Nature 425, 288–290.

Singer, F.J., Harting, A., Symonds, K.K., Coughenour, M.B., 1997. Density-dependence, compensation, and environmental effects on elk calf mortality in Yellowstone National Park. J. Wildl. Manage. 61, 12–25.

Spitze, K., 1985. Functional response of an ambush predator: Chaoborus americanus predation on Daphnia pulex. Ecology 66, 938–949.

Stenseth, N.C., Falck, W., Bjørnstad, O.N., Krebs, C.J., 1997. Population regulation in snowshoe hare and Canadian lynx: asymmetric food web configurations between hare and lynx. Proc. Natl. Acad. Sci. U.S.A. 94, 5147–5152.

Sundell, J., 2003. Population dynamics of microtine rodents: an experimental test of the predation hypothesis. Oikos 101, 416–427.

Sutherland, W.J., Ens, B.J., Goss-Custard, J.D., Hulscher, J.B., 1996. Specialization. In: Goss-Custard, J.D. (Ed.), The Oystercatcher. Oxford University Press, New York, pp. 105–132.

Sweitzer, R.A., Berger, J., 1992. Size-related effects of predation on habitat use and behaviour of porcupines Erethizon dorsatum. Ecology 73, 867–875.

Thiemann, G.W., Iverson, S.J., Stirling, I., Obbard, M.E., 2011. Individual patterns of prey selection and dietary specialisation in an Arctic marine carnivore. Oikos 120, 1469–1478.

Thompson, I.D., Peterson, R., 1988. Does wolf predation alone limit the moose population in Pukaskwa park? J. Wildl. Manage. 52, 556–559.

Toscano, B.J., Gatto, J., Griffen, B.D., 2014. Effect of predation threat on repeatability of individual crab behaviour revealed by mark-recapture. Behav. Ecol. Sociobiol. 68, 519–527.

Turner Jr., J.W., Morrison, M.L., 2001. Influence of predation by mountain lions on numbers and survivorship of a feral horse population. Southwest. Nat. 46, 183–190.

Valière, N., Fumagalli, L., Gielly, L., Miquel, C., Lequette, B., Poulle, M.-L., Weber, J.-M., Arlettaz, R., Taberlet, P., 2003. Long-distance wolf recolonization of France and Switzerland inferred from non-invasive genetic sampling over a period of 10 years. Anim. Conserv. 6, 83–92.

Van Ballenberghe, V., Ballard, W.B., 1994. Limitation and regulation of moose populations: the role of predation. Can. J. Zool. 72, 2071–2077.

Van Langevelde, F., Drescher, M., Heitkönig, I.M.A., Prins, H.H.T., 2008. Modelling instantaneous intake rate of herbivores as function of forage quality and mass: effects on facilitative and competitive interactions between differently sized herbivores. Ecol. Modell. 213, 273–284.

Varley, N., Boyce, M.S., 2006. Adaptive management for reintroductions: updating a wolf recovery model for Yellowstone National Park. Ecol. Modell. 193, 315–339.

Vaupel, J.W., Manton, K.G., Stallard, E., 1979. The impact of heterogeneity in individual frailty on the dynamics of mortality. Demography 16, 439–454.

Vucetich, J.A., Smith, D.W., Stahler, D.R., 2005. Influence of harvest, climate and wolf predation on Yellowstone elk, 1961-2004. Oikos 111, 259–270.

Vucetich, J.A., Hebblewhite, M., Smith, D.W., Peterson, R.O., 2011. Predicting prey population dynamics from kill rate, predation rate and predator–prey ratios in three wolf-ungulate systems. J. Anim. Ecol. 80, 1236–1245.

Vucic-Pestic, O., Ehnes, R.B., Rall, B.C., Brose, U., 2011. Warming up the system: higher predator feeding rates but lower energetic efficiencies. Glob. Change Biol. 17, 1301–1310.

Warburton, K., Retif, S., Hume, D., 1998. Generalists as sequential specialists: diets and prey switching in juvenile silver perch. Environ. Biol. Fishes 51, 445–454.

Werner, E.E., Hall, D.J., 1974. Optimal foraging and the size selection of prey by the bluegill sunfish. Ecology 55, 1042–1052.

Wilmers, C.C., Getz, W.M., 2005. Gray wolves as climate change buffers in Yellowstone. PLoS Biol. 3, e92.

Wittmer, H.U., Sinclair, A.R.E., McLellan, B.N., 2005. The role of predation in the decline and extirpation of woodland caribou populations. Oecologia 144, 257–267.

Wright, R.G., 1999. Wildlife management in the national parks: questions in search of answers. Ecol. Appl. 9, 30–36.

Wright, G.J., Peterson, R.O., Smith, D.W., Lemke, T.O., 2006. Selection of northern Yellowstone elk by gray wolves and hunters. J. Wildl. Manage. 70, 1070–1078.

Xia, J.Y., Rabbinge, R., Van Der Werf, W., 2003. Multistage functional responses in a ladybeetle-aphid system: scaling up from the laboratory to the field. Environ. Entomol. 32, 151–162.

Yeakel, J.D., Pires, M.M., Rudolf, L., Dominy, N.J., Kock, P.L., Guimarães Jr., P.R., Gross, T., 2014. Collapse of an ecological network in Ancient Egypt. Proc. Nat. Acad. Sci. U.S.A. 111 (40), 14472–14477.

> CHAPTER THREE

Individual Variation Decreases Interference Competition but Increases Species Persistence

Jean P. Gibert[1], John P. DeLong

School of Biological Sciences, University of Nebraska-Lincoln, Lincoln, Nebraska, USA
[1]Corresponding author: e-mail address: jeanphisth@gmail.com

Contents

Abstract

Interference competition is thought to stabilize consumer–resource systems. The magnitude of interference is linked to that of attack efficiency: when both levels are intermediate, populations are maximally stable and have high competitive ability. Individual variation can affect ecological dynamics through its effect on attack efficiency and handling time. Because interference has a non-linear effect on consumer foraging rates, individual variation in mutual interference can strongly affect ecological dynamics. Here, we explicitly incorporate individual variation in attack efficiency, handling time and interference into a dynamic consumer–resource model and show that variation increases species coexistence by depressing attack efficiency to a greater extent than predator interference. We argue that this differential effect of variation affects the

Advances in Ecological Research, Volume 52
ISSN 0065-2504
http://dx.doi.org/10.1016/bs.aecr.2015.01.002

45

equilibrium densities of consumers and their prey, thus altering their competitive ability. Intermediate levels of variation can maximize both consumer persistence and competitive ability. Our results show the importance of quantifying individual variation in natural populations for understanding the persistence and stability of species within communities.

1. INTRODUCTION

A major goal of ecology is to understand the factors underpinning species coexistence and stability in complex ecosystems (Allesina and Tang, 2012; May, 1972, 1973; McCann et al., 1998). Seminal work by Tilman showed that when two competing species share a common resource, the one that can reduce resource density the most will outcompete the other (Tilman, 1982, 1986). However, the ability to reduce resource density and persist may depend upon the factors controlling interaction strengths and consumer–resource interactions. A number of these factors have received a lot of attention, including foraging behaviour (Abrams and Matsuda, 2004; Schmitz et al., 1997), consumer and resource body sizes (Vucic-Pestic et al., 2010) and relative velocities (DeLong, 2014; Pawar et al., 2012), prey defence mechanisms (Hammill et al., 2010; Yoshida et al., 2004), and environmental temperature (Dell et al., 2014; Gibert and DeLong, 2014; O'Connor, 2009). And while all these factors are important, the underlying assumption in ecology has historically been that populations are homogeneous collections of individuals and that mean trait values are sufficient for understanding ecological processes (Lomnicki, 1988). Unfortunately, whenever non-linear relationships between underlying traits and ecological processes of interest occur, using mean trait values can be misleading (Bolnick et al., 2011; Inouye, 2005). Because non-linearities are common in consumer–resource interactions, overlooking individual phenotypic variation may impair our capacity to fully understand species persistence and competitive ability in natural communities.

Populations often show individual-level phenotypic variation in antipredator defences (Duffy, 2010), competitive ability (Lankau and Strauss, 2007), or resource utilization (e.g. Bolnick et al., 2003; Estes et al., 2003). Because interspecific interactions ultimately occur between individuals, individual phenotypic variation can affect interspecific interactions in multiple ways (Pettorelli et al., 2011). For instance, individual-level dietary specialization among southern sea otters (*Enhydra lutris nereis*) induces changes in the structure of the population-level resource utilization network, which in turn can alter the structure and dynamics of the food webs in which these organisms are embedded (Tinker et al., 2012). Individual variation also can affect the strength of consumer–resource interactions by changing the parameters of the functional response connecting species pairs (Bolnick et al., 2011; Gibert and Brassil, 2014; Schreiber et al., 2011; also see Doebeli, 1996 in an evolutionary context). In particular, increasing individual variation in attack efficiency (or attack rate) and handling time decreases interaction strengths, which in turn increases species persistence and stability (Gibert and Brassil, 2014). Together, these results underscore the need to understand how individual-level phenotypic variation affects ecological processes and, through that, the structure and dynamics of entire communities.

Interaction strengths can be influenced by 'mutual' interference competition among predators by dampening resource uptake at higher consumer densities (Arditi et al., 2004). Therefore, mutual interference is thought to stabilize the dynamics of consumer–resource interactions (Arditi et al., 2004; DeLong and Vasseur, 2011, 2013; Forrester et al., 2006). Interference is often thought to occur through behavioural mechanisms associated with territoriality and aggressiveness (Connell, 1961; Forrester et al., 2006; Kennedy and White, 1996), but more generally, interference competition is any form of interaction among consumers that inhibits foraging. Because interference is widespread among many different taxa, it may play an important role in stabilizing natural communities (DeLong and Vasseur, 2011, 2013; Skalski and Gilliam, 2001).

The parameters of the functional response, including mutual interference, are driven by organism traits, and these traits may influence more than one parameter at a time. For example, body size influences both attack efficiency and handling time in several taxa (DeLong and Vasseur, 2012a). Also, variation in different parameters can have opposite effects on foraging rates (Bolnick et al., 2011), so it may be important to link variation in underlying controlling traits to multiple parameters simultaneously (Gibert and Brassil, 2014; Pettorelli et al., 2011). Recently, a positive trait-based link between attack efficiency and mutual interference was discovered for predatory protists, where predator velocity was thought to increase the magnitude of attack efficiency and interference competition simultaneously (DeLong and Vasseur, 2013). Thus, while increasing individual variation can increase stability by lowering interaction strengths through attack efficiency, individual variation might also lower interference competition, potentially undermining the overall stabilizing effect. Because of this, the challenge now is to understand how individual variation in both mutual interference and attack efficiency influences the fate of interacting populations among natural communities.

Our goal is to extend recent work about how individual variation alters consumer–resource dynamics by studying its impact upon linked ecological attributes such as attack efficiency, handling time and interference competition. Schreiber et al. (2011) explored the effect of individual-level heritable variation in attack efficiencies in eco-evolutionary dynamics, while Gibert and Brassil (2014) explored the simultaneous effect of non-heritable variation in the attack efficiency and the handling time of a consumer–resource system. Here, we incorporate non-heritable individual variation in mutual interference by taking into account its functional relationship with attack efficiency (DeLong and Vasseur, 2013) and then we assess its effect upon the persistence and competitive ability of consumer–resource systems.

2. METHODS

2.1 The general model

To include mutual interference among consumers, we used a Rosenzweig–MacArthur consumer–resource model with a Hassell–Varley functional response (Hassell and Varley, 1969; Rosenzweig and MacArthur, 1963). The Hassell–Varley functional response introduces interference competition as a negative exponent, m, on the consumer density in both the numerator and denominator of the functional response (e.g. Arditi and Akçakaya, 1990; DeLong and Vasseur, 2011; Hassell and Varley, 1969). The dynamic model is thus:

$$\frac{dR}{dt} = rR\left(1 - \frac{R}{K}\right) - C\frac{\alpha RC^m}{1 + \alpha\eta RC^m},$$

$$\frac{dC}{dt} = \varepsilon C\frac{\alpha RC^m}{1 + \alpha\eta RC^m} - \beta C \tag{1}$$

where r is the maximal growth rate of the prey, K its carrying capacity, ε is the conversion efficiency, β is the mortality rate of the consumer, α its attack efficiency, η its handling time and m is the parameter that represents interference competition.

If $m = 0$, the model reduces to the classic Rosenzweig–MacArthur formulation, and if $m = -1$, it reduces to the ratio-dependent formulation (e.g. Arditi and Ginzburg, 1989). The level of interference, m, varies continuously in nature from 0 to -2.5, although it frequently takes intermediate values (Abrams and Ginzburg, 2000; DeLong and Vasseur, 2011, 2013). In the case of the predatory protist *Didinium nasutum* preying upon *Paramecium aurelia*, the magnitude of m is linked to that of attack efficiency (α) by:

$$m = -0.26 \ln(\alpha) - 0.67, \tag{2}$$

which was determined by estimating the functional response of the consumer across replicate populations (DeLong and Vasseur, 2013) (Fig. 1A). This relationship will later be used to introduce individual variation in interference.

2.2 Individual variation

Following previous theoretical studies we incorporated individual variation by assuming that both attack efficiency and handling time depend on the value of a normally distributed trait (Gibert and Brassil, 2014; Rall et al., 2012; Schreiber et al., 2011), x, with mean $\bar{x}$, variance σ^2 and probability density:

$$p(x, \bar{x}) = \frac{1}{\sqrt{2\pi\sigma^2}} \exp\left[-\frac{(x-\bar{x})^2}{2\sigma^2}\right]. \tag{3}$$

We assumed that the consumer's attack efficiency $\alpha(x)$ is maximal at a given optimal trait value $x = \theta_\alpha$ and decreases away from that maximum as:

$$\alpha(x) = \alpha_{\max} \exp\left[-\frac{(x-\theta_\alpha)^2}{2\tau^2}\right], \tag{4}$$

where $\alpha_{\max}$ is the maximal attack efficiency and τ^2 determines how steeply the attack efficiency declines away from θ_α (Fig. 1B). The handling time, $\eta(x)$, was assumed to be minimal at a given optimal value $x = \theta_\eta$ and to increase away from that minimum as:

$$\eta(x) = \eta_{\max} - (\eta_{\max} - \eta_{\min}) \exp\left[-\frac{(x-\theta_\eta)^2}{2\nu^2}\right], \tag{5}$$

where $\eta_{\max}$ and $\eta_{\min}$ are the maximal and minimal handling time, respectively, and ν^2 determines how steeply the handling time increases away from θ_η (Fig. 1C).

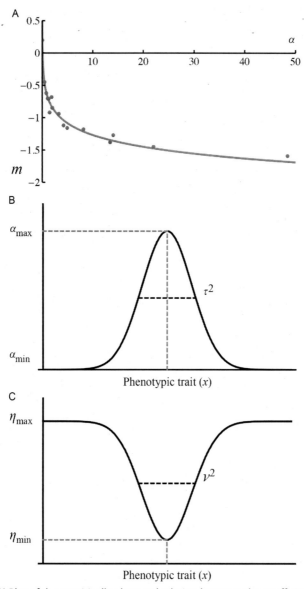

Figure 1 (A) Plot of the empirically observed relation between the coefficient of mutual interference (m) and the attack efficiency (α) in replicate populations of *Didinium nasutum* preying upon *Paramecium aurelia*. As attack efficiency increases, mutual interference becomes stronger. (B) Plot of the assumed relation between the attack efficiency (α) and the underlying phenotypic trait (x). (C) Plot of the assumed relation between the handling time (η) and the underlying phenotypic trait. *Panel (A): modified from DeLong and Vasseur (2013); Panels (B) and (C): modified from Gibert and Brassil (2014).*

We defined the quantities $d_\alpha^2 = (\bar{x} - \theta_\alpha)^2$ and $d_\eta^2 = (\bar{x} - \theta_\eta)^2$ as the distance between the mean trait in the population and the optimal value at which attack efficiency is maximal and handling time is minimal (referred to as phenotypic mismatch; see Raimundo et al., 2014; Schreiber et al., 2011 for similar definitions). Because the optimal value is set by past and existing selective pressures (Anderson et al., 2010), the phenotypic mismatch can be seen as a measure of how well adapted the consumer species is at attacking and handling a particular resource (Gibert and Brassil, 2014). The larger the mismatch is, the smaller the attack rate and the larger the handling time.

We explored three scenarios. We first recapitulated some of the results of Gibert and Brassil (2014) as a baseline for comparison, by including variation in attack efficiency and handling time only. Second, we included only variation in mutual interference, and, third, we included individual variation in all three parameters simultaneously. For the first scenario (variation in attack efficiency and handling time), the consumer–resource model is:

$$
\begin{aligned}
\frac{dR}{dt} &= rR\left(1 - \frac{R}{K}\right) - C\int_{-\infty}^{+\infty} \frac{\alpha(x)RC^m}{1 + \alpha(x)\eta(x)RC^m} p(x, \bar{x})\,dx, \\
\frac{dC}{dt} &= \varepsilon C\int_{-\infty}^{+\infty} \frac{\alpha(x)RC^m}{1 + \alpha(x)\eta(x)RC^m} p(x, \bar{x})\,dx - \beta C
\end{aligned}
\tag{6}
$$

where m is constant. For the second scenario (variation in interference only), the model is:

$$
\begin{aligned}
\frac{dR}{dt} &= rR\left(1 - \frac{R}{K}\right) - C\int_{-\infty}^{+\infty} \frac{\alpha RC^{m(\alpha(x))}}{1 + \alpha\eta RC^{m(\alpha(x))}} p(x, \bar{x})\,dx, \\
\frac{dC}{dt} &= \varepsilon C\int_{-\infty}^{+\infty} \frac{\alpha RC^{m(\alpha(x))}}{1 + \alpha\eta RC^{m(\alpha(x))}} p(x, \bar{x})\,dx - \beta C
\end{aligned}
\tag{7}
$$

where $m(\alpha(x)) = -0.26\ln(\alpha(x)) - 0.67$ and all other parameters are as in Eq. (1). Notice that in this model, α is only allowed to change with variation in the underlying trait x inside function $m(\alpha(x))$, but not outside it. Because this is not realistic, we only do it as a way of understanding variation in mutual interference alone while acknowledging that variation ought to be considered in multiple parameters at the same time. This leads to the third scenario, where variation is now being considered in all three parameters simultaneously (variation in attack efficiency, handling time and interference):

$$\frac{\mathrm{d}R}{\mathrm{d}t} = rR\left(1 - \frac{R}{K}\right) - C\int_{-\infty}^{+\infty} \frac{\alpha(x)RC^{m(\alpha(x))}}{1 + \alpha(x)\eta(x)RC^{m(\alpha(x))}} p(x, \bar{x})\,\mathrm{d}x$$

$$\frac{\mathrm{d}C}{\mathrm{d}t} = \varepsilon C\int_{-\infty}^{+\infty} \frac{\alpha(x)RC^{m(\alpha(x))}}{1 + \alpha(x)\eta(x)RC^{m(\alpha(x))}} p(x, \bar{x})\,\mathrm{d}x - \beta C. \tag{8}$$

We analysed these three scenarios for varying levels of phenotypic mismatch using intermediate values for the maximal attack efficiency and mutual interference, as this combination of parameters is thought to be the most likely in nature (DeLong and Vasseur, 2013). Considering different combination of parameters does not qualitatively affect our results.

The objective of our simulations was to assess the effect of individual variation on the equilibrium of the system (i.e. the intersection of the consumer and resource isoclines). Because of the way we incorporated individual variation in Eqs. (6)–(8), solving for these isoclines (the conditions at which dR/d$t = 0$ for the prey isocline and the conditions at which dC/d$t = 0$ for the predator isocline) and their intersection is now impossible analytically, so it was done numerically. The farther away this equilibrium is from the axes, the less likely consumers and resources are to go extinct due to random fluctuations. Finally, to assess the effect of variation upon community structure, we investigated its effect upon the persistence of consumers through their equilibrium density, C^*, as well as their competitive ability, through the equilibrium density of the resource, R^*. Low equilibrium resource densities (R^*) are associated with strong competitive ability of the consumers (Tilman, 1982, 1986). We therefore define the quantity $1/R^*$ as a measure of competitive ability: the larger the quantity, the larger the competitive ability of the consumer and vice versa.

3. RESULTS

Overall, individual variation can have a strong effect on equilibrium densities and species persistence when interference competition is considered. The effect of individual variation on interference competition depends on the levels of phenotypic mismatch in the trait that controls the consumer–resource interaction (Fig. 2). This effect seems to be mediated mainly by the interplay between attack efficiency and interference competition and ultimately affects the equilibrium densities of the interacting pair, resulting in differential persistence and competitive ability for the consumer at different levels of individual variation (Fig. 3).

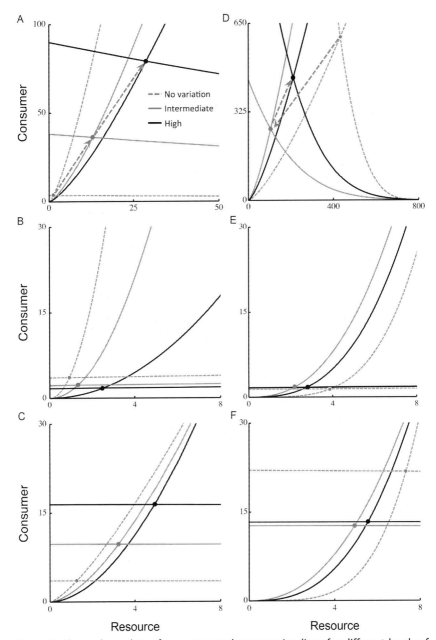

Figure 2 Phase-plane plots of consumer and resource isoclines for different levels of individual variation where the isoclines (values at which a species does not grow or decline) for consumers and resources are represented for different levels of individual variation. The intersection of the isoclines marks the equilibrium densities. Panels in the

(Continued)

3.1 Low phenotypic mismatch

When phenotypic mismatch is low ($d_\alpha \sim 0$ and $d_\eta \sim 0$), individual variation in attack efficiency and handling time increases equilibrium densities of both the consumer and the resource, moving them away from extinction thresholds (Fig. 2A). By doing so, individual variation potentially increases species persistence, as extinction due to demographic stochasticity is less likely to occur.

When it occurs only in interference, individual variation reduces the equilibrium density of the consumer but increases that of the resource (Fig. 2B). This makes consumers simultaneously less competitive due to a high R^* and more prone to extinction due to a low C^*. The change in equilibrium abundance for a given change in individual variation, however, is less pronounced than that observed when variation in both attack efficiency and handling time is considered (Fig. 2A and B).

The net effect of individual variation in interference competition, attack efficiency and handling time combined is intermediate to the effect produced when individual variation is included only in interference competition or in both the attack efficiency and the handling time. This is because the effects are opposite of each other. Individual variation increases the equilibrium density of consumers and resources, moving them away from the extinction threshold (Fig. 2C), which is qualitatively different from what happened when variation only in interference was included (Fig. 2B). However, this effect is also less pronounced than in a scenario with variation only in attack efficiency and handling time (notice the magnitude of the change in

---

Figure 2—Cont'd   left column refer to cases with low phenotypic mismatch, and panels in the right column to cases with large phenotypic mismatch. For the panels in the top row, individual variation was only considered in attack efficiency and handling time. In the second row, individual variation in interference competition only was considered. In the third row, individual variation in attack efficiency, handling time and interference is included. Variation in attack efficiency and handling time increases equilibrium densities (intersection moves away from axes) whenever mismatch is small, and decreases then increases densities whenever mismatch is large. Variation in mutual interference results in a small effect. The latter explains why variation in attack efficiency, handling time and interference results in a tempered version of the first case. Parameter values kept constant across all plots: $\alpha_{max} = 1.38$, $\eta_{max} = 0.08$, $\eta_{min} = 0$, $e = 0.15$, $r = 1.9$, $K = 841$, $\beta = 0.1$, $\tau = 1$, $\nu = 1$, $d_\eta^2 = 0$. Parameters that changed: (A) $d_\alpha^2 = 0$, $\sigma^2 = 0$ (grey, dashed), $\sigma^2 = 2.26$ (grey), $\sigma^2 = 14.19$ (black); (D) $d_\alpha^2 = 2$, $\sigma^2 = 0$ (grey, dashed), $\sigma^2 = 1.31509$ (grey), $\sigma^2 = 16.7242$ (black); (B), (C), (E), and (F) as in (A) but for $d_\alpha^2 = 0$ for (B) and (C) and $d_\alpha^2 = 2$ for (E) and (F).

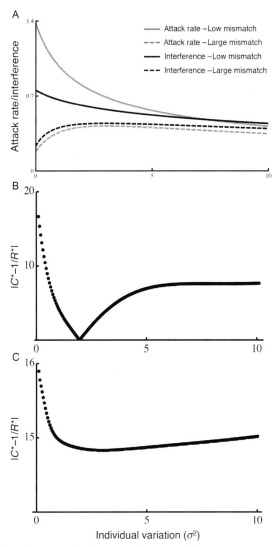

Figure 3 (A) Plot of the attack efficiency (grey) and interference competition (black) against individual variation under low phenotypic mismatch (solid) and large phenotypic mismatch (dashed). Variation decreases attack rates to a larger extent than interference competition when mismatch is small, and the effect on both parameters is comparable when mismatch is large. Parameter values as in Fig. 2. (B) Plot of the absolute value of the difference between consumer equilibrium density, C^*, and consumer competitive ability $1/R^*$, as a function of individual variation under low phenotypic mismatch. Variation maximizes both simultaneously whenever the curve is at its lowest point. (C) Same as in (B) but for large phenotypic mismatch. Parameter values as in Fig. 2.

Fig. 2A) and more similar in magnitude to a scenario with variation only in interference (notice the magnitude of the change in Fig. 2B).

3.2 Large phenotypic mismatch

When phenotypic mismatch is large ($|d_\alpha| \gg 0$) in a scenario with variation in both attack efficiency and handling time, low levels of individual variation decrease equilibrium densities, but high levels of variation increase equilibrium densities for both consumers and prey (Fig. 2D). These changes in equilibrium densities occur much farther away from extinction thresholds than in a scenario with small phenotypic mismatch (Fig. 1A), but are of larger magnitude.

Individual variation in interference decreases the equilibrium density of the resource at first, but it then increases as variation gets larger (Fig. 2E). This practically has no effect on consumer equilibrium densities and the overall effect of variation is comparatively small in magnitude.

The net effect of individual variation in interference competition, attack efficiency and handling time is, again, intermediate to the effect produced in the previous scenarios. Indeed, the densities for both resource and consumers behave as if only variation in attack efficiency and handling time was considered (Fig. 2F and D), but these fluctuations are of a much smaller magnitude, as in a scenario with only variation in interference (Fig. 2F and B).

3.3 Interference, attack efficiency consumer persistence and competitive ability

Individual variation has the same overall effect on interference competition as it has on attack efficiency (Fig. 3). If phenotypic mismatch is low, both attack efficiency and interference competition decrease with individual variation, but the effect seems to be more pronounced on attack efficiency than on interference (Fig. 3A). When phenotypic mismatch is large, however, both attack efficiency and interference increase with variation at first, and then decrease (Fig. 3A). The magnitude of this effect is comparable for both parameters.

Because variation on attack efficiency and interference alters the equilibrium densities of both consumers and resources (Fig. 2), it will ultimately affect consumer persistence as well as its overall competitive ability. For the full model (Eq. 8), when phenotypic mismatch is low, the consumer equilibrium density, C^*, increases with variation but its competitive ability, measured as $1/R^*$, decreases. Because variation maximizes C^* and $1/R^*$ simultaneously whenever the absolute value of that difference is small, our results suggest that intermediate levels of variation maximize the consumer's ability to persist (C^*) and to compete ($1/R^*$) (Fig. 3B). When phenotypic mismatch is large, the consumer equilibrium density decreases with individual variation at first and then increases slowly. Consumer competitive ability, however, increases with variation and then decreases. Interestingly, intermediate levels of variation maximize the consumer's ability to persist and to compete (Fig. 3C), despite the larger phenotypic mismatch.

4. DISCUSSION
4.1 Variation and interference

Individual variation in traits controlling ecological attributes such as attack efficiency and handling time can increase species persistence in consumer–resource interactions (Bolnick et al., 2011; Gibert and Brassil, 2014). In classic consumer–resource models, an increase in the attack efficiency increases interaction strengths, resulting in a decrease of species persistence and overall stability (e.g. Rosenzweig and MacArthur, 1963). Individual variation weakens interaction strengths by decreasing attack efficiencies, which in turn increases species persistence and stability (Gibert and Brassil, 2014). Our results suggest that this effect also occurs when consumer interference is considered. Interference is generally stabilizing (Ginzburg and Jensen, 2008), so it might be expected that individual variation in interference alone could have destabilizing effects, potentially leading to species extinctions. However, it seems to either decrease consumer equilibrium densities and increase resource equilibrium densities, or have negligible effects on both. When we consider variation in attack efficiency and link that to mutual interference through their empirically determined negative relationship (Fig. 1A), the net effect of variation is to increase species persistence. This may be due to a larger effect of individual variation in attack efficiency than in interference that would overcome the negative effect of variation in interference only. These results highlight the importance of considering possible functional relationships between dynamic parameters such as attack efficiency and interference as well as the importance of considering individual variation in the traits controlling these parameters in order to fully understand population dynamics and stability (DeLong and Vasseur, 2012b, 2013; Yodzis and Innes, 1992).

4.2 Variation and competitive ability

Our results also have important consequences for understanding community assembly. If phenotypic mismatch is low, the equilibrium resource density increases with individual variation, which decreases consumer competitive ability. If phenotypic mismatch is large, some variation can reduce resource density at first, momentarily increasing competitive ability. However, large phenotypic mismatch generally decreases competitive ability regardless of variation, meaning that poorly adapted species are in general poor competitors and populations that are already well adapted to their niche become less competitive when they become more variable. In the case of the *Didinium–Paramecium* system, after which our model is parameterized, if interference is too large, consumer uptake is heavily impaired, resulting in deterministic extinction (DeLong and Vasseur, 2013). If interference is low, however, equilibrium densities increase up to a point where the competitive ability of the populations is reduced (DeLong and Vasseur, 2013; Tilman, 1982, 1986). A similar rule might apply to individual variation when it affects both attack efficiency and interference. If variation is too small, populations are close to their extinction threshold. If variation is too large, their equilibrium densities increase to a point where it may impair their competitive abilities, threatening their persistence in the community. Variation in interference thus seem to counter the effect of variation in attack efficiency, naturally leading to the existence of an intermediate amount of variation that both minimizes the chance of extinction and maximizes the competitive ability of populations in a community.

4.3 Eco-evolutionary feedbacks

Individual variation can be important for ecological dynamics, but it also is the raw material upon which natural selection acts (Dobzhansky, 1937). In addition, evolutionary processes have been increasingly recognized to occur at ecological timescales, altering ecological processes and dynamics as they unfold (Grant and Grant, 2002; Hairston et al., 2005; Thompson, 1998). The interplay between ecological and evolutionary processes, or eco-evolutionary feedbacks, thus needs to be considered in future work. In this sense, individual variation has been recognized to increase species coexistence in eco-evolutionary dynamics (Schreiber et al., 2011; Vasseur et al., 2011), but variation has been assumed to be constant through time. However, phenotypic variation generally scales with mean trait values, a pattern known as Taylor's power law (Taylor, 1961) and prevalent across systems and taxa (DeLong, 2012). Thus, individual variation in a given trait will track the evolution of the mean trait value, potentially leading to changes in community structure due to alterations in competitive ability that are a consequence of changes in individual variation that track the evolution of underlying traits. This makes it paramount to also track the evolution of variation over time to understand eco-evolutionary and the stability and persistence of ecological systems in nature.

The effect of individual variation may also depend on the strength of selection acting on the traits that control the consumer–resource interaction (Gibert and Brassil, 2014; Yoshida et al., 2003). Strong stabilizing selection may reduce individual variation through time, with consequences for population stability and competitive ability. Unstable and uncompetitive populations will not fare well in communities, which implies that selection that reduces variation and increases mean fitness within populations may have negative effects for the population in a community. Populations may thus be the subject of antagonistic effects of natural selection (Raimundo et al., 2014). Together, this suggests that the interplay between ecological and evolutionary processes is central to understanding how communities are structured in nature (Bolnick et al., 2011; Fontaine et al., 2011; Guimarães et al., 2011; Thompson, 2005). Individual variation may be the key to bridging the gap between ecology and evolution.

4.4 Underlying controlling traits

Considering what traits operate as 'controlling' traits that influence parameters such as attack efficiency, handling time and interference is also important. For instance, the amount of variation in the controlling trait is linked to mutation rates, the amount of phenotypic plasticity and the strength of selective forces operating on the trait. Thus, by identifying probable controlling traits, we may have a deeper grasp of the processes controlling the variation ultimately affecting consumer–resource dynamics. It is possible that some traits, such as body size, might act as ecological 'magic traits'. In an evolutionary context, magic traits are involved in both mating and ecological activities, and when they experience disruptive selection they can lead to adaptive speciation (Gavrilets, 2004; Raimundo et al., 2014). Ecological magic traits would be traits influencing many dynamic parameters at once (e.g. DeLong and Vasseur, 2012b), while other traits only influence a limited set of parameters, if any. Specific links between traits, their optima and dynamic population parameters are needed to fully understand how individual variation influences consumer–resource dynamics. Identifying such traits and quantifying their distribution and their effect upon ecological processes of interest should be a major goal of ecology in the upcoming future (Gibert and Brassil, 2014; Pettorelli et al., 2011; Violle et al., 2012a,b).

4.5 Testable predictions from the theory of individual variation

To help move toward that goal, we can make some simple testable predictions as to how individual variation can affect interaction strengths in a system with interference competition. If the per-capita foraging rate of a consumer preying upon a resource is:

$$f(R, C) = \frac{\alpha R C^m}{1 + \alpha \eta R C^m}, \tag{9}$$

then, we can find an expression for the average foraging rate that explicitly depends upon individual variation by integrating over the functional response and the underlying trait distribution. We thus get:

$$\overline{f(R, C)} = \int_{-\infty}^{+\infty} \frac{\alpha(x) R C^m}{1 + \alpha(x)\eta(x) R C^m} p(x, \overline{x}) \, dx, \tag{10}$$

where m can be a function of the attack efficiency or a constant. In this case, we can see that while under some conditions increasing individual variation reduces foraging rates and thus, interaction strengths, this effect increases with resource density (Fig. 4A) and decreases with consumer density (Fig. 4B and C). These predictions can be tested in foraging experiments where the resource and consumer densities are manipulated in the same way it would be done for quantifying the parameter of mutual interference, m (DeLong and Vasseur, 2013). A measure of individual variation across treatments and one or several traits such as body size would need to be quantified as well. The latter is particularly doable in microcosms with protists (DeLong, 2012; DeLong and Vasseur, 2013) or mesocosm experiments with metazoan grazers and algae (Fussmann et al., 2003; Yoshida et al., 2003, 2004).

5. CONCLUSION

Because of their effect on population persistence and stability, understanding the interplay between individual variation and interference competition is central in ecology. Using dynamic models that explicitly take into account individual variation, we have shown that increasing individual

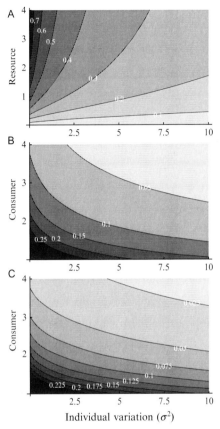

Figure 4 Plots of the effect of individual variation and either resource (A) or consumer density ((A) and (B): foraging rate as in Eq. 10, (C): foraging rate as in Eq. 8)) upon foraging rates (grey scale). Individual variation and consumer and resource densities have a joint effect upon foraging rates and should thus not be studied separately: foraging rates increase with resource density, decrease with consumer density and decrease with individual variation. Parameter values as in Fig. 2. R and C where kept constant and equal to 1 whenever the other quantity was varied.

variation simultaneously affecting attack efficiency, handling time and mutual interference can increase species persistence and stability as well as consumer competitive ability. Moreover, as variation is affected by selection, we argue that evolutionary processes may deeply affect the way communities are structured. Finally, our results underscore the need for comprehensive studies that quantify the level of individual variation in natural populations, making specific testable hypotheses as to how individual variation can interact with resource and consumer densities to alter foraging rates and through that, interaction strengths.

ACKNOWLEDGEMENTS

J.P.G. was supported by the School of Biological Sciences Special Funds and an Othmer Fellowship. We thank two anonymous reviewers for insightful suggestions.

REFERENCES

Abrams, P.A., Ginzburg, L., 2000. The nature of predation: prey dependent, ratio dependent or neither? Trends Ecol. Evol. 15, 337–341.

Abrams, P.A., Matsuda, H., 2004. Consequences of behavioral dynamics for the population dynamics of predator–prey systems with switching. Popul. Ecol. 46, 13–25.

Allesina, S., Tang, S., 2012. Stability criteria for complex ecosystems. Nature 483, 205–208.

Anderson, B., Terblanche, J.S., Ellis, A.G., 2010. Predictable patterns of trait mismatches between interacting plants and insects. BMC Evol. Biol. 10, 204.

Arditi, R., Akçakaya, H.R., 1990. Underestimation of mutual interference of predators. Oecologia 83, 358–361.

Arditi, R., Ginzburg, L.R., 1989. Coupling in predator–prey dynamics: ratio-dependence. J. Theor. Biol. 139, 311–326.

Arditi, R., Callois, J.-M., Tyutyunov, Y., Jost, C., 2004. Does mutual interference always stabilize predator–prey dynamics? A comparison of models. C.R. Biol. 327, 1037–1057.

Bolnick, D.I., Svanbäck, R., Fordyce, J.A., Yang, L.H., Davis, J.M., Hulsey, C.D., Forister, M.L., 2003. The ecology of individuals: incidence and implications of individual specialization. Am. Nat. 161, 1–28.

Bolnick, D.I., Amarasekare, P., Araújo, M.S., Bürger, R., Levine, J.M., Novak, M., Rudolf, V.H.W., Schreiber, S.J., Urban, M.C., Vasseur, D.A., 2011. Why intraspecific trait variation matters in community ecology. Trends Ecol. Evol. 26, 183–192.

Connell, J.H., 1961. The influence of interspecific competition and other factors on the distribution of the barnacle Chthamalus stellatus. Ecology 42, 710–723.

Dell, A.I., Pawar, S., Savage, V.M., 2014. Temperature dependence of trophic interactions are driven by asymmetry of species responses and foraging strategy. J. Anim. Ecol. 83, 70–84.

DeLong, J.P., 2012. Experimental demonstration of a "rate–size" trade-off governing body size optimization. Evol. Ecol. Res. 14, 343–352.

DeLong, J.P., 2014. The body-size dependence of mutual interference. Biol. Lett. 10, 20140261.

DeLong, J.P., Vasseur, D.A., 2011. Mutual interference is common and mostly intermediate in magnitude. BMC Ecol. 11 (1), 8.

DeLong, J.P., Vasseur, D.A., 2012a. A dynamic explanation of size–density scaling in carnivores. Ecology 93, 470–476.

DeLong, J.P., Vasseur, D.A., 2012b. Size-density scaling in protists and the links between consumer-resource interaction parameters. J. Anim. Ecol. 81, 1193–1201.

DeLong, J.P., Vasseur, D.A., 2013. Linked exploitation and interference competition drives the variable behavior of a classic predator–prey system. Oikos 122, 1393–1400.

Dobzhansky, T., 1937. Genetics and the Origin of Species. Columbia University Press, New York, p. 364.

Doebeli, M., 1996. An explicit genetic model for ecological character displacement. Ecology 77, 510–520.

Duffy, M.A., 2010. Ecological consequences of intraspecific variation in lake Daphnia. Freshw. Biol. 55, 995–1004.

Estes, J.A., Riedman, M.L., Staedler, M.M., Tinker, M.T., Lyon, B.E., 2003. Individual variation in prey selection by sea otters: patterns, causes and implications. J. Anim. Ecol. 72, 144–155.

Fontaine, C., Guimarães Jr., P.R., Kéfi, S., Loeuille, N., Memmott, J., van der Putten, W.H., van Veen, F.J.F., Thébault, E., 2011. The ecological and evolutionary implications of merging different types of networks. Ecol. Lett. 14, 1170–1181.

Forrester, G.E., Evans, B., Steele, M.A., Vance, R.R., 2006. Assessing the magnitude of intra- and interspecific competition in two coral reef fishes. Oecologia 148, 632–640.

Fussmann, G.F., Ellner, S.P., Hairston Jr., N.G., 2003. Evolution as a critical component of plankton dynamics. Proc. Biol. Sci. 270, 1015–1022.

Gavrilets, S., 2004. Fitness Landscapes and the Origin of Species. Princeton University Press, Princeton.

Gibert, J.P., Brassil, C.E., 2014. Individual phenotypic variation reduces interaction strengths in a consumer-resource system. Ecol. Evol. 4, 3703–3713.

Gibert, J.P., DeLong, J.P., 2014. Temperature alters food web body-size structure. Biol. Lett. 10, 20140573.

Ginzburg, L.R., Jensen, C.X.J., 2008. From controversy to consensus: the indirect interference functional response. Verh. Int. Ver. Limnol. 30, 297–301.

Grant, P.R., Grant, B.R., 2002. Unpredictable evolution in a 30-year study of Darwin's finches. Science 296, 707–711.

Guimarães Jr., P.R., Jordano, P., Thompson, J.N., 2011. Evolution and coevolution in mutualistic networks. Ecol. Lett. 14, 877–885.

Hairston Jr., N.G., Ellner, S.P., Geber, M.A., Yoshida, T., Fox, J.A., 2005. Rapid evolution and the convergence of ecological and evolutionary time. Ecol. Lett. 8, 1114–1127.

Hammill, E., Petchey, O.L., Anholt, B.R., 2010. Predator functional response changed by induced defenses in prey. Am. Nat. 176, 723–731.

Hassell, M.P., Varley, G.C., 1969. New inductive population model for parasites and its bearing on biological control. Nature 223, 1133–1137.

Inouye, B.D., 2005. The importance of the variance around the mean effect size of ecological processes: comment. Ecology 86, 262–265.

Kennedy, E.D., White, D.W., 1996. Interference competition from house wrens as a factor in the decline of bewick's wrens. Conserv. Biol. 10, 281–284.

Lankau, R.A., Strauss, S.Y., 2007. Mutual feedbacks maintain both genetic and species diversity in a plant community. Science 317, 1561–1563.

Lomnicki, A., 1988. Population Ecology of Individuals. Princeton University Press, Princeton.

May, R.M., 1972. Will a large complex system be stable? Nature 238, 413–414.

May, R.M., 1973. Qualitative stability in model ecosystems. Ecology 54, 638–641.

McCann, K.S., Hastings, A., Huxel, G.R., 1998. Weak trophic interactions and the balance of nature. Nature 395, 794–798.

O'Connor, M.I., 2009. Warming strengthens an herbivore-plant interaction. Ecology 90, 388–398.

Pawar, S., Dell, A.I., Savage, V.M., 2012. Dimensionality of consumer search space drives trophic interaction strengths. Nature 486, 485–489.

Pettorelli, N., Coulson, T., Durant, S.M., Gaillard, J.-M., 2011. Predation, individual variability and vertebrate population dynamics. Oecologia 167, 305–314.

Raimundo, R.L.G., Gibert, J.P., Hembry, D.H., Guimarães, P.R., 2014. Conflicting selection in the course of adaptive diversification: the interplay between mutualism and intraspecific competition. Am. Nat. 183, 363–375.

Rall, B.C., Brose, U., Hartvig, M., Kalinkat, G., Schwarzmüller, F., Vucic-Pestic, O., Petchey, O.L., 2012. Universal temperature and body-mass scaling of feeding rates. Philos. Trans. R. Soc. Lond. B Biol. Sci. 367, 2923–2934.

Rosenzweig, M.L., MacArthur, R.H., 1963. Graphical representation and stability conditions of predator–prey interactions. Am. Nat. 97, 209–223.

Schmitz, O.J., Beckerman, A.P., O'Brien, K.M., 1997. Behaviorally mediated trophic cascades: effects of predation risk on food web interactions. Ecology 78, 1388–1399.

Schreiber, S.J., Bürger, R., Bolnick, D.I., 2011. The community effects of phenotypic and genetic variation within a predator population. Ecology 92, 1582–1593.

Skalski, G.T., Gilliam, J.F., 2001. Functional responses with predator interference: viable alternatives to the holling type II model. Ecology 82, 3083–3092.

Taylor, L.R., 1961. Aggregation, variance and the mean. Nature 189, 732–735.

Thompson, J.N., 1998. Rapid evolution as an ecological process. Trends Ecol. Evol. 13, 329–332.

Thompson, J.N., 2005. The Geographic Mosaic Theory of Coevolution. The University of Chicago Press, Chicago, p. 443.

Tilman, D., 1982. Resource Competition and Community Structure. Princeton University Press, New Haven, p. 296.

Tilman, D., 1986. Resources, competition and the dynamics of plant communities. In: Crawley, M.J. (Ed.), Plant Ecology. Blackwell scientific publications, Oxford, pp. 51–75.

Tinker, M.T., Guimarães Jr., P.R., Novak, M., Marquitti, F.M.D., Bodkin, J.L., Staedler, M., Bentall, G., Estes, J.A., 2012. Structure and mechanism of diet specialisation: testing models of individual variation in resource use with sea otters. Ecol. Lett. 15, 475–483.

Vasseur, D.A., Amarasekare, P., Rudolf, V.H.W., Levine, J.M., 2011. Eco-evolutionary dynamics enable coexistence via neighbor-dependent selection. Am. Nat. 178, E96–E109.

Violle, C., Enquist, B.J., McGill, B.J., Jiang, L., Albert, C.H., Hulshof, C., Jung, V., Messier, J., 2012a. The return of the variance: intraspecific variability in community ecology. Trends Ecol. Evol. 27, 244–252.

Violle, C., Enquist, B.J., McGill, B.J., Jiang, L., Albert, C.H., Hulshof, C., Jung, V., Messier, J., 2012b. Viva la variance! a reply to Nakagawa & Schielzeth. Trends Ecol. Evol. 27, 475–476.

Vucic-Pestic, O., Rall, B.C., Kalinkat, G., Brose, U., 2010. Allometric functional response model: body masses constrain interaction strengths. J. Anim. Ecol. 79, 249–256.

Yodzis, P., Innes, S., 1992. Body size and consumer-resource dynamics. Am. Nat. 139, 1151–1175.

Yoshida, T., Jones, L.E., Ellner, S.P., Fussmann, G.F., Hairston Jr., N.G., 2003. Rapid evolution drives ecological dynamics in a predator–prey system. Nature 424, 303–306.

Yoshida, T., Hairston Jr., N.G., Ellner, S.P., 2004. Evolutionary trade-off between defence against grazing and competitive ability in a simple unicellular alga, Chlorella vulgaris. Proc. Biol. Sci. 271, 1947–1953.

CHAPTER FOUR

Predictors of Individual Variation in Movement in a Natural Population of Threespine Stickleback (*Gasterosteus aculeatus*)

Kate L. Laskowski[1,2,3], Simon Pearish[2], Miles Bensky, Alison M. Bell

School of Integrative Biology, University of Illinois, Champaign, Illinois, USA
[3]Corresponding author: e-mail address: kate.laskowski@gmail.com

Contents

Abstract

Species abundances and distributions are inherently tied to individuals' decisions about movement within their habitat. Therefore, integrating individual phenotypic variation within a larger ecological framework may provide better insight into how populations structure themselves. Recent evidence for consistent individual differences in behaviour prompts the hypothesis that variation in behavioural types might be related to variation in movement in natural environments. In a multiyear mark–recapture study, we found that individual sticklebacks exhibited consistent individual differences in behaviour both

---

[1] Current address: Department of Biology & Ecology of Fishes, Leibniz Institute of Freshwater Ecology and Inland Fisheries, Berlin, Germany.
[2] These authors contributed equally.

Advances in Ecological Research, Volume 52
ISSN 0065-2504
http://dx.doi.org/10.1016/bs.aecr.2015.01.004

within a standardized testing arena designed to measure exploratory behaviour and within a river. Therefore, we asked whether individual differences in movement in a natural river were related to an individual's exploratory behavioural type. We also considered whether body condition and/or the individual's habitat or social environment use was related to movement. There was no evidence that an individual's exploratory behavioural type was related to movement within the river. Instead, an individual's habitat use and body condition interacted to influence natural movement patterns. Individuals in good condition were more likely to move further in the river, but only if they inhabited a vegetated complex part of the river; body condition had no influence on movement in those individuals inhabiting open areas of the river. Our results suggest that individual traits could help improve predictions about how populations may distribute themselves within patchy and complex environments.

1. INTRODUCTION

Understanding how populations structure themselves is a fundamental goal of ecology. Population processes are largely mediated by the individual animals within populations. For example, individual activity and movement patterns influence population abundance and distribution (Clobert et al., 2001). Intraspecific variation in traits related to these processes is therefore inherent to the study of population dynamics, yet classic ecological theory frequently simplifies this variation by describing populations by average traits. Recently, there has been a call for explicit consideration of individual variation in ecologically relevant traits in population and community ecology (Bolnick et al., 2011; Dall et al., 2012; Violle et al., 2012). Individual traits are broadly defined as any 'measurable feature of an individual organism, including size, morphology, behaviour and physiology' (Bolnick et al., 2011). Here, we consider traits to be any feature of an individual that is measurable and repeatable over time. There is growing appreciation that variation in individual traits can influence diverse population-level processes, from generating variation in predation pressures (Miller et al., 2014; Sherratt and Macdougall, 1995) to the non-random dispersal of individuals to novel environments (Cote et al., 2010; Duckworth and Badyaev, 2007). Therefore, there is a growing need to understand whether and how individual trait variation may scale up to influence larger scale dynamics in populations and communities.

One growing area of research on intraspecific variation is the study of animal personality. Consistent individual differences in behaviour, or behavioural types, are widespread across taxa (Bell et al., 2009), influenced by important ecological factors, such as predation and competition (Bell and Sih, 2007; Laskowski and Bell, 2013) and may have fitness consequences

(Smith and Blumstein, 2008). In particular, there is evidence that individuals consistently differ in general activity levels and willingness to explore novel and potentially risky environments (Chapman et al., 2011; Cote et al., 2010; Dingemanse et al., 2003; Fraser et al., 2001; Lindström et al., 2013), behaviours which have been hypothesized to influence movement decisions and therefore population structure. These behaviours are frequently measured using standardized open-field assays to rapidly assess individual behavioural types. Behavioural types have been implicated in the spread of invasive species (Phillips and Suarez, 2012) and as a mechanism for species range expansion (Duckworth and Badyaev, 2007). Behaviourally mediated movement patterns are interesting because if certain behavioural types are more likely to move, the individuals that arrive at new resources or habitats will be a particular subset of the population, the most active or exploratory individuals, for example (Cote et al., 2010). Therefore, consistent individual differences in behaviour, such as activity and exploration, hold the potential to help explain species ranges and abundances.

It is likely, though, that movement is influenced not just by behavioural types, but also by other traits (Bowler and Benton, 2005; Clobert et al., 2001; Ronce, 2007). For example, differences in body condition have been related to differences in movement across a wide array of taxa with the general trend being that individuals in high condition move further than low condition individuals (Bonte et al., 2012). Condition is an index that is frequently used to approximate physiological wellness (Pope and Kruse, 2007) and usually does not change abruptly (Robb et al., 1992; Tosh et al., 2010). Biotic and abiotic environmental factors can also influence movement and often differ consistently between individuals. For example, consistent differences in habitat use are well known in threespine stickleback (Bentzen and McPhail, 1984). Those individuals using complex, vegetated habitats might be more likely to move greater distances if the abundance of refuges offer safety from predation (Gilliam and Fraser, 2001; Patrick et al., 2001). Similarly, movement decisions might also be related to the type of social environment that an individual uses (Magnhagen and Bunnefeld, 2009; Ward et al., 2013) and individuals often consistently occur in certain types of social environments (Brown and Brown, 2000; Saltz, 2011). For example, some lizards consistently avoid conspecifics, resulting in their further dispersal from groups (Cote and Clobert, 2007).

Each of these traits (body condition, habitat and social environment use) may by themselves influence on individual movement decisions and potentially interact with each other to influence movement. Body condition and habitat use may both influence movement decisions if individuals in poorer body condition are willing to move between refuges in a complex habitat because it is safe, compared to an open, unstructured environment, for example (Krause et al., 1998). Another possibility is that individuals in poor condition might instead rely on public information provided by the social group, which would limit their need to travel to sample patches (Templeton and Giraldeau, 1996; Valone, 1989; Valone and Templeton, 2002). A better understanding of the relative influence and interaction of these traits on an individual's movement patterns can generate insight into larger scale ecological processes.

Therefore in this chapter, we present the results of two studies designed to answer a series of questions about the causes of intraspecific variation in movement within a natural population of threespine sticklebacks, a species well known for its consistent individual variation in behaviour (Bell, 2005; Dingemanse et al., 2007; Huntingford, 1976; Laskowski and Bell, 2014). This species is also a model for investigating several population-level processes, such as colonization of new habitats (Bell and Foster, 1994) and structured habitat use by particular individuals (Bolnick et al., 2009; Schluter and McPhail, 1992). In the first study, we ask (1) whether individual exploratory behaviour in an open-field arena is repeatable and (2) whether exploratory behaviour in an open-field predicts movement within a stream. In the second study, we expand our study to (3) compare the influence of exploratory behavioural type and other factors (condition, habitat and social environment) on individual movement patterns. Specifically, we compared seven *a priori* models: models with only the effect of either behavioural type, body condition, habitat use or social environment use; a model with the interaction between body condition and habitat use; a model with the interaction between body condition and social environment use; and finally, a null model with no fixed effects. We chose the open-field arena as our standardized behavioural assay because it is designed to measure exploratory behaviour and/or general activity levels, which are plausibly related to movement (Chapman et al., 2011; Cote et al., 2010; Dingemanse et al., 2003; Fraser et al., 2001; Lindström et al., 2013).

2. METHODS

2.1 Overview

The two studies were conducted over 3 years in the Navarro River in Mendocino County, California, USA. We used mark–recapture techniques to test whether exploratory behaviour in a standardized open-field arena predicted individual movement within the river. The study sites were divided into transects. We sampled each transect and measured the exploratory behaviour of each individual in an open-field assay. Then, the individuals were marked, released and recaptured. We estimated movement as the number of transects moved between captures. In the first study (2010 and 2011), individuals were remeasured for exploratory behaviour in the open-field assay after recapture. In the second study (2013), we expanded the scope by considering other traits, such as body condition, habitat use and social environment use as possible predictors of movement.

2.2 Population location and characteristics

The sites for both studies were fairly narrow (4–10 m wide) and shallow (maximum 2 m depth) stretches of the river. The site for Study 1 was characterized by clear water, a stone and pebble bottom with considerable vegetation (azalea and eucalyptus) overhanging the edges. The second study was conducted at a different but nearby (<3 km away) stretch of the river where there was a clearer demarcation between different habitat types. The Navarro River is not dammed and experiences regular seasonal variation in flow and depth. Both studies took place during the summer low point. Small riffles bounded each study site and extensive sampling up and downstream of the riffles recaptured no marked individuals so we assume that dispersal outside of the study sites was negligible. At each site, the river was divided into 15 m transects (total 240 m section of river for each site). All transects were the same size and there were no barriers to movement between any of transects.

2.3 Animal marking and capture techniques

2.3.1 Study 1

At the beginning of the capture period, we arbitrarily chose a transect near the centre of the study section of river where we used six silver wire minnow traps to capture adult sticklebacks (age 1 year). Any sticklebacks that were found in these traps were measured in a standardized open-field assay designed to measure exploratory behaviour (described below). After measuring each individual's exploratory behaviour, we recorded standard length (to the nearest mm) and transect at capture, and then marked each individual with a unique identification using subcutaneous UV elastomer (Northwest Marine Technology, Inc.). Each individual was marked at up to six sites on their dorsal side using a combination of three different colours. The fish were then allowed to recover in a dark, aerated bucket until the end of the day and released at their point of capture.

We repeated this process until all transects had been sampled (one transect per day). We alternated sampling an upstream and a downstream transect each day from our starting transect in order to minimize disturbance in the same area. After all transects had been sampled once, we waited 2 days and then began our recapture efforts following the same schedule (average time between capture and recapture: 8.6 days; range 1–27).

During the recapture period, in addition to using minnow traps, we also seined each transect in an attempt to recapture as many individuals as possible and to avoid bias associated with particular capture methods, e.g., 'trappability' (Biro and Dingemanse, 2009; Wilson et al., 1993). A logistic regression also demonstrated that trapping technique had no effect on probability of recapture (Appendix), so we assume that our recaptured individuals were not biassed towards certain behavioural types. We remeasured the exploratory behaviour of recaptured, marked individuals. After the behavioural assay, we determined each individual's identity and individuals were placed in dark aerated bucket to rest until the end of the day. The fish were then released where captured. Some individuals ($n = 64$ out of 246 recaptured individuals) were recaptured more than once. We counted the number of transects between consecutive captures as our measure of movement, hereafter referred to as 'movement in the river'. For example, an individual that was initially caught in transect 3 and then recaptured in transect 6 was given a score of 3. If that individual was recaptured again in transect 4, it received a score of 2 (transects 6–4) for its second recapture.

2.3.2 Study 2

Each transect in Study 2 included a shallow bank that lacked vegetation (open habitat) and a relatively deep bank that was covered by patches of low hanging tree branches and grasses (cover habitat). These two habitat types were usually separated by a deep central area that stickleback did not occupy. The open habitat was a bed of sand and fine gravel that stretched uninterrupted from the furthest upstream transect to the furthest downstream transect. Stickleback were evenly distributed throughout this habitat type. The cover habitat was dominated by the submerged branches of the trees that lined the deep bank. Stickleback tended to aggregate at tree branches that were separated by as much as 10 m, i.e., the cover habitat was patchier. Thus, stickleback distribution was less evenly distributed across space in the cover habitat compared to the open habitat. Mark–recapture techniques were similar to those employed in Study 1 with a few modifications. We used snorkelling surveys and hand nets to capture individuals in order to assess each individual's habitat and social environment use at the time of capture (Pearish et al., 2013). Starting at the edge of a randomly selected transect, we collected one, haphazardly selected individual at a time with a hand net, alternating between habitats (open vs. cover). Before approaching, we noted the focal individual's social environment use by recording whether the focal fish was alone or in a shoal (<10 cm or four body lengths from another fish; Pitcher, 1993). No fish escaped capture so we assume our sample was not biassed towards more 'catchable' individuals (Biro and Dingemanse, 2009; Wilson et al., 1993). Each individual was placed into a separate 500-mL opaque container of river water and held overnight. The following day, individuals were observed in the standardized open-field assay to measure exploratory behaviour.

Following the behavioural assay, fish were weighed, measured for standard length and given unique markings using fluorescent visible implant elastomer. Each fish was released back into the river at the transect from which it was collected. After all transects were sampled once, we began the recapture period (average time between capture and recapture; 16.6 days; range 7–26 days). We started at the transect furthest downstream and moved methodically up the river using block nets to isolate each transect. This prevented fish from moving up or downstream in response to our activity. We used three methods for recapturing marked individuals (seining, snorkelling and electrofishing) in an effort to avoid biassing recapture towards particular behavioural types. We spent 3 h in each transect.

2.4 Standardized open-field assay to measure exploratory behaviour

Our open-field arena consisted of a 1.5-m diameter circular pool. The pool was divided into nine equally sized sections: eight sections around the outside with one central section. Rocks were placed within each perimeter section to provide cover. The arena was placed in the shade and we monitored the water temperature and changed it as needed to maintain the same temperature throughout the day.

An individual stickleback was placed into an acclimation chamber that was placed in the centre of the open-field arena. The chamber was made of PVC pipe end caps (height 8 cm, diameter 10 cm) with a hole that was plugged with a stopper on one side. After 3 min, we remotely pulled the stopper from the chamber and removed the stopper from the arena.

We recorded latency to emerge from the chamber, the total number of sections crossed, and the number of unique sections (out of 9) that an individual swam through for 3 min after emergence. Results were similar for all three behaviours that were recorded during the open-field assay. Here, we focus on total number of sections as it exhibited the most interindividual variation (results for the other behaviours are in Appendix). We interpret the total number of sections as activity level and/or exploratory behaviour which has been hypothesized to influence natural movement behaviour (e.g., Cote et al., 2010; Fraser et al., 2001). Hereafter, we refer to this as 'exploratory behaviour in the pool'. Trials where the individual did not exit the chamber after 5 min were terminated (occurred in 19% of trials). In Study 2, we increased the cutoff for emergence time from 5 to 10 min in order to maximize the number of individuals that emerged. If an individual did not emerge, we gently poured it from the chamber into the arena (occurred in 3% of trials). In Study 2, we did not remeasure behaviour at recapture as Study 1 indicated behaviour in the open-field arena was significantly repeatable (see Section 3, see also Pearish et al., 2013).

2.5 Data analysis

Exploratory behavioural data were normally distributed. Movement in the field was non-normal so we specified a Poisson error distribution and confirmed that this distribution was a good fit with Q–Q plots. We corrected for the anticonservative nature of the Poisson distribution by using quasi-likelihood estimation of parameters (see below).

2.5.1 Study 1

Our first question was whether there was evidence for consistent individual differences in behaviour, both in the open-field arena and in the river (for those individuals that were recaptured multiple times). To test this, we estimated the repeatability of exploratory behaviour in the pool and movement in the river separately. Because movement in the river was Poisson distributed, we corrected for the estimation of the residual error according to Nakagawa and Schielzeth (2010). We used (generalized) linear mixed models with exploratory behaviour in the pool or river as the response variable and individual (nested within year) as a random effect. Because we were interested in a general characterization of among individual variance in these behaviours, we did not include any other fixed effects. For exploratory behaviour in the pool we used data on all individuals in these models, even those that were not recaptured, as they still provide information about the behavioural variation present in this population (Martin et al., 2011). For movement in the river only, individuals that were recaptured at least once ($N = 246$) were included as we could not estimate movement in the river for nonrecaptured individuals. To determine whether any fixed effects could explain interindividual variation in exploratory behaviour in the pool, we ran a second linear mixed model for exploratory behaviour in the pool including the following fixed effects: time of day, individual standard length, number of capture (first, second, etc., capture) and year. We then reestimated the repeatability of exploratory behaviour in the pool using the variance components from the full model. We used (generalized) linear mixed models with the 'lme' command in the 'nlme' package (for total sections) and the 'glmmPQL' command in the 'MASS' package (distance) in Rv2.15 (http://www.r-project.org/). For exploratory behaviour in the pool, we tested for the significance of the individual effects using a log-likelihood ratio test comparing a model fit with the random individual effect to one without the individual effect. However, because of the nature of movement in the river (Poisson distributed) we could not accurately assess the significance of the individual effect (Nakagawa and Schielzeth, 2010).

Our second question was whether exploratory behaviour in the pool was related to natural movement in the river. To test this, we used a generalized linear mixed model with movement in the river as the response variable and exploratory behaviour in the pool at the previous capture as a predictor. We also included standard length and days since previous capture as predictors. We then reestimated the repeatability of distance moved from this full model to determine if the fixed effects could explain interindividual variation in behaviour in the river. Individual was included as a random effect as some individuals were recaptured more than once. Models were constructed using the 'glmmPQL' command in the lme4 package in R. Additionally, we tested whether exploratory behaviour in the pool predicted the probability of recapture. To test this, we used a binary regression with recapture (yes/no) as the response variable. We included exploratory behaviour in the pool, sex, standard length and year as fixed effects. For this model, we only used behavioural data from the initial capture.

2.5.2 Study 2
In Study 2, we measured exploratory behaviour in the pool as well as several other traits that might influence movement (condition, habitat and social environment at time of capture). These data allowed us to test the relative

importance of different factors on movement within the river. This analysis was conducted in two steps. First, we developed a set of *a priori* models corresponding to the predictions that were developed from our knowledge of the stickleback system and relevant literature (Table 5). Average movement was upstream (Fig. 1) and fish in upstream transects were limited in how far upstream they could travel. To account for this, we used generalized linear mixed models with transect included as a random factor in all models. Models were constructed using the 'lme4' package in R.

We tested seven models corresponding to our six *a priori* models and one model that contained only the random effect of transect (Table 5). Fixed factors included in models were body condition, habitat use, social

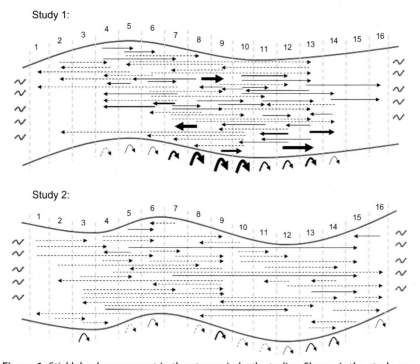

Figure 1 Stickleback movement in the stream in both studies. Shown is the study section of river divided into 15 m transects (vertical grey dashed lines) with a riffle at either end. Transect number is above the river. The direction of the arrowed line indicates the direction in which the fish moved between initial and subsequent captures. The thickness of the arrowed lines indicates the number of individuals that moved between those two transects (dashed line = 1 individual). The curved arrows below the river indicate the number of individuals that were recaptured within that same transect.

environment use and exploratory behaviour in the pool. We used Fulton's K calculated from body measurements taken during the initial capture as the body condition index. Fulton's K is calculated by

$$K = \frac{W}{L^3} \times 100,000$$

where W is weight (g), L is length (mm) and values are multiplied by 100,000 to achieve an index with values close to 1 (Pope and Kruse, 2007). Fulton's K is usually avoided in cases where comparisons across age classes, populations or species are desired but is appropriate for this application as we limited our study to threespine stickleback of a single population. The models that included two fixed factors also contained the interaction term between them (body condition $\times$ habitat use, body condition $\times$ social environment use).

The second step of our analysis was to use multimodel inferencing to compare our *a priori* models and calculated model-averaged estimates. This method is preferred over step-wise model selection because model selection uncertainty is accounted for in the model-averaged parameter estimates (Burnham et al., 2011). The ability of each model to predict movement was assessed using the second order, quasi-Akaike's information criteria (QAICc) to account for overdispersion in the data and small sample size (Burnham and Anderson, 2002). For each model, we calculated the number of parameters estimated (K), QAICc, delta (Δ, difference in QAICc between the focal model and the model with the lowest QAICc), Akaike's weight (w) and a conditional R^2. Models with deltas less than 2 are considered to have 'substantial support' while models with deltas of 4 or more receive 'considerably less support' (Burnham and Anderson 2002). Akaike's weight is the relative probabilities of each model given the data and sums to one over the set of models (Burnham et al., 2011; Johnson and Omland, 2004). Conditional R^2 represents the variance explained by fixed and random factors (Nakagawa and Schielzeth, 2013). When no single model can be specified as the best model, a 95% confidence set can be constructed by summing Akaike's weights from the largest to smallest until the sum is ≥ 0.95. This set can be used to calculate model-averaged estimates and 95% confidence intervals for fixed effects (Burnham and Anderson, 2002). We interpreted effects with estimates with confidence intervals that did not overlap zero as statistically significant. We used the 'MuMIn' package in R for analysis.

3. RESULTS
3.1 Study 1
3.1.1 General characteristics of captured fish
We marked 422 individual sticklebacks in the first study. Of these, we recaptured 246 individuals at least once (Table 1). Most individuals that were recaptured in the same or adjacent transect (distance = 0 or 1, Table 1), though one individual travelled a maximum of nine transects. Table 2 lists the general characteristics of all fish captured in each year.

Table 1 Summary of sample sizes for each year in each study

	Study 1		Study 2
	2010	2011	2013
	Number of individuals		
Marked	182	240	431
Recaptured	64	118	118
$x2$	10	39	0
$x3$	1	12	0
$x4$	0	2	0
Movement	Number of recaptures		
0	24	48	45
1	27	68	24
2	9	29	11
3	10	16	6
4	3	5	6
5	0	0	8
6	0	2	6
7	1	1	3
≥ 8	1	2	9

'Marked' lists the total number of individuals initially marked in each year and 'Recaptured' lists the number of individuals recaptured at least once; of those individuals that were recaptured, a number were recaptured two, three or four ($x2$, $x3$, $x4$) times. Movement indicates the number of transects an individual moved in the river between the previous and current captures.

Table 2 Average values (±SE) for traits measured in both studies

Measure	Study 1		Study 2
	2010	2011	2013
SL (mm)	42.31 (±0.33)	40.13 (±0.29)	25.12 (±0.16)
Number of transects	1.28 (±0.15)	1.28 (±0.09)	2.43 (±0.28)
Days between captures	8.54 (±0.55)	8.71 (±0.45)	16.60 (±0.51)
Total sections in pool	15.91 (±0.93)	20.63 (±0.87)	13.19 (±0.43)
Latency to emerge(s)	98.29 (±8.31)	61.04 (±5.62)	131.88 (±7.96)
Unique sections	5.54 (±0.25)	6.71 (±0.19)	6.09 (±0.13)

3.1.2 Consistent individual differences in behaviour

In general, we found evidence for consistent individual variation in exploratory behaviour in the pool and in the river. Repeatability of exploratory behaviour in the pool was $r = 0.29$ and inclusion of the individual effect in the model was strongly supported by the log-likelihood ratio test (log-likelihood ratio $= 23.95$, $p < 0.001$). The repeatability of movement in the river was $r = 0.26$.

Exploratory behaviour in the pool depended on time of day, size, number of recaptures and year (Table 3). For example, individuals that were measured later in the day and were relatively large were more active in the arena. An individual's exploratory behaviour in the pool tended to decrease with increasing number of captures (Table 3). Including these fixed effects did not change the estimate of the total variation among individuals in exploratory behaviour in the pool (conditional repeatability estimate: $r = 0.24$ log-likelihood ratio $= 16.65$, $p < 0.001$) or movement in the river (conditional repeatability: $r = 0.23$).

3.1.3 Relationship between exploratory behaviour in the pool and movement in the river

Among those individuals that were recaptured, we found no evidence that exploratory behaviour in the pool predicted movement between capture and recapture (Fig. 1 and Table 4). Interestingly, we also found that there was no effect of days since previous capture on movement in the river (Table 4), suggesting that even if given more time, individuals would not have moved further in the river.

There was no influence of exploratory behaviour in the pool on probability of recapture (Appendix). Using the other two behaviours, latency to

Table 3 Summary of the effects on exploratory behaviour in the pool in Study 1

Factor	Estimate	F value	p value
Time of day	0.11	2.95	0.043*
Standard length	0.58	16.48	<0.001*
Capture #	−1.61	1.13	0.069
Year: 2011	5.77	26.75	<0.001*

Effects significant at the $p < 0.05$ level are marked with '*'.

Table 4 Summary of the effects on movement in the river between captures in Study 1

Factor	Estimate	95% CI	p value
Total sections in the pool	0.006	(−0.004, 0.017)	0.218
Standard length	0.01	(−0.018, 0.010)	0.469
Days since capture	0.003	(−0.020, 0.027)	0.765

emerge from the acclimation chamber or unique sections as predictors did not change our results: there was no relationship between these two behaviours and movement in the river or probability of recapture (Appendix).

3.2 Study 2

3.2.1 General characteristics of captured fish

We marked 431 individual sticklebacks in the second study. Of these, we recaptured 118 individuals (Table 1). As in the first study, there was considerable variation in movement in the river (range 0–12 transects). Although movement in the river tended to be in the upstream direction, visual inspection of the data did not suggest that fish were converging on particular transects (Fig. 1).

3.2.2 Relative influence of different traits on movement in the river

Several of the *a priori* models received substantial support (delta < 2), including models that contained the main effects of body condition, social environment use and habitat use, and the model that contained only the random effect of transect (Table 5). Only the 'body condition × social environment use' model received substantially less support (delta ≥ 4). Since there was not a single 'best' model, we created a 95% confidence set of models (which excluded the 'body condition × social environment use'

Table 5 Results of information theoretic analysis comparing seven different models from Study 2 ordered from most to least informative

Model	K	QAICc	Δ	w	R²
1. Random effect only (transect)	2	93.8	0	0.28	0.10
2. Body condition	3	94.3	0.45	0.23	0.13
3. Social environment use	3	95.3	1.51	0.13	0.11
4. Exploratory behaviour in the pool	3	95.7	1.84	0.11	0.11
5. Habitat use	3	95.7	1.85	0.11	0.11
6. Condition × habitat use	5	95.9	2.07	0.10	0.18
7. Condition × social environment use	5	97.8	4.00	0.04	0.13

Models 6 and 7 include both the main effects and the interaction term between the two factors. K is the number of parameters estimated. QAICc is a smaller-is-better measure of goodness of fit. Delta (Δ) is the difference between the 'best' model with the lowest QAICc and all other models. Akaike's weight (w) is the relative probabilities of each model given the data. R^2 is the conditional R^2 that shows the variance explained by fixed and random effects.

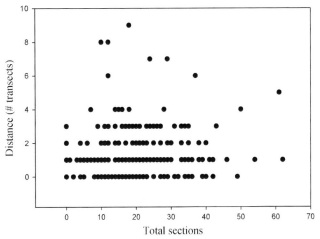

Figure 2 Relationship between exploratory behaviour in the pool and behaviour in the river in Study 1.

model) from which we calculated model-averaged estimates and 95% confidence intervals (Fig. 2).

Based on the model-averaged estimates, as in the first study, we found no evidence that exploratory behaviour in the pool influenced behaviour in the stream (Table 6). Instead, we found a significant interaction between habitat use and body condition: fish in better condition moved further but only

Table 6 Model-averaged estimates and 95% confidence intervals of factors predicting movement in Study 2 calculated from the 95% confidence set of models

Factor	Estimate	95% CI
Body condition	0.41*	0.12, 0.71
Social environment	0.21*	0.13, 0.29
Open–field behaviour	0.01	0.00, 0.03
Habitat use	−1.23	−3.46, 1.00
Condition × habitat use	1.42*	1.16, 1.69

Asterisks highlight significant factors where confidence intervals did not overlap zero. $N=82$.

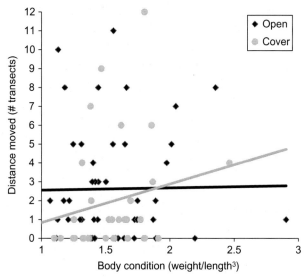

Figure 3 Interaction between body condition and movement in Study 2. Black diamonds and line represent individuals from open habitats. Grey circles and line represent fish from cover habitats. Among fish in cover habitats, individuals that were in better condition moved further. Condition was not related to distance moved among fish in the open. N, open $= 48$, cover $= 34$.

among individuals that occurred in the cover habitat (effect of condition × habitat type interaction $= 1.42$ [1.16, 1.69], Fig. 3). There was also a main effect of social environment use, where individuals that occurred in shoals tended to be more likely to be recaptured further away than fish that occurred alone (Table 6). We detected a non-significant trend for fish in the open to move further than fish that occurred in the cover habitat (Table 6).

As in the first study, the amount of time between capture and recapture was not correlated with movement in the river (Kendall's tau $= -0.02$, $p = 0.75$, $n = 118$).

Similar to the first study, these results did not depend upon the behavioural measure from the open-field arena: neither latency to emerge nor unique sections influenced an individual's movement within the river (Appendix).

4. DISCUSSION

Our studies tested how intraspecific phenotypic variation influences natural movement within a river population of sticklebacks. Consistent with the growing literature on animal personalities, we found strong evidence for consistent individual differences in behaviour in a standardized assay designed to measure exploratory behaviour and in behaviour in their natural environment. However, despite previous demonstrations that an individual's behavioural type can influence movement patterns in the wild in other species (Chapman et al., 2011; Cote et al., 2010; Dingemanse et al., 2003; Duckworth and Badyaev, 2007; Fraser et al., 2001; Herborn et al., 2010), we found no evidence that a stickleback's exploratory behaviour in the pool predicted how they moved within a river. Instead, we found that movement in the river was jointly influenced by an individual's body condition and its habitat use.

Individuals that were in better condition moved further in the river, but only if they occurred in the cover habitat. The cover habitat in this study consisted of distinct patches of submerged grass and tree branches separated by areas that were not inhabited by stickleback. The open habitat, consisting of relatively homogeneous gravel beds, was much more continuous and stickleback inhabiting this habitat were more evenly distributed. This difference in space between patches (i.e., patch isolation *sensu*; Bowler and Benton, 2005) in the open versus cover habitats might explain why we only detected condition-dependent movement in the cover habitat. High condition individuals might have been better able than low condition individuals to afford the costs associated with moving between patches in the cover habitat (e.g., lost foraging time, energetic requirements of locomotion, risk of predation; Hanski, 1998). This is further supported by the fact that individual sticklebacks were found to consistently occupy the same habitat (cover vs. open) between capture and recapture (S. Pearish, Unpublished manuscript). An individual's habitat use could be driven by a number of factors, such as food resource availability, predation risk or competitive exclusion and future work determining the relative influence of these could shed light on the mechanisms underlying this condition-dependent movement.

Our data support the hypothesis that group living encourages individuals to move further: fish that had been captured in shoals moved further in the river than fish that had been captured by themselves. Previous work has shown that fish in groups are less inhibited in the presence of a predator than fish that are alone (Magnhagen and Bunnefeld, 2009), perhaps due to the added safety from predation provided by group living (Krause and Ruxton, 2002). A plausible explanation for why shoaling individuals in our study moved further is that their perception of the predation risk of movement was lower compared to fish that occurred alone. Including the influence of the social environment and an individual's propensity to shoal might then help improve predictions about dispersal and migration on a larger scale.

We found no evidence that exploratory behaviour in the pool predicted movement in the river at two different field sites, in three different years and for neither juvenile nor adult sticklebacks. The resounding failure to detect a relationship between exploratory behaviour in the pool and movement in the river urges us to be cautious about assuming that an individual's behaviour in a standardized assay reflects natural behaviour in the field (Niemelä and Dingemanse, 2014). While some studies have shown that individual behaviour in a standardized open-field arena assay predicts behaviour and movement in the field (Chapman et al., 2011; Cote and Clobert, 2007; Dingemanse et al., 2003; Fraser et al., 2001; Herborn et al., 2010; Kobler et al., 2011; Wilson and McLaughlin, 2007), others have found no, or mixed, relationships between behaviour in standardized assays and natural behaviours in the field (Coleman and Wilson, 1998; Minderman et al., 2010; Wilson et al., 1993), a finding that is probably even more common but underreported due to publication bias.

One potential reason for the failure to detect a relationship between exploratory behaviour in the pool and movement in the river is because exploratory behaviour in the pool reflected a response to a novel and risky environment, while movement in the river reflected behaviour in a familiar environment. By switching to juveniles in the second study, we attempted to increase the likelihood that the movement we observed represented exploration of novel areas, but it is possible that even the youngest individuals had ample time to explore the entire study site prior to capture. This possibility is supported by the fact that fish that experienced more time between capture and recapture did not move further. A different behavioural assay, one designed to measure behaviour in a familiar environment, might have better success predicting movement in the river. It is also conceivable that exploratory behaviour in the pool might have better success predicting longer-range movement during the winter, when dispersal from the river is not limited by shallow riffles (Chapman et al., 2011). Instead, our studies were carried out during the summer, when male sticklebacks defend breeding territories. Finally, perhaps our measure of movement in the field (number of transects) is too coarse, i.e., it does not detect nonlinear movement, or movement over a smaller spatial scale. Studies utilizing internal acoustic or radio tags should be fruitful by providing more detailed and nuanced measures of individuals' natural movement behaviours (e.g., Kobler et al., 2009; Taylor and Cooke, 2014).

Altogether our studies demonstrate that individual phenotypic variation has the potential to impact processes that occur on larger, population-level scales. However, we also show that not all measures of individual variation are informative about natural processes and we urge caution when making the assumption that behavioural types as assessed using standardized behavioural assays are predictive of behaviour in the wild. Further investigation of the potential influence of individual variation on larger scale processes, such as population structure and distribution, can only provide a more integrative understanding of ecological dynamics.

APPENDIX. SUPPLEMENTARY RESULTS WITH OTHER BEHAVIOURAL VARIABLES

1.1 Total sections

1.1.1 Study 1: Effect of total sections on probability of recapture

Table A.1 Summary of factors influencing the probability of recapture in both years as estimated by a binary logistic regression

Factor	Odds ratio	p value
Total sections	1.01	0.487
Standard length	0.99	0.506
Technique: seine	1.02	0.551

1.2 Latency to emerge from the chamber

1.2.1 Study 1

Latency to emerge was log transformed to approximate a Gaussian distribution. We performed all the same analyses using the same methods as with 'total sections' (see Section 2).

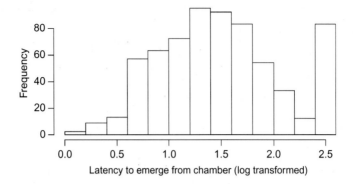

i. Repeatability of latency to emerge
ii. Conditional repeatability of latency to emerge
Similar to the results with total sections in the open-field arena, we found evidence for consistent individual differences in latency to emerge from the acclimation chamber (repeatability = 0.25; log-likelihood ratio: 21.18, $p < 0.001$).

Latency to emerge was influenced by an individual's standard length, capture number and year of capture (Table A.2). Individuals that were larger emerged more quickly from the acclimation chamber (larger scores indicate longer latencies). Additionally, individuals were quicker to emerge with subsequent captures and generally emerged more quickly in 2011. Inclusion of these fixed effects did not alter the amount of consistent individual variation (conditional repeatability = 0.24; log-likelihood ratio: 19.92, $p < 0.001$).

Table A.2 Summary of the effects on latency to emerge (log transformed) from the acclimation chamber into the open-field arena in Study 1

Factor	Estimate	F value	p value
Study 1: 2010 and 2011			
Time of day	−0.003	1.46	0.228
Standard length	−0.019	8.54	0.001*
Capture #	−0.063	5.40	0.021*
Year: 2011	−0.264	29.43	<0.001*

Effects significant at the $p < 0.05$ level are marked with '*'.

iii. Effect of latency on distance moved within stream
iv. Effect of latency on probability of recapture

We found no indication that variation in latency to emerge predicted an individual's movement in the stream (Table A.3) or probability of recapture (Table A.4).

Table A.3 Summary of the effects on distance moved in the stream between captures in Study 1 as estimated by a GLMM with quasi-Poisson distribution

Factor	Estimate	95% CI	p value
Latency to emerge	−0.148	(−0.35, 0.05)	0.163
Standard length	0.017	(−0.01, 0.05)	0.233
Days since capture	0.008	(−0.01, 0.03)	0.446

Table A.4 Summary of factors influencing the probability of recapture in both years as estimated by a binary logistic regression

Factor	Odds ratio	p value
Latency to emerge	1.04	0.833
Standard length	1.00	0.878
Technique: seine	1.01	0.562

Males were less likely to be recaptured.

1.2.2 Study 2
i. Comparison of six models predicting movement
ii. Model-averaged estimates of predictors

Table A.5 Results of information theoretic analysis from Study 2 using an alternative behavioural trait, latency to emerge, in order from most to least informative

Model	K	QAICc	Δ	w	R^2
Latency to emerge	3	91.5	0	0.39	0.11
Random effect only (transect)	2	92.9	1.38	0.20	0.10
Body condition	3	93.3	1.85	0.15	0.13
Social environment	3	94.4	2.90	0.09	0.11
Habitat use	3	94.7	3.24	0.08	0.11
Condition and habitat use	5	95.0	3.51	0.07	0.18
Condition and social environment	5	96.9	5.41	0.03	0.13

K is the number of parameters estimated. QAICc is a smaller-is-better measure of goodness of fit. Delta (Δ) is the difference between the 'best' model with the lowest QAICc and all other models. Akaike's weight (w) is the relative probabilities of each model given the data. R^2 is the conditional R^2 that shows the variance explained by fixed and random effects.

Table A.6 Model-averaged estimates and 95% confidence intervals of factors predicting movement in Study 2 calculated from the 95% confidence set of models that included an alternative behavioral trait, latency to emerge

Factor	Estimate	95% CI
Body condition	0.41*	0.12, 0.71
Social environment	0.21*	0.13, 0.29
Latency to emerge	0.00	0.00, 0.00
Habitat use	−1.21	−3.44, 1.02
Condition × habitat use	1.42*	1.16, 1.69

Asterisks highlight significant factors where confidence intervals did not overlap zero. $N = 82$.

1.3 Unique sections

1.3.1 Study 1

i. Repeatability of latency to emerge

As a bounded count variable (0–9), unique sections were difficult to analyse. A majority (58%) of individuals explored at least eight out of the nine sections generating a high amount of skew in the data that could not be correctly by transformation, or adequately modelled with alternative error distributions. However, we attempted to estimate the repeatability of unique sections using a GLMM with quasi-Poisson error. If we did this, the estimated repeatability was $r = 0.009$. Therefore, given the lack of consistent individual variation in this behaviour we did not find it justified to use this variable as an explanatory variable for Study 1.

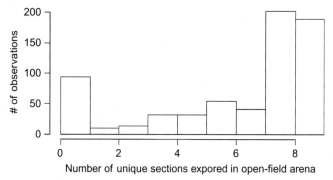

1.3.2 Study 2

i. Comparison of six models predicting movement

ii. Model-averaged estimates of predictors

Table A.7 Results of information theoretic analysis from Study 2 using an alternative behavioural trait, number of unique sections, in order from most to least informative

Model	K	QAICc	Δ	w	R²
Random effect only (transect)	2	93.8	0	0.29	0.10
Body condition	3	94.3	0.45	0.23	0.13
Social environment	3	95.3	1.51	0.13	0.11
Habitat use	3	95.7	1.85	0.11	0.11
Condition and habitat use	5	95.9	2.07	0.10	0.18
Number of unique sections	3	96.0	2.15	0.10	0.11
Condition and social environment	5	97.8	4.00	0.04	0.13

K is the number of parameters estimated. QAICc is a smaller-is-better measure of goodness of fit. Delta (Δ) is the difference between the 'best' model with the lowest QAICc and all other models. Akaike's weight (w) is the relative probabilities of each model given the data. R^2 is the conditional R^2 that shows the variance explained by fixed and random effects.

Table A.8 Model-averaged estimates and 95% confidence intervals of factors predicting movement in Study 2 calculated from the 95% confidence set of models that included an alternative behavioural trait, number of unique sections

Factor	Estimate	95% CI
Body condition	0.41*	0.12, 0.71
Social environment	0.21*	0.13, 0.29
Number of unique sections	0.02	0.00, 0.03
Habitat use	−1.23	−3.46, 1.00
Condition and habitat use	1.42*	1.16, 1.69

Asterisks highlight significant factors where confidence intervals did not overlap zero. $N=82$.

REFERENCES

Bell, A., 2005. Behavioural differences between individuals and two populations of stickleback (*Gasterosteus aculeatus*). J. Evol. Biol. 18, 464–473.

Bell, M.A., Foster, S.A., 1994. The Evolutionary Biology of the Threespine Stickleback. Oxford University Press, Oxford, UK.

Bell, A.M., Sih, A., 2007. Exposure to predation generates personality in threespined sticklebacks (*Gasterosteus aculeatus*). Ecol. Lett. 10, 828–834.

Bell, A.M., Hankison, S.J., Laskowski, K.L., 2009. The repeatability of behaviour: a meta-analysis. Anim. Behav. 77, 771–783.

Bentzen, P., McPhail, J., 1984. Ecology and evolution of sympatric sticklebacks (*Gasterosteus*): specialization for alternative trophic niches in the Enos Lake species pair. Canadian J. Zool. 62, 2280–2286.

Biro, P.A., Dingemanse, N.J., 2009. Sampling bias resulting from animal personality. Trends Ecol. Evol. 24, 66–67.

Bolnick, D.I., Snowberg, L.K., Patenia, C., Stutz, W.E., Ingram, T., Lau, O.L., 2009. Phenotype-dependent native habitat preference facilitates divergence between parapatric lake and stream stickleback. Evolution 63, 2004–2016.

Bolnick, D.I., Amarasekare, P., Araújo, M.S., Bürger, R., Levine, J.M., Novak, M., Rudolf, V.H., Schreiber, S.J., Urban, M.C., Vasseur, D.A., 2011. Why intraspecific trait variation matters in community ecology. Trends Ecol. Evol. 26, 183–192.

Bonte, D., Van Dyck, H., Bullock, J.M., Coulon, A., Delgado, M., Gibbs, M., Lehouck, V., Matthysen, E., Mustin, K., Saastamoinen, M., 2012. Costs of dispersal. Biol. Rev. 87, 290–312.

Bowler, D.E., Benton, T.G., 2005. Causes and consequences of animal dispersal strategies: relating individual behaviour to spatial dynamics. Biol. Rev. 80, 205–225.

Brown, C.R., Brown, M.B., 2000. Heritable basis for choice of group size in a colonial bird. Proc. Natl. Acad. Sci. 97, 14825–14830.

Burnham, K.P., Anderson, D.R., 2002. Model Selection and Multimodel Inference: A Practical Information-Theoretic Approach. Springer, New York, NY.

Burnham, K.P., Anderson, D.R., Huyvaert, K.P., 2011. AIC model selection and multi-model inference in behavioral ecology: some background, observations, and comparisons. Behav. Ecol. Sociobiol. 65, 23–35.

Chapman, B.B., Hulthén, K., Blomqvist, D.R., Hansson, L.A., Nilsson, J.Å., Brodersen, J., Anders Nilsson, P., Skov, C., Brönmark, C., 2011. To boldly go: individual differences in boldness influence migratory tendency. Ecol. Lett. 14, 871–876.

Clobert, J., Danchin, E., Dhondt, A.A., Nichols, J.D., 2001. Dispersal. Oxford University Press, Oxford.

Coleman, K., Wilson, D.S., 1998. Shyness and boldness in pumpkinseed sunfish: individual differences are context-specific. Anim. Behav. 56, 927–936.

Cote, J., Clobert, J., 2007. Social personalities influence natal dispersal in a lizard. Proc. R. Soc. B: Biol. Sci. 274, 383–390.

Cote, J., Fogarty, S., Weinersmith, K., Brodin, T., Sih, A., 2010. Personality traits and dispersal tendency in the invasive mosquitofish (Gambusia affinis). Proc. R. Soc. B: Biol. Sci. 277, 1571–1579.

Dall, S.R., Bell, A.M., Bolnick, D.I., Ratnieks, F.L., 2012. An evolutionary ecology of individual differences. Ecol. Lett. 15, 1189–1198.

Dingemanse, N.J., Both, C., Van Noordwijk, A.J., Rutten, A.L., Drent, P.J., 2003. Natal dispersal and personalities in great tits (Parus major). Proc. R. Soc. London Ser. B: Biol. Sci. 270, 741–747.

Dingemanse, N.J., Wright, J., Kazem, A.J., Thomas, D.K., Hickling, R., Dawnay, N., 2007. Behavioural syndromes differ predictably between 12 populations of three-spined stickleback. J. Anim. Ecol. 76, 1128–1138.

Duckworth, R.A., Badyaev, A.V., 2007. Coupling of dispersal and aggression facilitates the rapid range expansion of a passerine bird. Proc. Natl. Acad. Sci. 104, 15017–15022.

Fraser, D.F., Gilliam, J.F., Daley, M.J., Le, A.N., Skalski, G.T., 2001. Explaining leptokurtic movement distributions: intrapopulation variation in boldness and exploration. Am. Nat. 158, 124–135.

Gilliam, J.F., Fraser, D.F., 2001. Movement in corridors: enhancement by predation threat, disturbance, and habitat structure. Ecology 82, 258–273.

Hanski, I., 1998. Metapopulation dynamics. Nature 396, 41–49.

Herborn, K.A., Macleod, R., Miles, W.T., Schofield, A.N., Alexander, L., Arnold, K.E., 2010. Personality in captivity reflects personality in the wild. Anim. Behav. 79, 835–843.

Huntingford, F.A., 1976. The relationship between anti-predator behaviour and aggression among conspecifics in the three-spined stickleback, Gasterosteus aculeatus. Anim. Behav. 24, 245–260.

Johnson, J.B., Omland, K.S., 2004. Model selection in ecology and evolution. Trends Ecol. Evol. 19, 101–108.

Kobler, A., Klefoth, T., Mehner, T., Arlinghaus, R., 2009. Coexistence of behavioural types in an aquatic top predator: a response to resource limitation? Oecologia 161, 837–847.

Kobler, A., Maes, G.E., Humblet, Y., Volckaert, F.A., Eens, M., 2011. Temperament traits and microhabitat use in bullhead, Cottus perifretum: fish associated with complex habitats are less aggressive. Behaviour 148, 5–6.

Krause, J., Ruxton, G.D., 2002. Living in Groups. Oxford University Press, Oxford.

Krause, J., Loader, S.P., McDermott, J., Ruxton, G.D., 1998. Refuge use by fish as a function of body length-related metabolic expenditure and predation risks. Proc. R. Soc. London Ser. B: Biol. Sci. 265, 2373–2379.

Laskowski, K.L., Bell, A.M., 2013. Competition avoidance drives individual differences in response to a changing food resource in sticklebacks. Ecol. Lett. 16, 746–753. http://dx. doi.org/10.1111/ele.12105.

Laskowski, K.L., Bell, A., 2014. Strong personalities, not social niches, drive individual differences in social behaviour in sticklebacks. Anim. Behav. 90, 287–295.

Lindström, T., Brown, G., Sisson, S., Phillips, B., Shine, R., 2013. Rapid shifts in dispersal behavior on an expanding range edge. Proc. Natl. Acad. Sci. 110, 13452–13456.

Magnhagen, C., Bunnefeld, N., 2009. Express your personality or go along with the group: what determines the behaviour of shoaling perch? Proc. R. Soc. B: Biol. Sci. 276, 3369–3375.

Martin, J.G., Nussey, D.H., Wilson, A.J., Reale, D., 2011. Measuring individual differences in reaction norms in field and experimental studies: a power analysis of random regression models. Methods Ecol. Evol. 2, 362–374.

Miller, J.R.B., Ament, J.M., Schmitz, O.J., 2014. Fear on the move: predator hunting mode predicts variation in prey mortality and plasticity in prey spatial response. J. Anim. Ecol. 83, 214–222. http://dx.doi.org/10.1111/1365-2656.12111.

Minderman, J., Reid, J.M., Hughes, M., Denny, M.J., Hogg, S., Evans, P.G., Whittingham, M.J., 2010. Novel environment exploration and home range size in starlings Sturnus vulgaris. Behav. Ecol. 21, 1321–1329.

Nakagawa, S., Schielzeth, H., 2010. Repeatability for Gaussian and non-Gaussian data: a practical guide for biologists. Biol. Rev. 85, 935–956. http://dx.doi.org/10.1111/ j.1469-185X.2010.00141.x.

Nakagawa, S., Schielzeth, H., 2013. A general and simple method for obtaining R^2 from generalized linear mixed-effects models. Methods Ecol. Evol. 4, 133–142.

Niemelä, P.T., Dingemanse, N.J., 2014. Artificial environments and the study of 'adaptive' personalities. Trends Ecol. Evol. 29, 245–247.

Patrick, D.C., Carlo, R., Paul, C., 2001. Field test for environmental correlates of dispersal in hedgehogs Erinaceus europaeus. J. Anim. Ecol. 70, 33–46.

Pearish, S., Hostert, L., Bell, A.M., 2013. Behavioral type–environment correlations in the field: a study of three-spined stickleback. Behav. Ecol. Sociobiol. 67, 765–774.

Phillips, B.L., Suarez, A.V., 2012. The role of behavioural variation in the invasion of new areas. In: Candolin, U., Wong, B. (Eds.), Behavioural Responses to a Changing World: Mechanisms and Consequences. Oxford Universty Press, Oxford, pp. 190–200.

Pitcher, T.J., 1993. Behaviour of Teleost Fishes. Springer, New York, NY.

Pope, K., Kruse, C., 2007. Condition. Analysis and Interpretation of Freshwater Fisheries Data. American Fisheries Society, Bethesda, MD, pp. 423–471.

Robb, L.A., Martin, K., Hannon, S.J., 1992. Spring body condition, fecundity and survival in female willow ptarmigan. J. Anim. Ecol. 61, 215–223.

Ronce, O., 2007. How does it feel to be like a rolling stone? Ten questions about dispersal evolution. Annu. Rev. Ecol. Evol. Systemat. 38, 231–253.

Saltz, J.B., 2011. Natural genetic variation in social environment choice: context-dependent gene–environment correlation in *Drosophila melanogaster*. Evolution 65, 2325–2334.

Schluter, D., McPhail, J.D., 1992. Ecological character displacement and speciation in sticklebacks. Am. Nat. 140, 85–108.

Sherratt, T.N., Macdougall, A.D., 1995. Some population consequences of variation in preference among individual predators. Biol. J. Linn. Soc. 55, 93–107.

Smith, B.R., Blumstein, D.T., 2008. Fitness consequences of personality: a meta-analysis. Behav. Ecol. 19, 448–455.

Taylor, M.K., Cooke, S.J., 2014. Repeatability of movement behaviour in a wild salmonid revealed by telemetry. J. Fish Biol. 84, 1240–1246. http://dx.doi.org/10.1111/jfb.12334.

Templeton, J.J., Giraldeau, L.-A., 1996. Vicarious sampling: the use of personal and public information by starlings foraging in a simple patchy environment. Behav. Ecol. Sociobiol. 38, 105–114.

Tosh, J., Garber, A., Trippel, E., Robinson, J., 2010. Genetic, maternal, and environmental variance components for body weight and length of Atlantic cod at 2 points in life. J. Anim. Sci. 88, 3513–3521.

Valone, T.J., 1989. Group foraging, public information, and patch estimation. Oikos 56, 357–363.

Valone, T.J., Templeton, J.J., 2002. Public information for the assessment of quality: a widespread social phenomenon. Philos. Trans. R. Soc. London Ser. B: Biol. Sci. 357, 1549–1557.

Violle, C., Enquist, B.J., McGill, B.J., Jiang, L., Albert, C.H., Hulshof, C., Jung, V., Messier, J., 2012. The return of the variance: intraspecific variability in community ecology. Trends Ecol. Evol. 27, 244–252.

Ward, A., James, R., Wilson, A., Webster, M., 2013. Site fidelity and localised homing behaviour in three-spined sticklebacks (*Gasterosteus aculeatus*). Behaviour 150, 1689–1708.

Wilson, A.D., McLaughlin, R.L., 2007. Behavioural syndromes in brook charr, *Salvelinus fontinalis*: prey-search in the field corresponds with space use in novel laboratory situations. Anim. Behav. 74, 689–698.

Wilson, D.S., Coleman, K., Clark, A.B., Biederman, L., 1993. Shy-bold continuum in pumpkinseed sunfish (*Lepomis gibbosus*): an ecological study of a psychological trait. J. Comp. Psychol. 107, 250.

CHAPTER FIVE

Eco-Evolutionary Dynamics of Plant–Insect Communities Facing Disturbances: Implications for Community Maintenance and Agricultural Management

Ewen Georgelin*,†,1, Grigoris Kylafis*, Nicolas Loeuille*

*Institute of Ecology and Environmental Sciences—Paris (UPMC-CNRS-IRD-INRA-UPEC-Paris Diderot), Université Pierre et Marie Curie, UMR 7618, Paris, France
†Ecology, Systematic and Evolution (UPSUD, CNRS, AgroParisTech), Université Paris Sud, UMR 8079, Orsay, France
1Corresponding author: e-mail address: ewen.georgelin@u-psud.fr

Contents

Abstract

Understanding the response of natural communities to current global changes is crucial for the conservation and management of ecosystems. While the ecological and evolutionary responses of antagonistic or mutualistic systems have been studied separately, few studies investigate the eco-evolutionary response of systems combining different interaction types. We build an evolutionary model of a plant–pollinator–herbivore community, where both pollinators and herbivores are confronted with the same external disturbance, insecticide use. Pollinators' and herbivores' response to disturbances is controlled by a trait (e.g. sensitivity trait) that incurs a cost in reproduction. Using Adaptive Dynamics Theory, we find that herbivore evolution lowers densities of species and may drive pollinators to extinction while pollinator evolution increases densities and enhances community maintenance. We then show that coevolution, by constraining the variability of coevolving species, produces qualitative dynamics that cannot be predicted from the mere addition of single-species evolution scenarios.

1. INTRODUCTION

Human activities generate disturbances that affect both the ecological and evolutionary dynamics of communities and ecosystems (Palumbi, 2001; Vitousek, 1997). Several examples of trait evolution in response to anthropogenic disturbances, as well as their consequences for the maintenance of populations and ecosystem services have been reported (Dieckmann and Ferrière, 2004). Fisheries, for example, lead to maturation of fishes at earlier age and smaller size (Olsen et al., 2004). These changes in phenotypic traits affect the growth rate of populations and compromise fish stocks (Olsen et al., 2004). In agricultural landscapes, evolution of resistances to pesticides has been often reported (Rex Consortium, 2013). The emergence of pesticide resistance reduces the efficacy of pest control measures, along with side effects of pesticides on natural populations, which compromises the durability of agricultural systems (Potts et al., 2010). These changes in life-history traits can affect population growth rates (Coltman et al., 2003; Olsen et al., 2004) by generating positive ("Evolutionary Rescue"; Gomulkiewicz and Holt, 1995) or negative effects ("Evolutionary Deterioration" and "Evolutionary Suicide"; Dieckmann and Ferrière, 2004; Gyllenberg and Parvinen, 2001; Matsuda and Abrams, 1994) on population survival. These kind of processes have been named eco-evolutionary effects (Fussmann et al., 2007; Pelletier et al., 2009; Schoener, 2011). Drops in fish densities associated with evolution of earlier maturation (i.e. a case of "Evolutionary Deterioration") for instance, compromise the durability of fisheries (Olsen et al., 2004). These evolution-driven changes in ecological densities (or eco-evolutionary effects) may then propagate to the community level through ecological indirect effects (*sensu* Wootton, 1994).

Unfortunately, models studying eco-evolutionary dynamics and species extinction in the face of disturbances usually account for just one species (e.g. Gomulkiewicz and Holt, 1995) or one interaction type (e.g. antagonistic interactions; Matsuda and Abrams, 1994). Recently, several studies have emphasized the importance of the interplay between antagonistic and mutualistic interactions for natural communities, because they produce density-dependent effects of different signs on species growth rates (Fontaine et al., 2011; Georgelin and Loeuille, 2014). In fact, mutualistic interactions affect positively the growth rates of partner species while antagonistic interactions affect negatively these growth rates. When the interplay between the two interaction types is accounted for, the sign of the emergent indirect effects between antagonistic and mutualistic species sharing a resource is highly constrained by population traits and abundances (Georgelin and Loeuille, 2014). Evolutionary dynamics of traits can also change the signs of these indirect effects. Evolution of traits in antagonistic interactions may lead to an increase in densities of the species (i.e. a case of "Hydra Effect"; Abrams and Matsuda, 2005), while in mutualistic interactions, it can lead to a decrease in densities and even extinctions (Bronstein et al., 2004).

Agricultural landscapes are confronted with insect pests (antagonistic interactions) and rely on pollination services (mutualistic interactions). The joint objective of pest control and pollination service therefore needs models that consider the two interaction types. Particularly, given that pesticides are often used to limit pests, eco-evolutionary dynamics associated with such a disturbance have to be understood within the complete system, to provide an integrative understanding of the propagation of indirect effects within the agricultural system. The interplay of phenotypic changes and ecological dynamics may affect these communities through two mechanisms. First, some phenotypic traits are simultaneously involved in antagonistic and mutualistic interactions (Fontaine et al., 2011). An obvious example is the existence of floral signals in plants that are detected not only by pollinators but also by herbivores (Strauss, 1997). Secondary metabolites that repel herbivores of plants can also affect interactions with mutualists, for example by changing the attractiveness to pollinator (Strauss et al., 1999). Second, antagonistic and mutualistic species may be linked through indirect density-dependent effects via a third species (Georgelin and Loeuille, 2014; Wootton, 1994). For instance, pollinators and herbivores are linked through the density of plants. Hence, changes in herbivore density due to either direct effects of disturbances (ecological effects) or associated evolutionary responses (eco-evolutionary effects) will change pollinator density and vice versa. In this work, we will focus on this second mechanism, how an evolutionary response to disturbance in pollinator and herbivore populations will affect the community dynamics through this indirect density-dependent effect. We model the eco-evolutionary consequences of pesticide application on plant–pollinator–herbivore communities in agricultural landscapes.

The use of pesticides in agricultural landscapes affects both herbivores and pollinators (Rex Consortium, 2013). From an ecological point of view, an important concern posed by pesticides is the threat to pollination services. Declines in pollinator density and diversity put the crop yields at risks, reducing agricultural sustainability (Klein et al., 2007). From an evolutionary perspective, several examples of pesticide resistance have now been reported, often in a few generations (Mallet, 1989; Rex Consortium, 2013). To date, most evolutionary studies have been interested in designing strategies that manage or delay the onset of resistance in pests (Rex Consortium, 2013). While these studies are crucial for the maintenance of crop yields, we still have little information on the consequences of pesticide uses for eco-evolutionary dynamics of communities and for the simultaneous management of pests and pollination services (Loeuille et al., 2013).

Our goal here is to understand how adaptive responses of herbivores and pollinators to pesticides may change the ecological dynamics of these communities in agricultural landscapes. We incorporate evolutionary dynamics in a plant–pollinator–herbivore model we previously used to study consequences of pesticides for ecological dynamics (Georgelin and Loeuille, 2014). We let species sensitivities to pesticides evolve and study the consequences of trait evolution dynamics for the maintenance of the community (species densities and community composition). We study two evolutionary scenarios separately. First, only one species is allowed to evolve (either the pollinators or the herbivores). Second, we tackle the co-evolutionary dynamics of the two sensitivity traits. We focus on three questions:

(i) What are the expected evolutionary dynamics of the sensitivity traits and how do selected strategies depend on the intensity of disturbance? Based on direct selective effects, we expect that any increase in disturbance will select for a lower level of sensitivity (higher resistance).

(ii) What are the ecological consequences of trait evolution in terms of densities and community composition? We previously showed that ecological dynamics under a low pesticide sensitivity of pollinators enhances the viability of the community (through effects on density, extinction threshold and resilience; Georgelin and Loeuille, 2014). We therefore make the hypothesis that evolution of pollinators towards less sensitivity will help the maintenance of the community. Conversely, we make the opposite prediction for the evolution of herbivores.

(iii) Can the co-evolutionary dynamics be predicted from the superposition of two previous cases, where only one species evolves? If, as we predict, antagonist and mutualist adaptive responses have opposite effects on the community, we expect that the consequences of coevolution on the viability of the community will depend on the evolutionary potentials of the two species, as well as on eco-evolutionary effects. Our main results suggest that single antagonistic or mutualistic trait evolution have contrary effects on community structure. While adaptive response of herbivores decreases the densities of the two other species and eventually leads to extinction of pollinators, adaptive response of pollinators can help the herbivores to remain in the system. Thus, contrary to other models (Abrams and Matsuda, 2005; Bronstein et al., 2004), evolution reinforces the consequences of ecological dynamics rather than limiting them. However, when co-evolutionary dynamics are considered, mutation rates heavily constrain the maintenance of the community. Co-evolutionary outcomes cannot be inferred from the simple addition of single-species evolutionary outcomes. We discuss how these results can help explain the current pollination crisis and provide lessons for the future management of agricultural landscapes.

2. MODEL AND METHODS

2.1 Ecological dynamics

The model represents a community of three species, one plant species P, one pollinator species M and one herbivore species H:

$$\frac{dP}{dt} = r_p P\left(1 - \frac{P}{K_p}\right) + k_p\left(\frac{a_m MP}{1 + a_m h_m P}\right) - \left(\frac{a_h HP}{1 + a_h h_h P}\right) \tag{1}$$

$$\frac{dM}{dt} = r_m M\left(1 - \frac{M}{K_m}\right) + k_m(s_m)\left(\frac{a_m MP}{1 + a_m h_m P}\right) - g_m(s_m) lM \tag{2}$$

$$\frac{dH}{dt} = k_h(s_h)\left(\frac{a_h HP}{1 + a_h h_h P}\right) - d_h H - g_h(s_h) lH \tag{3}$$

Variables and parameters are defined in Table 1.

Interspecific interactions are modelled with Holling type II functional responses (Holling, 1959). Hence, herbivores and pollinators consuming plants are constrained by handling times h_h and h_m, respectively. Similar functional responses have been used and discussed extensively both for predator–prey (Oksanen et al., 1981; Rosenzweig and MacArthur, 1963) and for mutualistic interactions (Holland et al., 2002; Jang, 2002; Thébault and Fontaine, 2010; Wright, 1989). Plants and pollinators, in the absence of

Table 1 Definitions and dimensions of variables and parameters
Variable/parameter description (dimension)

P	Plant density (ind m^{-2})
M	Pollinator density (ind m^{-2})
H	Herbivore density (ind m^{-2})
r_p	Plant intrinsic growth rate (t^{-1})
r_m	Pollinator intrinsic growth rate (t^{-1})
K_p	Plant carrying capacity (ind m^{-2})
K_m	Pollinator carrying capacity (ind m^{-2})
a_m	Plant–pollinator encounter rate (ind^{-1} m^2 t^{-1})
a_h	Plant–herbivore encounter rate (ind^{-1} m^2 t^{-1})
h_m	Pollinator handling time (t)
h_h	Herbivore handling time (t)
k_p	Conversion efficiency of pollination by plant (dimensionless)
$k_m(s_m)$	Conversion efficiency function of pollinator (dimensionless)
$k_h(s_h)$	Conversion efficiency function of herbivore (dimensionless)
d_h	Herbivore intrinsic death rate (t^{-1})
l	Mortality rate due to disturbance (t^{-1})
$g_m(s_m)$	Sensitivity function of the pollinator (dimensionless)
$g_h(s_h)$	Sensitivity function of the herbivore (dimensionless)
s_m	Sensitivity trait of the pollinator (dimensionless)
s_h	Sensitivity trait of the herbivore (dimensionless)

interacting species, have independent growths, limited by carrying capacity (K_p and K_m, respectively). In other words, the mutualistic interaction is considered to be facultative: plants partly rely on anemophilous pollination or pollination by non-modelled pollinator species and pollinators rely also on other types of resources, such as non-modelled plant species. On the contrary, we consider herbivores as being specialist of the focal plant species. Herbivores, especially agricultural pests, are often reported to be more specialized than pollinators (Bernays and Graham, 1988; Fontaine et al., 2009). In the absence of plants, herbivore density decays exponentially due to intrinsic mortality rate d_h. The parameter l expresses the extra-mortality rate due to the disturbance (pesticide use). It expresses the quantity of pesticides that is released and affects both pollinators and herbivores. Organisms show different sensitivities to pesticides (Desneux et al., 2007; Goulson, 2013; Pelosi et al., 2013). Functions g_h and g_m, and their dependency on traits s_m and s_h, account for this variability. Traits s_h and s_m combine all relevant traits that have an impact on the sensitivity of the organisms to insecticides, such as detoxification traits or changes in patch or resource exploitation. In the example of peach-potato aphids, seven types of resistances are described, each associated with a modification of pesticide target enzymes (Bass et al., 2014). Some of these modifications provide resistance synergistically against the same insecticides. Hence, s_h and s_m, can be thought as the combination of these different traits, that modify the overall sensitivity of organisms to pesticides.

The ecological dynamics of this system and the impacts of disturbances on the structure of the community are detailed in Georgelin and Loeuille (2014). At the coexistence equilibrium, plant density is only dependent on herbivore parameters (top-down control) while pollinator and herbivore densities depend on both species parameters:

$$P^0 = \frac{d_h + g_h(s_h)l}{a_h(k_h(s_h) - h_h(d_h + g_h(s_h)l))} \tag{4}$$

$$M^0 = \frac{K_m(r_m - g_m(s_m)l + a_m P^0(k_m(s_m) + h_m(r_m - g_m(s_m)l)))}{r_m + a_m h_m P^0 r_m} \tag{5}$$

$$H^0 = \frac{(1 + a_h h_h P^0)((K_p - P^0)(1 + a_m h_m P^0)r_m + a_m K_p M^0 k_p)}{a_h K_p(1 + a_m h_m P^0)} \tag{6}$$

2.2 Evolutionary dynamics

To study the adaptive responses of both pollinators and herbivores to the disturbance, we first study how sensitivity traits affect the growth rates of each species. Low sensitivity to disturbance, for instance via detoxification mechanisms (Cresswell et al., 2012; Després et al., 2007), often incur allocation costs, reducing growth or reproduction (Carrière et al., 1994; Rex Consortium, 2013). To account for this trade-off, we assume that sensitivity traits negatively affect the conversion efficiency functions, $k_h(s_h)$ and $k_m(s_m)$. While allocation costs are documented, we have no information on the shape of the trade-off functions, which are crucial for evolutionary dynamics (de Mazancourt and Dieckmann, 2004). To investigate a wide range of evolutionary possibilities, we use exponential functions for k_m, k_h, g_m and g_h:

$$k_m(s_m) = k_{m_0}\exp(c_m s_m) \text{ and } k_h(s_h) = k_{h_0}\exp(c_h s_h)$$
$$g_m(s_m) = g_{m_0}\exp(z_m s_m) \text{ and } g_h(s_h) = g_{h_0}\exp(z_h s_h)$$

This formulation allows sensitivity traits s_h and s_m to take any real value. For an evolving species i, parameters c_i and z_i permit to control the shape of the g_i versus k_i trade-off. If $c_i > z_i$, the trade-off between reproduction and survival is convex, while when $c_i < z_i$, the trade-off is concave. Figure 1 depicts the functioning of this trade-off, based on the different values of c_i and z_i.

We use Adaptive Dynamics methods to analyse the eco-evolutionary dynamics of the community confronted with disturbances (Dieckmann and Law, 1996; Geritz et al., 1998; Metz et al., 1992). We first study the case when only one species evolves (pollinators or herbivores) then we study the coevolution of the two species. When only one species evolves, variations of sensitivity trait can be expressed by the following equations:

$$\frac{ds_h}{dt} = C_H \mu_H \sigma_H^2 H^0(s_h)\left(\frac{\partial \omega_{H_{mut}}(s_{h_{mut}}, s_h)}{\partial s_{h_{mut}}}\right)_{s_{h_{mut}} \to s_h} \tag{7}$$

for herbivores and

$$\frac{ds_m}{dt} = C_M \mu_M \sigma_M^2 M^0(s_m)\left(\frac{\partial \omega_{M_{mut}}(s_{m_{mut}}, s_m)}{\partial s_{m_{mut}}}\right)_{s_{m_{mut}} \to s_m} \tag{8}$$

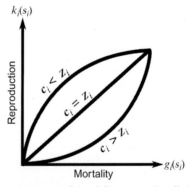

Figure 1 Schematic representation of the different trade-off shapes, based on the relative values of z_i and c_i (i being either h or m). An increase in reproduction due to an increase in s_h or s_m entails an increase in mortality (or a decrease in survival). According to z_i and c_i, the trade-off can be convex, concave or linear.

for pollinators, where C_i (i being either H for herbivores or M for pollinators) is a scaling parameter, μ_i is the per individual mutation rate, σ_i^2 is the variance of mutation effects on the sensitivity traits s_m or s_h. H^0 and M^0 are the herbivore and pollinator population densities at the ecological equilibrium (Eqs. 4–6) and the fitness gradient term corresponds to the selective pressures acting on the trait (Dieckmann and Law, 1996). Equations (7) and (8) imply that the phenotypic change is the product of two terms: (i) the phenotypic variability produced by mutations (products $\mu_H \sigma_H^2 H^0(s_h)$ and $\mu_M \sigma_M^2 M^0(s_m)$ in Eqs. 7 and 8, respectively), hereafter referred to as "evolutionary potential" and (ii) a term that embodies the natural selection process acting on this evolutionary potential. The fitness of a mutant is defined by its growth rate when rare in the environment set by the resident populations (Geritz et al., 1998; Metz et al., 1992). Invasion fitness equation, in the case of herbivores, is expressed by:

$$\omega_{H_{mut}}\left(s_{h_{mut}}, s_h\right) = k_h\left(s_{h_{mut}}\right) \frac{a_h P^0(s_h)}{1 + a_h h_h P^0(s_h)} - d_h - g_h\left(s_{h_{mut}}\right) l \qquad (9)$$

and in the case of pollinators by:

$$\omega_{M_{mut}}\left(s_{m_{mut}}, s_m\right) = r_m \left(1 - \frac{M^0(s_m)}{K_m}\right) + k_m\left(s_{m_{mut}}\right) \frac{a_m P^0(s_h)}{1 + a_m h_m P^0(s_h)}$$
$$- g_m\left(s_{m_{mut}}\right) l \qquad (10)$$

Evolution ends when the fitness gradient vanishes (Eqs. 7 and 8). Such points are called singular strategies (Geritz et al., 1998). Singular strategies can be classified according to two stability properties: invasibility and convergence. A singular strategy is non-invasible when no mutant can invade such a strategy (Maynard Smith, 1982). A singular strategy is convergent if only mutants with closer trait value to the singular strategy can invade (Christiansen, 1991; Eshel, 1983). Convergence and invasibility criteria can be studied by differentiating fitness functions twice according to sensitivity traits s_h or s_m (Geritz et al. 1998 and see Supplementary material on https://www.dropbox.com/s/qa7g6usf3hpsejr/appendix_georgelin_aer.pdf?dl=0).

In the case of coevolution, the trait variation depends on the traits of both resident populations (pollinators and herbivores):

$$\frac{ds_h}{dt} = C_H \mu_H \sigma_H^2 H^0(s_h, s_m) \left(\frac{\partial \omega_{H_{mut}}\left(s_{h_{mut}}, s_h\right)}{\partial s_{h_{mut}}}\right)_{s_{h_{mut}} \to s_h} \qquad (11)$$

$$\frac{\mathrm{d}s_m}{\mathrm{d}t} = C_M \mu_M \sigma_M^2 M^0(s_h, s_m) \left(\frac{\partial \omega_{M_{mut}}(s_{m_{mut}}, s_h, s_m)}{\partial s_{m_{mut}}} \right)_{s_{m_{mut}} \to s_m} \tag{12}$$

The two fitness equations become:

$$\omega_{H_{mut}}(s_{h_{mut}}, s_h) = k_h(s_{h_{mut}}) \frac{a_h P^0(s_h)}{1 + a_h h_h P^0(s_h)} - d_h - g_h(s_{h_{mut}})l \tag{13}$$

$$\omega_{M_{mut}}(s_{m_{mut}}, s_m, s_h) = r_m \left(1 - \frac{M^0(s_m, s_h)}{K_m} \right) + k_m(s_{m_{mut}}) \frac{a_m P^0(s_h)}{1 + a_m h_m P^0(s_h)}$$
$$- g_m(s_{m_{mut}})l \tag{14}$$

Sets of strategies (s_m^*, s_h^*) that make fitness Eqs. (13) and (14) vanishing are called Singular Coalitions. Invasibility and convergence criteria of such coalitions can again be studied by differentiating fitness functions (Kisdi, 2006).

2.3 Numerical simulations of eco-evolutionary dynamics

Analytical tools of Adaptive Dynamics (Eqs. 7–10) were used when possible. In addition to this mathematical analysis, we undertook extensive numerical simulations, for two reasons. First, the analytical tools of Adaptive Dynamics are based on restrictive hypotheses (stable ecological equilibrium, rare mutations with weak phenotypic effects and clonal reproduction) (Dieckmann and Law, 1996; Geritz et al. 1998; Metz et al., 1992). Numerical simulations allow us to relax these hypotheses. Second, in the case of coevolution, only sufficient (but not necessary) conditions of convergence can be obtained analytically (Kisdi, 2006). When such conditions are violated, it cannot be guaranteed whether the system will converge towards the coalition or not. Details on the simulation procedures can be found in Supplementary material on https://www.dropbox.com/s/qa7g6usf3hpsejr/appendix_georgelin_aer.pdf?dl=0.

3. RESULTS

Below, we summarize the main results of our analysis. Full details of the mathematical computations and numerical simulations as well as details on the dynamics of plant–herbivore and plant–pollinator subsystems can be found in Supplementary material on https://www.dropbox.com/s/qa7g6usf3hpsejr/appendix_georgelin_aer.pdf?dl=0.

3.1 Evolution of herbivore sensitivity trait only

Depending on the trade-off shape, two qualitative dynamics are possible. For a convex or linear trade-off ($z_h \leq c_h$), the fitness gradient is always positive and then, ever-increasing sensitivities are selected (Fig. 2A). For a concave trade-off ($z_h > c_h$), sensitivity will reach a Continuously Stable Strategy (CSS), a singular strategy that is convergent and non-invasible. Evolution

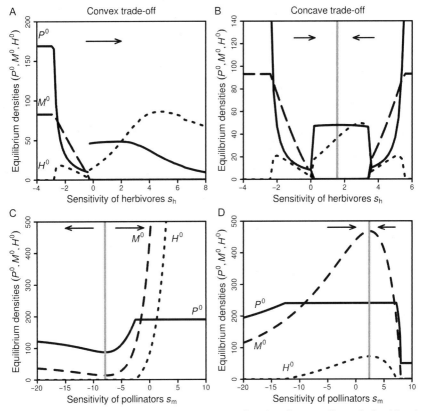

Figure 2 Density variations of plants (continuous lines), pollinators (long-dashed lines) and herbivores (dashed lines) at the ecological equilibrium during the evolution of herbivores (A and B) or of pollinators (C and D). Two types of trade-offs are depicted, convex (A and C) and concave (B and D). Vertical lines correspond to the sensitivity singular strategies s_h^* or s_m^* and arrows depict the direction of evolution. Parameters values are $r_p = 0.5$ t^{-1}; $r_m = 0.5$ t^{-1}; $K_p = 50$ ind m^{-2}; $K_m = 50$ ind m^{-2}; $a_m = 1$ ind^{-1} m^2 t^{-1}; $a_h = 1$ ind^{-1} m^2 t^{-1}; $h_m = 0.2$ t; $h_h = 0.1$ t; $k_p = 0.5$; $d_h = 3.5$ t^{-1}; $l = 1$ t^{-1}; $k_{m_0} = 1$; $g_{m_0} = 1$; $k_{h_0} = 1$; $g_{h_0} = 1$; $s_m = 2$ in graphs (A) and (B); $s_h = 1.25$ in graphs (C) and (D); $c_h = 0.3$ and $z_h = 0.2$ in (A); $c_h = 0.4$ and $z_h = 0.8$ in (B); $c_m = 0.5$ and $z_m = 0.1$ in (C); $c_m = 0.1$ and $z_m = 0.3$ in (D).

ceases at this point (Fig. 2B). The singular strategy is described by the following analytical formula:

$$s_h = \frac{1}{z_h} \left(\ln \frac{c_h}{g_{h_0}(z_h - c_h)} + \ln \frac{d_h}{l} \right) \tag{15}$$

Equation (15) is composed of two terms. The first term represents the evolutionary constraints associated with the trade-off shape (relative values of c_h and z_h). The second part stresses the importance of ecological conditions, represented by the ratio of natural mortality (d_h) to pesticide-driven mortality (l). Adaptive response to disturbance therefore strongly depends on background mortality. A low (high) level of natural mortality yields a low (high) equilibrium sensitivity.

Regardless of the trade-off shape, evolution of herbivore sensitivity always leads to a decrease in the densities of both plants and pollinators, as long as all species are present (Fig. 2A and B). Herbivore sensitivity evolution affects negatively the density of plants by increasing top-down pressures. Consequently, pollinator density decreases as well and given a high sensitivity to disturbance (high sensitivity s_m), pollinator population may even go extinct (Fig. 2A and B). In such instances, the disturbance leads to an Evolutionary Murder of pollinators. Once pollinators disappear, plant and herbivore densities may display cyclic dynamics (Fig. 2B). When pollinator sensitivity is sufficiently low to avoid extinction, however, herbivore evolution leads to an eco-evolutionary equilibrium where the three species coexist at reduced densities.

3.2 Evolution of pollinator sensitivity trait only

When only pollinators evolve, the existence of a singular strategy s_m^* is constrained by the trade-off shape. When the trade-off is linear ($z_m = c_m$), no such singularity exists. In this case, pollinators evolve towards ever-decreasing sensitivities when disturbance intensity parameter l is high and towards ever-increasing sensitivities when l is low. When the trade-off is not linear, a singular strategy exists, described by the following analytical formula:

$$s_m = \frac{1}{c_m - z_m} \left(\ln \left(\frac{z_m g_{m_0}}{c_m k_{m_0}} \right) + \ln \left(\frac{l}{(a_m P^0(s_h))/(1 + a_m h_m P^0(s_h))} \right) \right) \tag{16}$$

Similar to the evolution of herbivores, the properties of the singularity depend on the trade-off shape. For a convex trade-off ($z_m < c_m$), the singular strategy is invasible and non-convergent (i.e. a Repeller; Fig. 2C).

Conversely, for a concave trade-off ($z_m > c_m$), the singular strategy is a CSS (Fig. 2D). Yet, the equilibrium value s_m^* is dependent not only on the trade-off shape (first term of Eq. 16) but also on the ecological conditions (second term of Eq. 16), in specific, the ratio of mortality due to pesticide to resource benefits stemming from plants. Therefore, a higher density of plants will typically reduce the evolved sensitivity of pollinators.

Evolution of pollinator sensitivity, in the absence of herbivores, always leads to an increase in the densities of plants and pollinators (Fig. 2C and D). Eventually, plant density becomes sufficiently high for the herbivore to invade the system in spite of the disturbance. We call this phenomenon Evolutionary Facilitation: the evolution of pollinator sensitivity creates the conditions for herbivores to establish in the community. Once herbivores are present pollinator and herbivore densities increase with pollinator sensitivity evolution while plant density remains constant, due to the top-down control exerted by herbivores (Eqs. 4–6).

3.3 Coevolution

Regarding the coevolution of pollinators and herbivores, we find that if the two trade-offs are concave, the Singular Coalition is convergent and non-invasible (Absolute Convergence Stability; Kisdi, 2006). Coevolution therefore ends at this point. In the three other cases (when one or both of the trade-offs are convex), numerical simulations are necessary to determine the behaviour of the Singular Coalitions (see Supplementary material on https://www.dropbox.com/s/qa7g6usf3hpsejr/appendix_georgelin_aer.pdf?dl=0).

The outcomes of co-evolutionary dynamics, however, are also highly dependent on the evolutionary potentials of species. In Fig. 3A, we study the case where the coalition is convergent and non-invasible. We choose one initial point (triangle in Fig. 3A) and then run three simulation scenarios. In one scenario, pollinators and herbivores have the same mutation rates, while in the other two, herbivores evolve faster or slower than pollinators. For each scenario, we run 10 simulations and calculate the mean trajectories of sensitivity traits and population densities. When pollinators evolve faster or when both species evolve at the same speed, evolution ends at a Singular Coalition, as expected (dashed and dotted lines, Fig. 3B1, 3B2, 3C1 and 3C2). However, during the co-evolutionary dynamics, trait values and densities largely depend on these relative mutation rates. The interplay between pollinator and herbivore evolutionary dynamics also affects the direction of density variations. Densities first increase then decrease, when pollinators

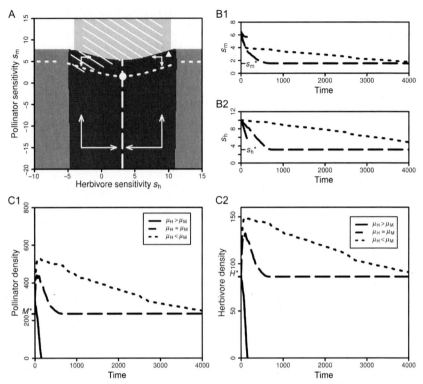

Figure 3 Numerical simulations of the eco-evolutionary dynamics of the community when the two trade-offs are concave and for different evolutionary potentials of the two species. (A) Community composition and evolutionary dynamics. Colours depict community composition. The dark region stands for the coexistence of the three species, dark grey for plant–pollinator community, light grey for plant–herbivore community and white for plant only. White hatching represents regions of cyclic dynamics. The dotted line depicts the singular strategy isocline of pollinators and the long-dashed line depicts the herbivore isocline. Arrows show the direction of evolution of sensitivity traits. Dot depicts the eco-evolutionary singularity. Triangle depicts the start of the simulations. (B) Sensitivity of pollinators (1) and herbivores (2) during the eco-evolutionary dynamics for the three different scenarios of evolutionary potentials. The two evolutionary equilibriums that are theoretically predicted (s_h^* and s_m^*) are depicted on the y-axis. (C) Density of pollinators (1) and herbivores (2) for the three same scenarios (μ_m is the pollinator mutation rate and μ_h is the herbivore one). Densities theoretically predicted at the eco-evolutionary equilibrium (H^* and M^*) are depicted on the y-axis. Parameters values are $r_p = 0.5\ t^{-1}$; $r_m = 0.5\ t^{-1}$; $K_p = 50\ \text{ind m}^{-2}$; $K_m = 50\ \text{ind m}^{-2}$; $a_m = 1\ \text{ind}^{-1}\ m^2\ t^{-1}$; $a_h = 1\ \text{ind}^{-1}\ m^2\ t^{-1}$; $h_m = 0.2\ t$; $h_h = 0.1\ t$; $k_p = 0.5$; $d_h = 3.5\ t^{-1}$; $l = 1\ t^{-1}$; $k_{m_0} = 1$; $g_{m_0} = 1$; $k_{h_0} = 1$; $g_{h_0} = 1$; $c_m = 0.2$; $c_h = 0.2$; $z_m = 0.4$; $z_h = 0.4$; $\mu_m = 10^{-6}\ \text{ind}^{-1}$ and $\mu_h = 10^{-5}, 10^{-6}$ or $10^{-7}\ \text{ind}^{-1}$.

evolve faster or at the same pace (Fig. 3C1 and 3C2). By contrast, when herbivores evolve faster than pollinators, densities always decrease and Evolutionary Murder of pollinators is observed (plain lines, Fig. 3C1 and 3C2).
Subsequently, the community collapses, as cyclic dynamics confine plants
and herbivores to low populations. In this case, pollinator evolution is
too slow to rescue the community from collapsing.

The evolutionary potential of species is however not simply defined by
mutation rates μ_i or mutation amplitudes σ_i^2. In Eqs. (7) and (8), evolutionary
potentials are also dependent on H^0 and M^0, respectively, the population
densities of herbivores and pollinators. Therefore, because we have observed
that the evolution of one partner can strongly influence the density of the
other, we expect that co-evolutionary dynamics can create strong
density-dependent effects constraining the evolutionary potentials of the
two species. In Fig. 4, we depict the case of a convex trade-off for herbivores

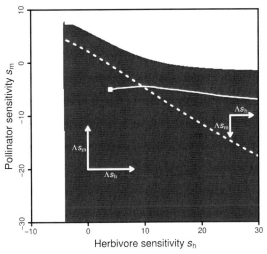

Figure 4 Eco-evolutionary dynamics of the community when the trade-off of herbivores is convex and trade-off of pollinators is concave. The dark zone depicts the coexistence of the three species. The dotted line represents the evolutionary isocline of
pollinators. There is no herbivore isocline, since the trade-off is convex so that their sensitivity trait increases continuously. Arrows show the direction of evolution of sensitivity
traits. In this case, herbivore sensitivity s_h always increases, while pollinator sensitivity s_m
theoretically converges towards the isocline. The continuous line shows a
co-evolutionary trajectory, in a scenario where herbivores evolve faster than pollinators.
The white square depicts the beginning of the simulation. Parameter values are:
$r_p = 0.5\ t^{-1}$; $r_m = 0.5\ t^{-1}$; $K_p = 50$ ind m^{-2}; $K_m = 50$ ind m^{-2}; $a_m = 1$; $a_h = 1$; $h_m = 0.2$;
$h_h = 0.1$; $k_p = 0.5$; $d_h = 3.5\ t^{-1}$; $l = 1\ t^{-1}$; $k_{m_0} = 1$; $g_{m_0} = 1$; $k_{h_0} = 1$; $g_{h_0} = 1$; $c_m = 0.2$;
$c_h = 0.2$; $z_m = 0.4$; $z_h = 0.1$; $\mu_m = 10^{-6}$ ind^{-1} and $\mu_h = 10^{-5}$ ind^{-1}.

and concave trade-off for pollinators. In this case, we expect that sensitivity trait of herbivores will always increase while trait of pollinators will stick to the isocline. However, when herbivores evolve faster than pollinators, herbivore evolution is sufficiently fast for pollinator sensitivity to cross its isocline, then to be brought further and further from it, therefore causing a large maladaptation in the pollinator population. Indeed, herbivore evolution strongly decreases the density of pollinators, thereby deteriorating their evolutionary potential. Evolution of pollinators towards their isocline is therefore very slow (Fig. 4). Coevolution here, and differences in evolutionary potentials, makes pollinator populations constantly maladapted to the disturbance.

4. DISCUSSION

Ecological dynamics of mixed antagonistic and mutualistic interactions create complex density-dependent effects (Georgelin and Loeuille, 2014). We show here that such indirect effects constrain the evolutionary and co-evolutionary dynamics that can be observed in such systems. These eco-evolutionary dynamics and their consequences for community composition and structure may have profound implications for the simultaneous management of pollination service and pest control in agricultural landscapes.

When only herbivore sensitivity evolves, we find that herbivore adaptive response is detrimental to the maintenance of the whole community. Here, we observe that herbivore evolution (1) lowers its own density and (2) also lowers the densities of the other species (plants and pollinators). Effect (1) is consistent with other theoretical works suggesting that adaptive response of a given population should not be expected to lead systematically to higher population levels (Dieckmann and Ferrière, 2004), and can even lead to population extinction (Evolutionary Suicide; Gyllenberg and Parvinen, 2001). In our case, evolution lowers the herbivore population making it vulnerable to demographic stochasticity (Evolutionary Deterioration; Matsuda and Abrams, 1994). We also note that such deteriorating effects are hierarchical. In the simulations we made, pollinators disappear first during antagonistic evolution. Jones et al. (2009), in a model of antagonistic–mutualistic evolving community, show that evolution of antagonists can lead the whole community to extinction. Their model assumes that the mutualism is obligatory and antagonists need the presence of mutualists to survive (Jones et al., 2009). Such obligatory links facilitate the propagation of deteriorating evolutionary effects in their community. Our results confirm the idea that mutualists may suffer from antagonistic evolution and are prone to extinction, even in the absence of obligatory links among species.

The evolutionary responses of herbivores to disturbances have large implications for the conservation and management of agricultural communities. In agricultural landscapes, direct mortality due to pesticides or habitat changes is often proposed to explain the decrease of pollinator densities or their extinctions (González-Varo et al., 2013; Potts et al., 2010). Our model suggests that the (often observed) adaptive response of herbivores to pesticides may create indirect ecological effects that decrease pollinator populations and therefore contribute to the loss of pollination services and to the destabilization of communities. While the occurrence of such effects needs to be tested, they illustrate the importance of an evolutionary perspective for the management of agricultural systems (Loeuille et al., 2013).

Contrary to herbivore evolution, we find that pollinator adaptive response to disturbance is beneficial to the community maintenance. Pollinator evolution increases plant density. This creates indirect bottom-up effects that enhance pollinator and herbivore densities. Also, pollinator evolution can ultimately create and maintain the conditions of the establishment and survival of herbivores in the system (a phenomenon we name Evolutionary Facilitation). A recent surge of interest exists in co-evolutionary models of mutualisms (Ferrière et al., 2002; Gomulkiewicz et al., 2003; Guimarães et al., 2011; Nuismer et al., 2012; Zhang et al., 2011), including in antagonistic–mutualistic systems (Jones et al., 2009). These models focus on the role of cheaters or exploiters in mutualistic interactions. Considering the evolution of birth rate, Jones et al. (2009) find that mutualism evolution can lead to an Evolutionary Purging (i.e. Evolutionary Murder of the exploiter by the mutualist), which contradicts our results. Jones et al. (2009) focus on mutualists that are pollinating seed parasites (as it is found in fig and yucca systems for example). Hence, under some conditions (weak intraspecific competition of mutualists and antagonists), the density of pollinating seed parasites is high and creates high parasitism pressure on host plants. This reduces plant density and the availability of non-parasited seeds that are necessary for antagonist survival. By contrast, mutualists, in our model, always have a positive effect on plant growth rate, which explains the discrepancy between the result generated by our and Jones' models. Positive effects of evolution on community structure have already been shown in other mutualistic network models. Evolution and coevolution between partners can lead to nestedness in mutualistic networks (Zhang et al., 2011), an architecture that increases the maintenance and resilience of such communities (Bascompte et al., 2006; Thébault and Fontaine, 2010). We note also that the possibility of Evolutionary Facilitation we observe here is not restricted to the evolution of mutualistic interactions and can also be observed in a food web context. Evolutionary changes in foraging behaviour of predators can thereby favour the maintenance of community diversity (Loeuille and Loreau, 2009; Urban, 2013).

Evolutionary Facilitation has important implications for the conservation of natural communities. The current concern about the loss of mutualistic interactions in natural communities needs to integrate such an evolutionary perspective (Kiers et al., 2010). Mutualistic interactions are not only important as a support for the ecosystem functioning or services they provide (e.g. pollination services; Georgelin and Loeuille, 2014; Potts et al., 2010) but also for their positive influence on the evolutionary dynamics of communities. The loss of pollinators may in turn decrease such positive evolutionary effects and threaten the different species of the community through the modification of evolutionary or co-evolutionary trajectories. The occurrence of Evolutionary Facilitation in our model strengthens the idea that pollinators may help the population of herbivores, in the presence of pesticide application. This result is found both on an ecological scale (Georgelin and Loeuille, 2014) and on an evolutionary scale. A pesticide that is highly selective (i.e. that does not affect pollinator populations too much) would not be efficient because, in this case, high pollinator populations will buffer herbivore populations against this disturbance (Georgelin and Loeuille, 2014).

The range of possible dynamics is much larger when coevolution occurs. We find that a combination of positive and negative density-dependent effects arise, respectively, due to mutualistic and antagonistic evolution, and the community outcomes are heavily constrained by the evolutionary potentials of pollinators and herbivores. When herbivores have a higher evolutionary potential, the phenomenon of Evolutionary Murder of pollinators may occur. However, if pollinator evolution is sufficiently fast, it can rescue the community from this Evolutionary Murder. Evolutionary potentials are constrained by several attributes, such as rate and amplitude of mutations, population densities and generation times of organisms (Dawkins and Krebs, 1979; Fisher, 1930). Higher densities lead to higher genetic and phenotypic variabilities in populations, enhancing their responses to environmental changes (Leimu et al., 2006). Also, the rate of appearance of new mutations is thought to be higher in larger populations, which facilitates adaptation (Lanfear et al., 2013). Considering agricultural landscapes, pests have by definition high densities. Furthermore, pesticide use is a common response to outbreaks when pest densities are even more important (Barbosa and Shultz, 1987). In such circumstances, herbivore pest evolutionary response is likely to be fast compared to pollinators, whose populations may be already deteriorated by other detrimental factors (González-Varo et al., 2013). Consistent with this view, pest resistances to new toxic compounds have been frequently observed, often in a few generations (Mallet, 1989; Rex Consortium, 2013), while the evolutionary responses of pollinators remain largely undocumented (but see Cresswell et al., 2012). In our model, the relative evolutionary potentials of the two species vary during the co-evolutionary dynamics, due to density changes that are evolutionary driven (Fussmann et al., 2007). In the last scenario we investigate, sensitivity of pollinator is maintained far from its optimum and pollinator population remains maladapted, because fast herbivore evolution strongly reduces pollinator density. This result suggests the need for the joined consideration of the community context and the interplay between different interaction types, when studying the evolutionary responses of populations to disturbances (Fussmann et al., 2007; Loeuille et al., 2013). While such an integrative approach is particularly difficult in the field, not accounting for coevolution and evolutionary potentials of the different species may lead to misconceptions about the dynamics of communities in the face of global changes (Dieckmann and Ferrière, 2004; Loeuille et al., 2013).

The question of trade-off is of particular interest, here. We link the evolution of sensitivity to disturbance to decreases in the growth rates of organisms, through allocation costs. We use exponential functions to depict the relationship between these two components. This allows us to model a wide variety of trade-off shapes, convex, concave or linear. We however acknowledge that other shapes are possible (e.g. saturating functions). We show that these shapes are crucial for the evolutionary dynamics of sensitivity traits, constraining the fate of the community. We have few empirical knowledge regarding the shapes of these trade-offs in natural systems, and we believe that describing such trade-offs is an important challenge for future insecticide resistance research. Recent works have stressed the importance of trade-off shapes for other evolutionary scenarios (defence and growth in predator–prey systems) (Kasada et al., 2014), and we hope this kind of work will develop in future years. Another limit is our focus on allocation costs. In some situations, such allocation costs associated to resistance to pesticides have not been found (Lopes et al., 2008). Resistance can even confer selective advantages in some environments without pesticides (McCart et al., 2005). Instead of direct impacts on reproduction rate, costs can be associated to traits that are difficult to measure, or acting in a different environment or at a different time in the life cycle of organisms (e.g. Gazave et al., 2001). Gazave et al. (2001) found that resistant phenotypes in the mosquito *Culex pipiens* have a lower survival during the wintering phase. Costs of resistance can also happen due to changes in interspecific interactions rather than from changes in life-history traits. Resistance to organophosphates in peach-potato aphids for instance reduces their response to alarm pheromones and may lead to a higher susceptibility to predation (Foster et al., 2003). In such instances, costs of resistance to insecticide can act in a different way than the ones we modelled here, and the analysis of associated evolutionary dynamics goes beyond the scope of the present chapter. Such an understanding of different cost structures is however an important step to fully understand the eco-evolutionary responses of communities to external disturbances.

Here, we discuss our results through the case of plant–pollinator–herbivore communities confronted with pesticides in agricultural landscapes. However, since we have little information on the eco-evolutionary dynamics of antagonistic–mutualistic communities confronted with external disturbances, we keep our model simple. Hence, these results may give insights for many multiple-interaction types communities, that are confronted with external disturbances, like climate changes (Parmesan, 2006), fragmentation and habitat destruction (Georgelin and Loeuille, 2014; González-Varo et al., 2013). The analysis shown here allows a first understanding of eco-evolutionary dynamics of antagonistic–mutualistic ecological communities, in a simple context. Additional complexities like space, or network architecture, would inevitably influence the results we report here. Another fruitful exercise would be to compare how adaptive processes differ from those found in other antagonistic/mutualistic networks. How positive and negative feedbacks (or activation/inhibition mechanisms) affect stability, and adaptation is for instance a current question in gene regulatory networks (Crombach and Hogeweg, 2008; Pinho et al., 2014). Social networks also contain positive and negative relationships, and Structural Balance Theory can then be used to study how the adaptive modification of social relationships can affect group stability (Doreian, 2002; Hummon and Doreian, 2003). Because adaptation is guided by different selective processes acting on units of different scales in social networks, gene networks and ecological networks, the effect of adaptation on stability and complexity is expected to differ among network types. However, comparing such differences could help us to understand how network dynamics *per se* matters for adaptation, and to establish interdisciplinary links among different scales of organization.

ACKNOWLEDGEMENTS

In addition to INRA, IRD, UPMC, UPEC, Paris Diderot and CNRS that routinely support our research, the authors acknowledge the financial support of the Region Ile de France (DimAstrea grant allocated to N. L. and G. K.). We also thank Nicolas Chazot and Élisa Thébault for comments on an early version of this chapter.

REFERENCES

Abrams, P.A., Matsuda, H., 2005. The effect of adaptive change in the prey on the dynamics of an exploited predator population. Can. J. Fish. Aquat. Sci. 62, 758–766.

Barbosa, P., Shultz, J.C., 1987. Insect Outbreaks. Academic Press, Waltham.

Bascompte, J., Jordano, P., Olesen, J.M., 2006. Asymmetric coevolutionary networks facilitate biodiveristy maintenance. Science 312, 431–433.

Bass, C., Puinean, A.M., Zimmer, C.T., Denholm, I., Field, L.M., Foster, S.P., Gutbrod, O., Nauen, R., Slater, R., Williamson, M.S., 2014. The evolution of insecticide resistance in the peach potato aphid, *Myzus persicae*. Insect Biochem. Molec. 51, 41–51.

Bernays, E., Graham, M., 1988. On the evolution of host specificity in phytophagous arthropods. Ecology 69, 886–892.

Bronstein, J.L., Dieckmann, U., Ferrière, R., 2004. Coevolutionary dynamics and the conservation of mutualisms. In: Ferrière, R., Dieckmann, U., Couvet, D. (Eds.), Evolutionary Conservation Biology. Cambridge University Press, Cambridge, pp. 305–326.

Carrière, Y., Deland, J.P., Roff, D.A., Vincent, C., 1994. Life-history costs associated with the evolution of insecticide resistance. Proc. R. Soc. Lond. B Biol. Sci. 258, 35–40.

Christiansen, F.B., 1991. On conditions for evolutionary stability for a continuously varying character. Am. Nat. 138, 37–50.

Coltman, D.W., O'Donoghue, P., Jorgenson, J.T., Hogg, T.H., Strobeck, C., Festa-Bianchet, M., 2003. Undesirable evolutionary consequences of trophy hunting. Nature 426, 655–658.

Cresswell, J.E., Page, C.J., Uygun, M.B., Holmbergh, M., Li, Y., Wheeler, J.G., Laycock, I., Pook, C.J., Hempel de Ibarra, N., Smirnoff, N., Tyler, C.R., 2012. Differential sensitivity of honey bees and bumble bees to a dietary insecticide (imidacloprid). Zoology 115, 365–371.

Crombach, A., Hogeweg, P., 2008. Evolution of evolvability in gene regulatory networks. PLoS Comput. Biol. 4, e1000112.

Dawkins, R., Krebs, J.R., 1979. Arms races between and within species. Proc. R. Soc. Lond. B Biol. Sci. 205, 489–511.

de Mazancourt, C., Dieckmann, U., 2004. Trade-off geometries and frequency-dependent selection. Am. Nat. 164, 765–778.

Desneux, N., Decourtye, A., Delpuech, J.M., 2007. The sublethal effects of pesticides on beneficial arthropods. Annu. Rev. Entomol. 52, 81–106.

Després, L., David, J.P., Gallet, C., 2007. The evolutionary ecology of insect resistance to plant chemicals. Trends Ecol. Evol. 22, 298–307.

Dieckmann, U., Ferrière, R., 2004. Adaptive dynamics and evolving biodiversity. In: Ferrière, R., Dieckmann, U., Couvet, D. (Eds.), Evolutionary Conservation Biology. Cambridge University Press, Cambridge, pp. 188–224.

Dieckmann, U., Law, R., 1996. The dynamical theory of coevolution: a derivation from stochastic process. J. Math. Biol. 34, 579–612.

Doreian, P., 2002. Event sequences as generators of social network evolution. Soc. Networks 24, 93–119.

Eshel, I., 1983. Evolutionary and continuous stability. J. Theor. Biol. 103, 99–111.

Ferrière, R., Bronstein, J.L., Rinaldi, S., Law, R., Gauduchon, M., 2002. Cheating and the evolutionary stability of mutualisms. Proc. R. Soc. Lond. B Biol. Sci. 269, 773–780.

Fisher, R.A., 1930. The Genetical Theory of Natural Selection. Clarendon Press, Oxford.

Fontaine, C., Thébault, E., Dajoz, I., 2009. Are insect pollinators more generalist than insect herbivores? Proc. R. Soc. Lond. B Biol. Sci. 276, 3027–3033.

Fontaine, C., Guimarães, P.R., Kéfi, S., Loeuille, N., Memmot, J., van der Putten, W.H., van Veen, F.J.F., Thébault, E., 2011. The ecological and evolutionary implications of merging different types of networks. Ecol. Lett. 14, 1170–1181.

Foster, S.P., Young, S., Williamson, M.S., Duce, I., Denholm, I., Devine, G.J., 2003. Analogous pleiotropic effects of insecticide resistance genotypes in peach-potato aphids and houseflies. Heredity 91, 98–106.

Fussmann, G.F., Loreau, M., Abrams, P.A., 2007. Eco-evolutionary dynamics of communities and ecosystems. Funct. Ecol. 21, 465–477.

Gazave, E., Chevillon, C., Lenormand, T., Marquine, M., Raymond, M., 2001. Dissecting the cost of insecticide resistance genes during the overwintering period of the mosquito *Culex pipiens*. Heredity 87, 441–448.

Georgelin, E., Loeuille, N., 2014. Dynamics of coupled mutualistic and antagonistic interactions, and their implications for ecosystem management. J. Theor. Biol. 346, 67–74.

Geritz, S.A.H., Kisdi, E., Meszéna, G., Metz, J.A.J., 1998. Evolutionary singular strategies and the adaptive growth and branching of the evolutionary tree. Evol. Ecol. 12, 35–57.

Gomulkiewicz, R., Holt, R.D., 1995. When does evolution by natural selection prevent extinction. Evolution 49, 201–207.

Gomulkiewicz, R., Nuismer, S.L., Thompson, J.N., 2003. Coevolution in variable mutualisms. Am. Nat. 162, S80–S93.

González-Varo, J.P., Biesmeijer, J.C., Bommarco, R., Potts, S.G., Schweiger, O., Smith, H.G., Steffen-Dewenter, I., Szentgyörgyi, H., Woyciechowski, M., Vilà, M., 2013. Combined effects of global change pressures on animal-mediated pollination. Trends Ecol. Evol. 28, 524–530.

Goulson, D., 2013. An overview of the environmental risks posed by neonicotinoid insecticides. J. Appl. Ecol. 50, 977–987.

Guimarães, P.R., Jordano, P., Thompson, J.N., 2011. Evolution and coevolution in mutualistic networks. Ecol. Lett. 14, 877–885.

Gyllenberg, M., Parvinen, K., 2001. Necessary and sufficient conditions for evolutionary suicide. Bull. Math. Biol. 63, 981–993.

Holland, J.N., DeAngelis, D.L., Bronstein, J.L., 2002. Population dynamics and mutualisms: functional responses of benefits and costs. Am. Nat. 159, 231–244.

Holling, C.S., 1959. Some characteristics of simple types of predation and parasitism. Can. Entomol. 91, 385–398.

Hummon, N.P., Doreian, P., 2003. Some dynamics of social balance processes: bringing Heider back into balance theory. Soc. Networks 25, 17–49.

Jang, S.R.J., 2002. Dynamics of herbivore-plant-pollinator models. J. Math. Biol. 44, 129–149.

Jones, E.I., Ferrière, R., Bronstein, J.L., 2009. Eco-evolutionary dynamics of mutualists and exploiters. Am. Nat. 174, 780–794.

Kasada, M., Yamamichi, M., Yoshida, T., 2014. Form of an evolutionary tradeoff affects eco-evolutionary dynamics in a predator-prey system. Proc. Natl. Acad. Sci. 111, 16035–16040.

Kiers, T.E., Palmer, T.M., Ives, A.R., Bruno, J.F., Bronstein, J.L., 2010. Mutualisms in a changing world: an evolutionary perspective. Ecol. Lett. 13, 1459–1474.

Kisdi, E., 2006. Trade-off geometries and the adaptive dynamics of two co-evolving species. Evol. Ecol. Res. 8, 959–973.

Klein, A.M., Vaissière, B.E., Cane, J.H., Steffan-Dewenter, I., Cunningham, S.A., Kremen, C., Tscharntke, T., 2007. Importance of pollinators in changing landscapes for world crops. Proc. R. Soc. Lond. B Biol. Sci. 274, 303–313.

Lanfear, R., Kokko, H., Eyre-Walker, A., 2013. Population size and the rate of evolution. Trends Ecol. Evol. 29, 33–41.

Leimu, R., Mutikainen, P., Koricheva, J., Fischer, M., 2006. How general are positive relationships between plant population size, fitness and genetic variation? J. Ecol. 94, 942–952.

Loeuille, N., Loreau, M., 2009. Emergence of complex food web structure in community evolution models. In: Verhoef, H.A., Morin, P.J. (Eds.), Community Ecology. Oxford University Press, Oxford, pp. 163–178.

Loeuille, N., Barot, S., Georgelin, E., Kylafis, G., Lavigne, C., 2013. Eco-evolutionary dynamics of agricultural networks: implications for a sustainable management. Adv. Ecol. Res. 48, 339–435.

Lopes, P.C., Sucena, E., Santos, M.E., Magalhães, S., 2008. Rapid experimental evolution of pesticide resistance in C. elegans entails no costs and affects the mating system. PLoS One 3, e3741.

Mallet, J., 1989. The evolution of insecticide resistance: have the insects won. Trends Ecol. Evol. 4, 336–340.

Matsuda, H., Abrams, P.A., 1994. Runaway evolution to self-extinction under asymmetrical competition. Evolution 48, 1764–1772.

Maynard Smith, J., 1982. Evolution and the Theory of Games. Cambridge University Press, Cambridge.

McCart, C., Buckling, A., ffrench-Constant, R.H., 2005. DDT resistance in flies carries no cost. Curr. Biol. 15, 587–589.

Metz, J.A.J., Nisbet, R.M., Geritz, S.A., 1992. How should we define "fitness" for general ecological scenarios. Trends Ecol. Evol. 7, 198–202.

Nuismer, S.L., Jordano, P., Bascompte, J., 2012. Coevolution and the architecture of mutualistic networks. Evolution 67, 338–354.

Oksanen, L., Fretwell, S.D., Arruda, J., Niemela, P., 1981. Exploitation ecosystems in gradients of primary production. Am. Nat. 118, 240–261.

Olsen, E.M., Heino, M., Lilly, G.R., Morgan, M.J., John, B., Ernande, B., Dieckmann, U., 2004. Maturation trends indicative of rapid evolution preceded the collapse of northern cod. Nature 428, 932–935.

Palumbi, S.R., 2001. Humans as the world's greatest evolutionary force. Science 293, 1786–1790.

Parmesan, C., 2006. Ecological and evolutionary responses to recent climate change. Annu. Rev. Ecol. Evol. Syst. 37, 637–669.

Pelletier, F., Garant, D., Hendry, A.P., 2009. Eco-evolutionary dynamics. Philos. Trans. R. Soc., B, Biol. Sci. 364, 1483–1489.

Pelosi, C., Barot, S., Capowiez, Y., Hedde, M., Vandenbulcke, F., 2013. Pesticides and earthworms. A review. Agron. Sustain. Dev. 34, 199–228.

Pinho, R., Garcia, V., Irimia, M., Feldman, M.W., 2014. Stability depends on positive autoregulation in Boolean gene regulatory networks. PLoS Comput. Biol. 10, e1003916.

Potts, S.G., Biesmeijer, J.C., Kremen, C., Neumann, P., Schweiger, O., Kunin, W.E., 2010. Global pollinator declines: trends, impacts and drivers. Trends Ecol. Evol. 25, 345–353.

Rex Consortium, 2013. Heterogeneity of selection and the evolution of resistance. Trends Ecol. Evol. 28, 110–118.

Rosenzweig, M.L., MacArthur, R.H., 1963. Graphical representation and stability conditions of predator-prey interactions. Am. Nat. 97, 209–223.

Schoener, T.W., 2011. The newest synthesis: understanding the interplay of evolutionary and ecological dynamics. Science 331, 426–429.

Strauss, S.Y., 1997. Floral characters link herbivores, pollinators and plant fitness. Ecology 78, 1640–1645.

Strauss, S.Y., Siemens, D.H., Decher, M.B., Mitchell-Olds, D., 1999. Ecological costs of plant resistance to herbivores in the currency of pollination. Evolution 53, 1105–1113.

Thébault, É., Fontaine, C., 2010. Stability of ecological communities and the architecture of mutualistic and trophic networks. Nature 329, 853–856.

Urban, M.C., 2013. Evolution mediates the effects of apex predation on aquatic food webs. Proc. R. Soc. Lond. B Biol. Sci. 280, 20130859.

Vitousek, P.M., 1997. Human domination of Earth's ecosystems. Science 277, 494–499.

Wootton, J.T., 1994. The nature and consequences of indirect effects in ecological communities. Annu. Rev. Ecol. Syst. 25, 443–466.

Wright, D.H., 1989. A simple, stable model of mutualism incorporating handling time. Am. Nat. 134, 664–667.

Zhang, F., Hui, C., Terblanche, J.S., 2011. An interaction switch predicts the nested architecture of mutualistic networks. Ecol. Lett. 14, 797–803.

CHAPTER SIX

Population and Community Body Size Structure Across a Complex Environmental Gradient

Anthony I. Dell[‡,1], Lei Zhao[§,¶], Ulrich Brose[†], Richard G. Pearson*, Ross A. Alford*

*College of Marine and Environmental Sciences and Centre for Tropical Biodiversity and Conservation, James Cook University, Townsville, Queensland, Australia
[†]Systemic Conservation Biology, Department of Biology, Georg-August University Göttingen, Göttingen, Germany
[‡]National Great Rivers Research and Education Center (NGRREC), East Alton, Illinois, USA
[§]Research Centre for Engineering Ecology and Nonlinear Science, North China Electric Power University, Beijing, China
[¶]Department of Life Sciences, Imperial College London, London, United Kingdom
[1]Corresponding author: e-mail address: adell@lc.edu

Contents

Advances in Ecological Research, Volume 52
ISSN 0065-2504
http://dx.doi.org/10.1016/bs.aecr.2015.02.002

Abstract

We monitored the invertebrate community of leaf litter in and around a drying inter-mittent pool bed to explore patterns of ecological organisation across a complex envi-ronmental gradient, with particular focus on population and community size structure. We measured the body size of 24,609 individuals from 313 taxa ranging over 6 orders of magnitude in size to explore how the functional properties of individuals, populations and communities are affected by moisture (aquatic vs. terrestrial) and light (diurnal vs. nocturnal), and how these properties change across the aquatic–terrestrial habitat tran-sition that occurs as the pool bed dried. We found strong effects of moisture on some population (size structure) and many community (species richness, abundance, even-ness, biomass and size structure) properties, with additional temporal effects across the aquatic–terrestrial ecotone. There was no difference between diurnal and nocturnal populations or communities. Our results facilitate understanding of how the physical environment influences functional attributes, and particularly the size structure, of natural populations and communities.

1. INTRODUCTION

Identification of ecological patterns that span habitats and domains of life, and that link attributes of individuals to communities and ecosystems, can offer considerable insight into the universal mechanisms that structure ecological systems (Dell et al., 2011; Nathan et al., 2008; Pawar et al., 2012; Schramski et al., 2015; Simpson et al., 2010). Similarly, identification of patterns and processes that are unique to particular habitats can shed light on the biological mechanisms that drive ecological complexity across land-scapes. Such insights are essential for answering basic questions about how natural ecosystems operate, how humans are affecting natural systems and how these effects can be managed (Petchey and Belgrano, 2010). This is par-ticularly true for predicting how ecological systems will respond to future environmental scenarios that are beyond the boundaries of those currently observed (Dell et al., 2014c; McGill et al., 2006).

Body size is a key trait of individuals, influencing many biological pro-cesses central to their ecology and evolution (Brown et al., 2004; Kalinkat et al., 2013b; Pawar et al., 2012; Peters, 1983; Schmidt-Nielsen, 1984).

Because size is so important to the ecology of individuals, patterns in the body size of co-occurring individuals have important implications for the functioning and dynamics of higher levels of ecological organisation, such as populations and local communities (Brose et al., 2006; Brown et al., 2004; Gaston et al., 2001; Kalinkat et al., 2013b; Petchey et al., 2008; White et al., 2007; Woodward et al., 2005a,b). Understanding the biological mechanisms that determine body size distributions across levels of ecological organisation (i.e. individuals, populations and communities) is central for testing and validating ecological theory (Pawar et al., 2012; Schramski et al., 2015) and for understanding and predicting effects of human activities that alter size distributions, such as hunting, fishing and conversion of native plant communities for agriculture or urbanisation (Achard et al., 2002; Angelsen and Kaimowitz, 2001; Estes et al., 2011; Jennings and Blanchard, 2004; Roman and Palumbi, 2003).

While there are universal biological and physical constraints on size structure that operate across environments (Brose et al., 2005, 2006; White et al., 2007; Woodward et al., 2005a), the physical environment can place additional constraints (Denny, 1990; Riede et al., 2011) and might affect size distributions in unique ways, such that different habitats should have unique signatures in their population and community size distributions (Brose et al., 2005, 2006; Yvon-Durocher et al., 2011). Frequently, there are substantial taxonomic differences between communities in different environments, such as between aquatic and terrestrial or nocturnal and diurnal environments, but how these differences influence the functional properties of populations and communities, such as their size structure, is not well understood (Chase, 2000; Link, 2002; Shurin et al., 2006; Yvon-Durocher et al., 2011). Because the physical environment affects properties of individuals—such as how they move, behave and interact—that influence higher levels of ecological organisation (Dell et al., 2011, 2014c; Denny, 1990; Kalinkat et al., 2013a; Pawar et al., 2012), effects of the physical environment should manifest as functional differences between populations and communities (Yvon-Durocher et al., 2011). Uncovering these patterns could help elucidate the drivers of structural and functional differences between communities from different habitats.

A key step in understanding the link between size structure and the physical environment is characterisation of patterns in real systems. However, little is known about (i) whether the physical environment influences these patterns in systematic ways, (ii) how size structures are related at different levels of ecological organisation (across individuals, populations and communities) and (iii) how patterns of size structure co-vary within sets of interacting populations in a local community. We address each of these issues in this chapter.

We used the unique physical environment in and around a drying intermittent pool bed to assess how the physical environment (moisture and light) influences the size structure of populations and local communities. Intermittent pool beds periodically and predictably cycle between aquatic and terrestrial habitat in the same location, so they provide an opportunity to examine functional differences in local communities as they naturally transition from aquatic to terrestrial habitat during pool bed drying. Thus, they permit exploration of ecological differences between habitats in a way that avoids confounding sources of variation, such as differences in spatial or temporal scale, taxonomic resolution or sampling and laboratory protocols (Dell et al., 2014a; Yvon-Durocher et al., 2011). Importantly, the limited spatial and temporal scale of our study means that individual organisms were able to colonise anywhere in the sampled area (i.e. local communities in and around pool beds were drawing from the same species pool). We expected that moisture and light would affect population and community size structure because they both influence how organisms move, behave and interact. For example, physical constraints on movement are very different in water than on land (Denny, 1990) and differences in ambient light can affect how predators and prey detect each other (Dieguez, 2003; Fraser and Metcalfe, 1997; Gilbert and Hampton, 2001). Nonetheless, we had no prior expectations about how these compositional changes would affect population and communities size distributions, due primarily to the complexity of processes that link the ecology of individuals to the trait distributions of higher levels of ecological organisation.

2. FIELD EXPERIMENT METHODS

2.1 Study Site

Our study was carried out in Goondaloo Creek, a small stream in northeastern Australia (Fig. A1). Annual surface flow within Goondaloo Creek is intermittent, resulting from the seasonal tropical climate of the region and the steep topography of the streambed. Surface flow normally commences with the onset of the wet season in January and ceases between March and May (depending on the extent of the wet season) after which numerous natural intermittent pools remain. Larger pools can persist for up to 5 months before drying, although the duration of each is variable and depends on their size, substrate, shading and groundwater seepage

(Dell et al., 2014a; Smith and Pearson, 1987). The experiment was undertaken within a small side channel of Goondaloo Creek that only experienced flooding during heavy rainfall (Fig. A1). The streambed at this site was composed of rocks interspersed with sand. Riparian plant species (Table A1) provided a canopy cover of about 50% across the creek bed. Under these conditions, most of the streambed receives at least a few hours of direct sunlight each day. Deciduous species provide a steady input of leaves into the stream bed from July to September (Smith and Pearson, 1987). During the wet season, flooding washes the majority of this litter downstream so that each year leaf packs within the streambed consist of recently abscised leaves. A diverse community of macroinvertebrates occupies the pools of Goondaloo Creek (Dell et al., 2014a; Smith and Pearson, 1987), and although fish have been recorded in its lower reaches, they rarely occur as far upstream as the study site (Dell et al., 2014a).

2.2 Experimental Pool Bed

An experimental pool bed was constructed in the dry season in 2001. We used this pool bed to control pool topography and its filling and drying regime. The pool bed consisted of a plywood frame embedded within the natural streambed, positioned where a natural intermittent pool bed normally formed (Fig. 1). Once embedded in the streambed, the frame was covered with rubber aquarium liner (JR'S Foam & Rubber Pty Ltd) and a 3-cm

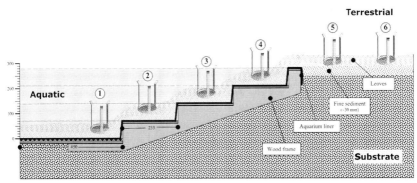

Figure 1 Profile view of one half of the experimental pool bed, which consisted of a wood frame embedded within the substrate of a natural streambed. Sampling trays were positioned at six levels (numbered), running from the centre of the pool bed into the surrounding terrestrial landscape. As pool depth decreased over time, levels 1–4 progressively dried (level 4 lost surface water on 7 October 2002, level 3 on 11 October 2002, level 2 on 15 October 2002 and level 1—and the entire pool bed—on 19 October 2002; see Table 1).

layer of natural sediment, excavated from the same location and passed through a 2.5-cm-mesh sieve. The wooden frame consisted of four flat concentric circular levels in a stepped design, so that the topography of the pool bed was radially symmetrical, with depth decreasing from the middle of the pool to its margins (Fig. 1). Burying the wooden frame within natural substrate removed any unnatural barrier to the movement of non-flying animals.

In June 2002—1 year following pool bed construction and when surface flow had ceased—accumulated leaf litter was removed and the sediment was evenly distributed over the wooden frame. Sampling trays (see section 2.4) were placed in and around the pool bed, and a disk of filter paper was placed in each tray to provide an estimate of substrate moisture content (see section 2.6). Sampling trays were distributed across six levels (18 trays per level). Four levels were within the inner margins of the pool bed, and therefore corresponded to pool depth, and two levels were outside the pool margins and represented more permanent terrestrial habitat (Fig. 1). Within the pool bed (level 4 and below), conditioned leaf litter (see below) was placed evenly across the substrate to a depth of approximately 3 cm. Dry leaves, which had previously been collected but not immersed in the natural pools, were laid around the margin of the experimental pool bed for a distance of about 1.5 m (levels 5 and 6). A ruler was placed vertically in the middle of the pool so that pool depth could be measured. Water level was maintained by slow dripping from a single irrigation nozzle (Pope™ Veriflow® Dripper) of rainwater gravity fed from a 9000-L polyethylene tank (Gough Plastics) located nearby. The tank was sealed against faunal colonisation, and water was filtered through a 0.005-mm sediment filter (Raindance™) prior to delivery into the pool bed.

This method created a pool bed with a complex litter layer whose topography, volume, shape and substrate were known, which mimicked nearby natural pool beds, and whose water level could be reduced when desired by piercing the rubber pond liner at the desired level (Fig. A2).

2.3 Leaf Packs

We focused on litter communities because leaf packs occur within nearby natural intermittent pool beds throughout the year (Dell et al., 2014a; Smith and Pearson, 1987), serve as food and microhabitat for many aquatic and terrestrial invertebrates (Davies and Boulton, 2009; Dudgeon and Wu, 1999; Murphy and Lugo, 1986; Reddy, 1995; Richardson, 1992) and are easily manipulated. Leaves of multiple species were collected from the streambed in the dry season prior to sampling. A garden vacuum (Flymo™ Garden Blower Vac) was used to collect recently fallen leaves, and no leaves were collected from existing dry pool beds. Leaves were shaken vigorously over a 2.5-cm-mesh sieve to remove fine detritus and animals and to standardise initial leaf fragment size. Decomposition is inhibited when litter is dry (Reddy, 1995), so leaves were sealed in lightproof plastic garbage bags and stored in an air-conditioned room until required. One month prior to the start of the experiment, approximately three quarters of the leaves were immersed in a nearby natural pool to leach soluble compounds and allow initial colonisation of leaves by microbes, flora and fauna (Webster and Benfield, 1986; Xiong and Nilsson, 1997).

2.4 Sampling Protocol

Sampling trays allowed effective sampling at day and night and in both aquatic and terrestrial environments and allowed sampling of the entire community. Sampling trays were constructed from a 16-cm length of 10-cm-diameter PVC pipe with three guide legs cut out at the top and 0.05-mm heavy duty nylon mesh (Australian Filter Specialists) attached to the base (Fig. 1). Each tray sat embedded in the substrate with the three legs visible from the surface. Sampling involved taking a core of habitat (surface area of 78.54 cm^2 and volume of 2356 cm^3) from within the trays, minimising escape of flying, burrowing and crawling animals (Figs. 1, A2 and A3). This involved pushing a 10-cm diameter × 30-cm high, sharpened galvanised steel cylinder with a clear plastic lid through the litter and sediment within the guide legs until it lay flush with the base of the sampling tray to form a seal (Fig. A3). The entire assembly was removed from the pool bed and was placed in 95% ethanol preservative, after first removing the disk of filter paper that was used to estimate substrate moisture levels (see section 2.6, Fig. A3).

2.5 Sampling Schedule

Once filled, the experimental pool bed remained full for 96 days before sampling to allow for colonisation of the inhabiting community (Table 1). Sampling was undertaken on nine occasions over 56 days, beginning when the pool bed was full and ceasing 32 days after total surface water loss. Sampling was more frequent during pool drying, to capture the rapid ecological changes that occur (Dell et al., 2014a). Separate diurnal and nocturnal

Table 1 Details of Sampling Schedule and Its Relationship to Pool Depth

Sample[a]	Date	Days Since Pool Filled	Days Until Pool Dried[b]	Max. Pool Depth[c]	Level					
					1	2	3	4	5	6
1	25 September 2002	96	−24	245	−24	−20	−16	−12		
2	5 October 2002	106	−14	249	−14	−10	−6	−2		
3	9 October 2002	110	−10	173	−10	−6	−2	2		
4	13 October 2002	114	−6	108	−6	−2	2	6		
5	17 October 2002	118	−2	34	−2	2	6	10		
6	21 October 2002	122	2	−	2	6	10	14		
7	29 October 2002	130	10	−	10	14	18	22		
8	9 November 2002	141	21	−	21	25	29	33		
9	20 November 2002	152	32	−	32	36	40	44		

Sampling occurred on nine occasions, each including both diurnal and nocturnal sampling. 'Level' shows surface water at each sampling level (1–6, see Fig. 1) in relation to sampling date: blue (black in the print version) means the community was submerged, and brown (dark grey in the print version) means no surface water was present. Values within 'Level' denote days until loss of surface water for all immersed communities (blue (black in the print version) cells) and days after loss of surface water for all dry but previously immersed communities (brown (dark grey in the print version) cells): day 0 is loss of surface water for that level.
[a]Each sample consisted of separate diurnal and nocturnal samples.
[b]With day 0 equating to total loss of surface water from the pool bed.
[c]Pool depth (mm) measured from the top of the ∼30 mm layer of sediment placed above the aquarium liner (see Fig. 1).

sampling was undertaken on each occasion, with diurnal samples always taken prior to nocturnal samples (Table 1). On each of the 18 (9 × 2) sampling occasions a single sampling tray was randomly selected and removed from each of the six levels in and around the pool bed (Fig. A3). Each set of six samples therefore represented a transect running from the middle of the pool bed laterally into the dry streambed, across the aquatic–terrestrial ecotone at the pool's margins (Fig. 1). Samples on the outside the pool bed (level 6) were taken first and those in the middle of the pool bed (level 1) taken last, to minimise disturbance effects.

Drying of the pool was initiated on October 6th (Table 1), about 5 days after the last of the nearby (∼500 m) natural pools had dried. Drying involved stopping the flow from the rainwater dripper and piercing the rubber pond liner at a lower level each day to prevent the pool from refilling with any rainfall. The rate of drying was approximately 2 cm/day, which corresponded to 4 days between each of the four levels within the pool (Fig. 1 and Table 1) and reflected drying rates of nearby natural pools (Dell et al., 2014a; Smith and Pearson, 1987).

2.6 Pool Depth and Relative Moisture

Daily pool depth was measured by reading from the ruler placed in the pool bed. A point estimate of the 'relative moisture' of the substrate within each community was obtained at the time of sampling. This was done gravimetrically, by measuring the amount of water retained in 3.2-cm-diameter disks of absorbent, non-biodegradable glass microfibre filter paper (GF/B Whatman®) placed within each sampling tray during initial construction of the pool bed (Dell et al., 2014a). The filter paper disk was placed into a small airtight plastic jar and returned to the laboratory for processing within 3 h. The filter paper was removed from the jar and placed onto a 4-cm square aluminium tray. Larger pieces of fine sediment and detritus stuck to the paper were carefully removed with forceps. The weight of the aluminium tray and the moist filter paper was measured on a balance accurate to 0.001 g; the tray and its contents were then placed in an oven at 60 °C until constant weight (>48 h). Following this, the weights of the aluminium tray and the dry filter paper were determined separately. Relative moisture was calculated by dividing the mass of water soaked up by the filter paper divided by the mass of the dry filter paper.

2.7 Sample Processing

In the laboratory, each sample was elutriated so that lighter material (animals, leaves, small sticks, sediment and fine organic matter) flowed onto a stack of four nylon mesh sieves (1, 0.5, 0.25 and 0.05 mm; Australian Filter Specialists). The remaining material (large sediment and sticks) was periodically examined for animals (none found) and was retained for later determination of organic and inorganic content.

Material retained in the 1-mm sieve was sorted under a magnifying lamp in a white plastic tray containing approximately 5 cm of water. Any organisms found were placed into a vial with 70% alcohol, together with other individuals that were apparently of the same taxon. The remaining contents of the sorting tray were again washed back through the four sieves and the material integrated with material remaining on the sieves after initial elutriation. Material retained on the 1-mm sieve was again sorted as above. Samples with a heavy detritus load were processed in this way several times until only a small amount of material remained in the 1-mm sieve after washing. When no additional organisms were apparent, the sediment and detritus remaining on the 1-mm sieve were combined with the other material from that sieve size collected at prior stages of processing, including during initial elutriation. This material was placed in an aluminium tray for measurement of organic and inorganic loads.

Material retained in the 0.5-mm sieve was extracted by flotation using Ludox™, a colloidal silica solution, diluted with distilled water to a specific gravity of 1.15. Flotation extraction involves placing the sample in the Ludox solution for 30 min, after which there was a clear separation of a scum containing individuals and other organic matter floating on top and heavier sediment at the bottom. The supernatant was poured back over the three remaining sieves (0.5, 0.25 and 0.05 mm) and the entire process repeated twice. Each time the inorganic material removed from the sample was added to an aluminium tray that held similar material from the 0.5-mm sieve. Following elutriation, material retained in the 0.5-mm sieve was placed into 70% ethanol for later processing in a Bogorov tray under a stereo-dissecting microscope. Non-animal waste material (inorganic and organic) was combined with similar material from this size class for measurement of organic and inorganic loads (see section 2.9).

Material retained in the 0.25-mm sieve was elutriated and sorted in the same way as the 0.5-mm material, except that every individual was mounted directly onto glass slides in groups of similar taxa. Specimens were cleared and mounted in Hoyer's medium (10% Gum Arabic, 16.7% distilled water, 66.6% chloral hydrate and 6.7% glycerine). The sediment and detritus that remained after individuals were removed from the Bogorov tray were placed into an aluminium tray for quantification of organic and non-organic loads (see section 2.9).

Material retained in the 0.05-mm sieve was processed as above to determine organic and inorganic loads, but due to the very long times required not all of the 108 samples were processed for fauna (data not shown here).

2.8 Taxonomic Identification and Body Size

The identity and body size of all organisms >0.250 mm (in the longest dimension) were determined. Unmounted specimens were identified, counted and measured in a glass dish under a stereo-dissecting microscope. Every individual was assigned to the lowest possible taxonomic level (most to species or genus, Fig. A4) using published keys and expert knowledge, and its life stage, sex (if possible) and maximum body length (excluding appendages) were determined. Length of unmounted specimens was estimated from a 0.5-mm grid pasted to the bottom of the glass dish. Mounted specimens were processed similarly, but were identified and measured under a high-power microscope with a graticule eyepiece. International experts confirmed identifications for most taxa, using preserved specimens or photographs.

Wet mass estimates were calculated for every individual collected. Because weighing all organisms was impractical, we converted individually measured body lengths to wet mass using published length-weight and dry-wet mass regressions (see Dell et al., 2011, 2013, 2014c; Pawar et al., 2012), using a two-step algorithm: first, body length was converted to body mass (ideally wet, otherwise dry or ash-free-dry) using 364 published size-mass regressions; second, all remaining dry masses (dry or ash-free-dry) were converted to wet mass using 10 published taxon-specific conversion ratios. This method of wet mass estimation is scalable to large numbers of individuals and was underpinned by a richer set of literature data and regressions than previous studies.

2.9 Organic and Inorganic Loads

We measured the amount of organic and inorganic material within each of the four size classes of material from each sieve (i.e. 1, 0.5, 0.25 and 0.05 mm). To do this, following removal of organisms from the sample, material from each sieve was placed separately in an aluminium tray and dried to constant mass at 45 °C (~5–6 days) before being weighed to 0.01 g. Samples were then placed in a muffle furnace at 550 °C until they attained constant mass, and then weighed. This process provided both organic and inorganic load estimates for the four size classes of material.

3. DATA PROCESSING AND ANALYSIS
3.1 Habitat Categories

We categorised communities by three environmental drivers: moisture, light and days until/after drying (Table 2). We used five moisture categories: 'aquatic' (surface water present), 'terrestrial' (never previously immersed), and 'wet', 'moist' and 'dry' (based on the relative moisture content of previously immersed samples that did not currently have surface water present) (Fig. 2 and Table 2). Communities were also categorised as either 'diurnal' or 'nocturnal', depending on the time of the day they were sampled. All 'aquatic' communities were also categorised by the number of days until loss of surface water to calculate 'time until drying', and all previously immersed communities were categorised by the number of days since loss of surface

Table 2 Environmental Categorisations Used to Group Communities for Analysis (See Main Text for More Detail)

Habitat	Description
Moisture	
Aquatic	Sediment layer covered by surface water
Wet	No surface water, relative moisture higher than lower 95% CI of 'aquatic' communities
Moist	No surface water, relative moisture in between lower 95% CI of 'aquatic' communities and higher 95% CI of 'terrestrial' communities
Dry	No surface water, relative moisture lower than higher 95% CI of 'terrestrial' communities
Terrestrial	Inorganic or organic substrate never covered by surface water, representing the permanent terrestrial environment surrounding intermittent pool beds
Light	
Diurnal	Sample collected between 10:00 and 14:00
Nocturnal	Sample collected between 22:00 and 02:00
Habitat transition	
Days until drying	The number of days until loss of surface water for that community, for communities that had surface water
Days after drying	The number of days since surface water loss for that community, for all previously immersed communities that currently did not have surface water

water to calculate 'time after drying'. These categorisations were defined for each level, not the entire pool bed (Table 1 and Fig. 1).

3.2 Functional Richness/Evenness/Divergence

Using individual body size as the functional trait, we calculated three components of each community: functional richness, functional evenness and functional divergence, following Mason et al. (2005). These measures provide a quantitative and continuous measure of the functional trait (i.e. body size) distribution within local ecological communities. Here, we used $\log_{10}$ transformation of body mass as the measure of body size. Functional richness is the proportion of niche space filled for a character and is expressed as

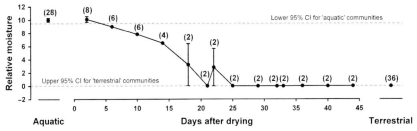

Figure 2 Changes in relative moisture ± SE within communities following surface water loss. For comparison, relative moisture values for 'aquatic' and 'terrestrial' communities are shown, with horizontal grey-dotted lines showing lower and upper 95 CI bounds, respectively. See Table 2 and main text for additional details of habitat categories. Error bars are SE. Values in parentheses are number of replicate communities within each category. Day = 0 represents loss of surface water for that community, not the entire pool bed.

$FR_i = \frac{SF_i}{R}$, where FR_i is the functional richness in community i, SF_i is the niche space filled by the individuals within the community and R is the relative body size range (i.e. the largest range in the set of communities). Functional richness is independent of abundance, since a section of niche space is considered occupied even if only very little abundance occurs within it. Functional evenness indicates the evenness of the distribution of individuals in niche space and applies only to the distribution of abundance in occupied niche space. Evenness is measured directly by dividing the occupied niche space into 100 narrow categories and applying Palou's evenness index (Pielou, 1966) to the abundance contained within each category. Functional divergence indicates the degree to which the abundance distribution in niche space maximises divergence in functional characters within the community and is calculated by the equation: $\frac{2}{\pi}\arctan\left\{5 \times \sum_{j=1}^{N}\left[\left(\ln M_j - \overline{\ln x}\right)^2 \times A_j\right]\right\}$, where M_j is the body mass value of the jth functional character category, A_j is the proportional abundance of the jth functional character category and $\overline{\ln x}$ is the abundance-weighted mean of the natural logarithm of character values for the categories. A community with high functional divergence will have the most abundant species occurring at the extremities of the functional character range, while a community with low functional divergence will have the most abundant species occurring towards the centre of the functional character range. Functional divergence can change without a change in either functional richness or functional evenness.

3.3 Statistical Analysis

We initially used two-way ANOVA to examine the influence of moisture and light on properties of populations (mean body size, range, standard deviation and skewness), communities (richness, abundance, evenness, biomass, functional richness, functional evenness and functional divergence) and inorganic/organic loads within samples. Because this preliminary exploratory analysis revealed only a single significant interaction between moisture and light (functional evenness), for clarity of interpreting results, we used a one-way ANOVA followed by Tukey post hoc tests to explore differences between the five moisture categories, while unpaired t-tests were used to compare differences between the two light categories. Analyses were undertaken in GraphPad Prism (version 6.0). In analyses where populations had different variances (i.e. the p-value of the Bartlett's test is small), we used the Geisser–Greenhouse correction prior to analyses.

Only minor differences were found between samples based on their organic or inorganic load (Fig. A5), so no attempt was made to standardise population and community attributes by organic or inorganic loads because it is difficult to know *a priori* what would be the best measure to standardise values by (i.e. organic, inorganic or some combination of both). We are exploring the effects of organic and inorganic loads on species richness, abundance and body size elsewhere.

4. RESULTS

A taxonomically and ecologically diverse suite of taxa was collected in and around the experimental pool bed, with a total of 24,609 individuals from 313 taxa ranging over 6 orders of magnitude in body size (Fig. 3 and Table A2). Arthropods were by far the most common taxa found, including 70 dipteran species, 53 beetle species, 39 hymenopteran species (including ants and wasps) and a large diversity of mites (Fig. 3 and Table A2). The body size distribution of the entire community showed three peaks, corresponding to a single cladoceran species at approximately 4.00×10^{-6} g (*Ceriodaphnia cornuta*), dipterans at approximately 3.55×10^{-5} g and a haplotaxida annelid at approximately 8.91×10^{-5} g (Fig. 3).

4.1 Richness, Abundance, Evenness and Biomass

When categorised by moisture, systematic and significant effects on species richness, abundance, evenness and biomass within local communities were

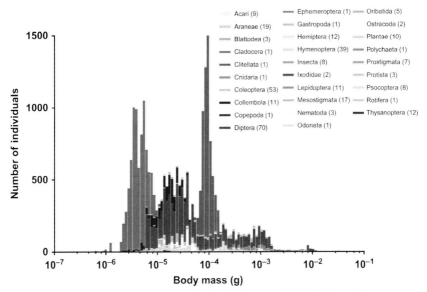

Figure 3 Body size frequency distribution for all 24,609 individuals from all 313 species recorded in the study. Values in parentheses are total number of species within that taxon.

evident (Fig. 4). Richness was highest in 'moist' communities and was significantly lower in both 'aquatic' and 'terrestrial' communities (Fig. 4A). 'Wet' and 'dry' communities, which had relative moisture levels intermediate between immersed (aquatic) and permanently dry (terrestrial) communities (Fig. 2 and Table 2), also had intermediate numbers of species (Fig. 4B). Whether communities were diurnal or nocturnal had no effect on species richness (middle panel in Fig. 4A, paired t-test, $t(53) = 1.394$, $p = 0.456$, two-tailed). Analysis of the full habitat transition from 25 days prior to drying until 44 days after drying revealed that richness increased from early on in 'aquatic' communities until after surface water loss (left panel in Fig. 4A). Across the temporal aquatic–terrestrial transition richness peaked at about 5–20 days following surface water loss, confirming that richness was highest in 'moist' communities. Patterns in family and order richness (data not shown) were qualitatively and quantitatively similar to species richness, suggesting that the physical environment did not systematically constrain the families or orders of species within the local communities.

Total abundance within local communities generally decreased as surface water was lost and the substrate dried, with 'terrestrial' communities having significantly lower total abundances than 'aquatic' and 'moist' communities

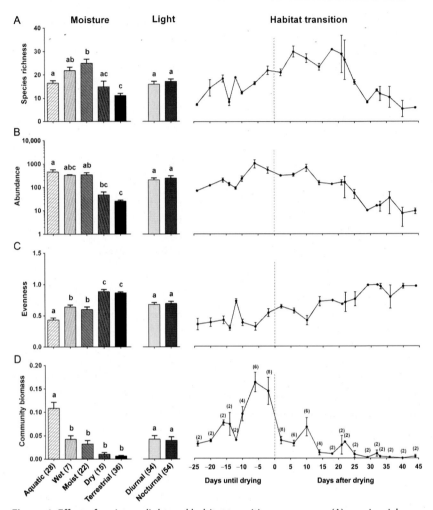

Figure 4 Effect of moisture, light and habitat transition on average (A) species richness, (B) total abundance of individuals, (C) Pielou's evenness index and (D) community biomass within local communities. Error bars are SE. See Table 2 for details of habitat categories. Values in parentheses are total number of replicate communities within each habitat category (habitat transition does not sum to 108 because it excludes communities from levels 5 and 6, which were permanently terrestrial). Grey-dotted line in habitat transition panel at day = 0 represents loss of surface water for that community, not the entire pool bed. Within each habitat category (i.e. moisture and light), different letters above the error bars denote significant differences between habitats ($p < 0.05$) as determined by ANOVA followed by post hoc Tukey test (moisture) or t-test (light).

(Fig. 4B). Whether communities were 'diurnal' or 'nocturnal' had no effect on their average abundances (Fig. 4B, paired t-test, $t(53) = 1.804$, $p = 0.598$, two-tailed). Abundance over the aquatic–terrestrial habitat transition largely mirrored patterns found for species richness, with abundance increasing in aquatic habitats as the habitat was ending (days until drying) and decreased in non-aquatic communities as the habitat developed (days after drying) (Fig. 4B). Unlike richness, average abundance within local communities peaked just prior to surface water loss and then decreased in non-aquatic communities as the habitat aged (Fig. 4B) and dried (Fig. 2).

Average evenness was lowest in 'aquatic' communities, intermediate in 'wet' and 'moist' communities and highest in 'dry' and 'terrestrial' communities (Fig. 4C). As with richness and abundance, whether communities were diurnal or nocturnal had no effect on evenness (Fig. 4C, paired t-test, $t(53) = 0.855$, $p = 0.726$, two-tailed). Across the habitat transition evenness increased in aquatic communities as the habitat dried (days until drying) and continued to increase in non-aquatic communities as the habitat aged (days after drying), confirming results when communities were grouped into moisture categories (Fig. 4C). Average community biomass, quantified as the total mass of all individuals within each local community, was also strongly affected by moisture (Fig. 4D), but not light (Fig. 4D, paired t-test, $t(53) = 0.525$, $p = 0.821$, two-tailed). Biomass was significantly higher in 'aquatic' communities than in other non-aquatic communities, where surface water was absent (Fig. 4D). This pattern was confirmed in the habitat transition panel, where biomass showed a distinct peak immediately prior to drying, and by a general decrease in biomass after surface water loss as the habitat dried (far right panel in Fig. 4D).

4.2 Population Size Structure

To examine effects of the environment on properties of the mean size structure of populations, we combined data for replicate communities within each habitat group (e.g. data from all 28 'aquatic' replicates were grouped prior to analysis). The mean of population average body size was not significantly affected by moisture (Fig. 5A, ANOVA, $F(4, 550) = 1.796$, $p = 0.128$) or light (Fig. 5A, unpaired t-test, $t(444) = 1.355$, $p = 0.176$, two-tailed), despite an apparent decrease in population average body size across the habitat transition following surface water loss (Fig. 5A). The average range of body sizes within populations was highest in 'aquatic' communities; lowest in 'dry' communities; and intermediate in 'wet', 'moist' and

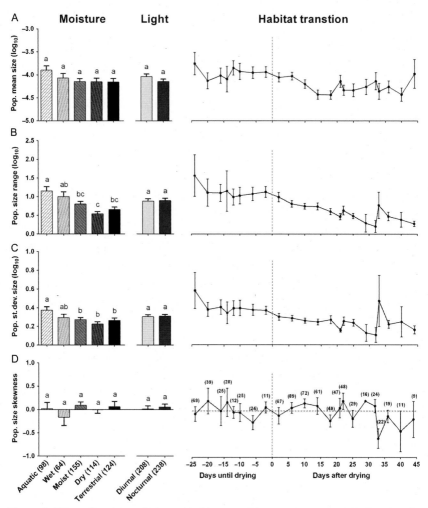

Figure 5 Effect of moisture, light and habitat transition on average population size distributions within local communities: (A) mean size, (B) standard deviation in size, (C) size range (i.e. maximum–minimum size) and (D) the skewness of the population distributions. Error bars are SE. See Table 2 for details of habitat categories. Values in parenthesis are total number of populations included within each habitat category (total number of replicate communities within each habitat category, whose data were combined, are the same as for Figure 4). Grey-dotted line in habitat transition panel at day = 0 represents loss of surface water for that community, not the entire pool bed. Within each habitat category (i.e. moisture and light), different letters above the error bars denote significant differences between habitats ($p < 0.05$) as determined by ANOVA followed by post hoc Tukey test (moisture) or t-test (light).

'terrestrial' communities (Fig. 5B). There was no significant difference between 'diurnal' and 'nocturnal' communities in the average population body size range (Fig. 5B, unpaired t-test, $t(253) = 0.148$, $p = 0.883$, two-tailed), and following surface water, the average size range decreased continuously, until a peak approximately 33 days following surface water loss (right panel in Fig. 5B). The standard deviation of average population body size (Fig. 5C) followed a similar pattern to average size range (Fig 5B), except that the standard deviations of means of population body sizes of 'dry' communities were not significantly lower than those of 'wet', 'moist' or 'terrestrial' communities. The average skewness of population distributions was centred around zero for all habitats, indicating that body sizes in all habitats were generally symmetrically distributed within populations, and we found no significant effect of moisture (Fig. 5D, ANOVA, $F(4, 285) = 0.679$, $p = 0.607$), light (Fig. 5D, unpaired t-test, $t(253) = 0.428$, $p = 0.669$, two-tailed) or habitat duration on skewness (Fig. 5D). The average skewness of population size distributions decreased slightly long after surface water loss, but was highly variable across the entire habitat transition (right panel in Fig. 5D).

4.3 Community Size Structure

Our examination of distributions of mean population size structure (Fig. 5) weighted all taxa equally regardless of abundance, so we also examined distributions of individual body size for all individuals across all local communities within each habitat type (Fig. 6), thus weighting abundant taxa more heavily than rare taxa. Strong effects of moisture on the shapes of the community-level size distributions were apparent (Fig. 6A). The size distribution of all 'aquatic' communities combined had a high peak at small sizes ($\sim 5.01 \times 10^{-6}$ g), while 'terrestrial' communities had a single broader peak centred around 4.47×10^{-5} g (Fig. 6A). Like 'aquatic' and 'terrestrial' communities, 'moist' communities also had a high peak, centred around a mass of approximately 8.91×10^{-5} g and caused by the appearance of many individuals of a single earthworm taxon (Fig. 3). The size distribution of 'wet' communities resembled a combination of 'aquatic' and 'moist' communities, with the highest peak that occurs at small sizes in the 'aquatic' distribution absent and the single peak at large sizes in the 'moist' communities not being as distinct (Fig. 6A). The 'dry' community size distribution resembled a combination of 'wet' and 'moist' communities, or possibly the disappearance of the very large single peak in the 'moist' communities. Although

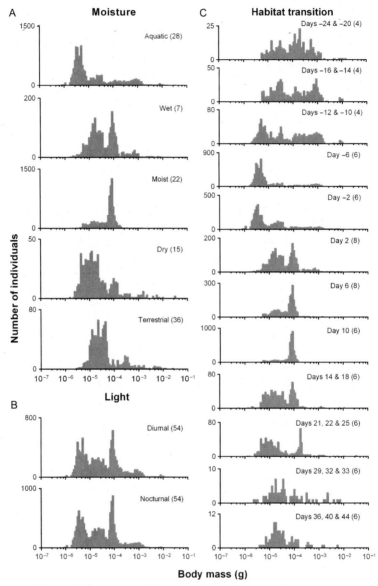

Figure 6 Effects of (A) moisture, (B) light and (C) habitat transition on community body size distributions. Data are for all local communities combined, for that habitat group, and data for some sampling dates in habitat transition are grouped due to lack of data.

the shape of 'dry' distribution looks similar to 'terrestrial' communities, the locations of the peaks do not match (Fig. 6A). There were no discernable differences between the size distributions of 'diurnal' and 'nocturnal' communities when data from all communities were combined (Fig. 6B), with three peaks apparent, corresponding to the three peaks observed when the data were categorised by moisture (Fig. 6A).

The habitat transition panel shows that the size frequency distributions of individuals in 'aquatic' communities did not all have equivalent shapes (Fig. 6C). Long before the habitat dried (i.e. 24–10 days prior to drying, top three panels in Fig. 6C), the communities had a broad and relatively even size distribution. As the disappearance of the aquatic habitat approached (days −6 to −2), a peak at lower sizes appeared and dominated the shape of the community size distribution, corresponding to the appearance of the cladoceran C. *cornuta* (Fig. 3). As surface water disappeared, so did the C. *cornuta* peak, and a second peak at greater size appeared at day 2 (due to the appearance of the annelid; Fig. 3), becoming the main feature of the size distribution on day 6 and day 10. As the non-aquatic habitat aged and dried (Fig. 2), the abundance of this earthworm declined until on the last few sampling days the distribution of individual sizes within the community was broad and relatively evenly distributed (bottom four panels of Fig. 6C).

4.4 Functional Richness/Evenness/Divergence

Functional richness (the amount of niche space filled by species in the community) and functional evenness (the evenness of abundance distribution in filled niche space) were highest in 'aquatic' and 'wet' communities, were lowest in 'dry' and 'terrestrial' communities and had intermediate values for 'moist' communities (Fig. 7A and B). Functional divergence, indicating the degree of niche differentiation within a community, was also highest in 'aquatic' communities, but unlike richness and evenness was lowest in 'moist' communities and again high in 'dry' and 'terrestrial' communities (Fig. 7C). Whether communities were 'diurnal' or 'nocturnal' did not significantly affect their functional richness (paired t-test, $t(53) = 0.570$, $p = 0.868$, two-tailed), functional evenness (paired t-test, $t(53) = 1.704$, $p = 0.390$, two-tailed) or functional divergence (paired t-test, $t(53) = 0.014$, $p = 0.990$, two-tailed) (Fig. 7A–C). Habitat transition impacted functional richness in local communities, showing patterns not evident in analysis of moisture categories alone: richness increased as the aquatic habitat came closer to disappearing (days until drying) and then decreased progressively in non-immersed communities (days after drying) (Fig. 7A). Functional evenness was not affected by habitat transition for immersed communities and decreased slowly with increasing habitat age once surface water was lost (Fig. 7B). There was substantial variation in the effect of habitat transition

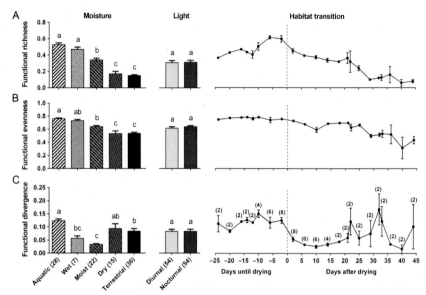

Figure 7 Effects of moisture, light and habitat transition on average (A) functional richness, (B) functional evenness and (C) functional divergence within local communities. Error bars are SE. See Table 2 for details of habitat categories. Values in parentheses are total number of replicate communities within each habitat category (habitat transition does not sum to 108 because it excludes communities from levels 5 and 6, which were permanently terrestrial). Grey-dotted line in habitat transition panel at day = 0 represents loss of surface water for that community, not the entire pool bed. Within each habitat category (i.e. moisture and light), different letters above the error bars denote significant differences between habitats ($p < 0.05$) as determined by ANOVA followed by post hoc Tukey test (moisture) or t-test (light).

on functional divergence, with 'days until drying' not appearing to have an important effect, and divergence being minimal in young non-aquatic communities (Fig. 7C).

5. DISCUSSION

The intermittent pool bed provided a complex environmental gradient at limited spatial and temporal scales, and we were able to sample effectively across this entire gradient. Our study is thus unlikely to include many of the possible confounding sources of variation of other comparative studies of ecological systems across diverse habitats (Dell et al., 2014a). It remains to be determined how closely our system mimics other habitats globally, but it is nonetheless clear that the unique environmental characteristics of intermittent pool beds, with distinct aquatic–terrestrial boundaries in both space and time, make them valuable study systems for important basic and applied questions in ecology (Dell et al., 2014a).

5.1 Moisture

We identified strong and systematic effects of moisture on a number of key attributes of population and community size structure, due probably to at least two mechanisms. First, within terrestrial habitats, differences in moisture can strongly influence the strength and outcome of ecological interactions and, subsequently, the functional structure of populations and communities (Allen et al., 2014; Hawkins et al., 2003; Lensing and Wise, 2006; McCluney and Sabo, 2009; McCluney et al., 2012; Melguizo-Ruiz et al., 2012; Spiller and Schoener, 2008; Verdeny-Vilalta and Moya-Laraño, 2014). Water is an essential resource, and limitations to water availability in terrestrial ecosystems can alter the movement and behaviour of individuals, ultimately influencing growth and mortality of populations (Hawkins et al., 2003; McCluney and Sabo, 2009; Spiller and Schoener, 2008; Verdeny-Vilalta and Moya-Laraño, 2014). Second, moisture levels determine whether a habitat is aquatic or terrestrial, and, therefore, the nature of the environmental and functional constraints for individuals, populations and communities. For example, trophic interactions in pelagic food webs tend to be more size structured than on land, due to differences in the mechanics of prey capture (Denny, 1990; Yvon-Durocher et al., 2011), and as pelagic environments have relatively little habitat heterogeneity, being much larger or smaller than predators is a common way to avoid predation (Chase, 1999; Denny, 1990; Yvon-Durocher et al., 2011). Such effects are likely to have important consequences for populations and community size structure (Hairston and Hairston, 1993; Pawar et al., 2012; Yvon-Durocher et al., 2011).

We failed to find significant effects of moisture on the distribution of the mean body sizes of populations, probably due to the large number of species of very different sizes within local communities. However, we did identify effects on the range and the standard deviation of average population body sizes within communities. The mean, range and standard deviation of population body sizes in aquatic communities were on average greater than those of terrestrial communities, possibly because of the greater size structure of feeding relationships in many aquatic systems (Chase, 1999; Denny, 1990; Yvon-Durocher et al., 2011). The mean skewness of population size distributions was centred around zero, indicating that the size distributions of most populations were symmetrical (Gouws et al., 2011). However, there was some evidence of temporal change in skewness across the habitat transition and during the terrestrial phase, which appears to be a novel result. Despite their importance for life history theory and macroecology, very few comparative studies of intraspecific size structure across multiple species exist (but see Gouws et al., 2011). We could find no studies that examine effects of the physical environment on animal population size structure across local landscapes, nor any that explore the size structure of multiple interacting species within local communities. This, therefore, is apparently the first analysis of such patterns, and we see this as a fruitful avenue of future research.

Moisture had significant effects on most aspects of the size structure of individuals within communities and on functional richness and evenness. A number of biological mechanisms could be responsible for these patterns, including those affecting the ecological interactions of the component species (see above). One potential mechanism is the effect of consumer search space dimensionality, which likely differs between aquatic and terrestrial habitats (Pawar et al., 2012, 2013). Whatever biological mechanisms are causing different peaks in our community size distributions, it is clear that within the transitional habitats (i.e. 'wet', 'moist' and 'dry'), both aquatic and terrestrial species were present. Unless 'aquatic' individuals alter their size as they transition from wet to dry, which appears unlikely, they should maintain body sizes (and size distributions) in these non-aquatic environments that resulted from selection in their preferred aquatic habitat. Thus, 'aquatic' taxa remaining in the terrestrial environment might be expected to have functional traits (i.e. size structure) that were optimised within the aquatic habitat. Counter to this argument, of course, is that traits that facilitate aquatic taxa to persist in the terrestrial environment following drying might be size-dependent. Either way, this is an issue that warrants further investigation.

5.2 Light

There were no functional differences between diurnal and nocturnal populations and communities in any of the attributes we measured. Most species in most habitats exhibit diel variations in their activity and location, including freshwater pond macroinvertebrates (Florencio et al., 2011; Gilbert and Hampton, 2001; Hampton and Duggan, 2003; Hampton and Friedenberg, 2002), and while a variety of ecological and evolutionary processes may drive these patterns, asymmetries in the detection capabilities of predators and prey appear to be a key driver (Brewer et al., 1999; Fraser and Metcalfe, 1997; Gergs et al., 2010; Gilbert and Hampton, 2001; Kronfeld-Schor and Dayan, 2003; Sana et al., 2008). Light effects on detection distance can even be size-dependent (Dieguez, 2003; Jara, 2007). Compositional shifts between day and night do not necessarily result in functional differences in the size structure of populations and communities, if species are replaced by similar-sized ones (our methods of sampling and analysis would not have identified such effects). Uncovering additional non-size-based functional effects would require monitoring of real functioning communities, for example, with modern non-invasive tracking methods (Dell et al., 2014b). If such a pattern were evident, it would have implications for a convergence of functional organisation between diurnal and nocturnal communities (Losos, 1992; Losos et al., 1998; Melville et al., 2006; Wiens et al., 2010). Alternatively, our spatial scale of sampling may have been too coarse to capture the diel movements of small litter invertebrates, which likely have small home ranges (Jetz et al., 2004). Our sampling method captured an entire volume of habitat, including all the benthic and pelagic/aerial microhabitats above the sampling tray. Traditional sampling methods that have identified diel variations, such as sweep nets or grab samples, generally only capture single microhabitats. Therefore, even if in our study animals were moving between these microhabitats over a 24-h period, we could not have captured these effects.

Direct observations throughout our study of large organisms (too large to be retained in our samples) suggested that the pool bed supported a different suite of species during the day (skinks, diurnal snakes) than at night (frogs, toads, large spiders, nocturnal snakes). Perhaps these larger species periodically move from shelters in the riparian zones into the streambed to forage, in contrast to leaf packs, which likely provided sufficient shelter for smaller invertebrates during both active and inactive times.

5.3 Habitat Transition

Although the ecology of the aquatic phase of intermittent pool beds has been extensively studied, relatively little is known about their ecology once surface water has disappeared or about the population and community dynamics that occur across the aquatic–terrestrial habitat transition during pool drying (Dell et al., 2014a; Steward et al., 2011). This is surprising considering the recognised human and ecological importance of these habitats globally

(Larned et al., 2010; Steward et al., 2012). The diverse invertebrate taxa we recorded in and around the pool bed following surface water loss (Fig. 3 and Table A2) are typical of similar habitats elsewhere (Dell et al., 2014a; Steward et al., 2011, 2012; Williams, 1987), showing that once surface water is lost, pool beds can support a rich and abundant community that may be as well adapted to the ephemeral nature of its terrestrial habitat as is the aquatic fauna is to its (Adis, 1992; Adis and Junk, 2002; Dell et al., 2014a; Lambeets et al., 2008; Steward et al., 2011, 2012; Tamm, 1984). Survival of this community is probably facilitated by the higher moisture content persisting in dry pool beds (for $\sim$20 days in this study) compared to the surrounding terrestrial landscape, and also the rich nutritional resource that previously immersed detritus and remaining aquatic organisms represent to new colonists (Dell et al., 2014a; McLachlan and Cantrell, 1980; Stehr and Branson, 1938; Williams, 1987).

We found strong impacts of moisture, and thus pool drying, on richness, abundance, biomass and evenness within local communities. Even within the aquatic community, as the pool dried local communities became richer, more abundant and had higher total biomass, as remaining aquatic fauna that did not aestivate or move elsewhere became concentrated into a decreasing area of habitat (Williams, 1987). Clearly, loss of surface water does not immediately negatively impact species richness as might be expected from previous studies of these habitats as ephemeral aquatic systems (Dell et al., 2014a). Instead, community richness, abundance and biomass peaked across the aquatic–terrestrial temporal ecotone, eventually declining as beds dried completely. This pattern is caused by the occurrence in the habitat transition of taxa from adjoining aquatic and terrestrial habitats (Dell et al., 2014a; Larned et al., 2010). The lower community evenness in aquatic than non-aquatic communities was due to greater dominance by a few taxa, but the biological mechanisms responsible for this are currently unclear.

5.4 Summary

Our data set is unusual in its comprehensive size-explicit description of individuals, populations and communities spanning a complex environmental gradient, albeit in a single pool bed. It offers a new perspective on how functional attributes of populations and communities might vary between habitats (Dell et al., 2014a), which to date has been hampered by a lack of high-quality empirical data (Yvon-Durocher et al., 2011) and potential confounding sources of variation from different sampling and analytical methods (Dell et al., 2014a). Using this data set, we identified strong effects of moisture on properties of populations (size structure) and communities (species richness, abundance, evenness, biomass and size structure), along with temporal effects across the aquatic–terrestrial ecotone. There were no functional differences between diurnal and nocturnal communities at the scale of our samples. Further studies of this nature are required to develop a functional understanding of ecological systems, of effects of human activities that are altering the size structure of natural ecosystems, for making quantitative predictions about such effects, and for developing management practices to ensure ecosystem survival in the face of global change.

ACKNOWLEDGEMENTS

Discussions with N. Connolly, J.B. Mackenzie, N.D. Martinez, S. Pawar and V.M. Savage greatly improved this manuscript. We thank A. Backer, C. Barry, J. Burden, A. Cairns, N. Connolly, L. Davis, I. Dell, C. Dudgeon, M. Edwards, G. Gilroy, G. Horner, P. Kiong, J. Madin and E. Potter-Madin for technical assistance in the field and laboratory. Many people graciously provided their expertise with taxonomic identification, but we especially thank R. Raven and C. Watts. We thank an anonymous reviewer for comments that greatly improved the quality and clarity of our manuscript. Field research was supported by an MRG grant from JCU to R.A.A. A.I.D. was supported by an Australian Postgraduate Award (Australian Government) and funds from the Department of Tropical Biology at James Cook University. L.Z. was supported by the National Special Water Programs (2009ZX07210-009), Department of Environmental Protection of Shandong Province (SDHBPJ-ZB-08) and the China Scholarship Council (201206730022).

APPENDIX

Table A1 Common Riparian Plants Found at the Study Site

Brachychiton australis
Cassine melanocarpa
Cochlospermum gillivraei
Cupaniopsis anacardioides
Diospyros geminata
Drypetes diplangia
Geijera salicifolia
Kailarsenia ochreata
Lophostemon grandiflorus
Mimusops elengi
Pleigynium timorense
Pongamia pinnata
Sterculia quadrifida

Table A2 Abundance and Body Size (Mean, Minimum and Maximum) of All Taxa Recorded

Kingdom	Phylum	Class	Family	Species	Abund.	Wet Body Mass (mg)		
						Mean	Min	Max
Protista	NA	NA	NA	Protista sp. Z	2	0.1438	0.1438	0.1438
Protista	Heterokontophyta	Bacillariophyceae	NA	Bacillariophyceae sp. A	1	0.0006		
Protista	Rhizopoda	NA	NA	Rhizopoda sp. A	2	0.0566	0.0477	0.0654
Plantae	Chlorophyta	NA	NA	Chlorophyta sp. A	2	0.0099	0.0093	0.0105
Plantae	Chlorophyta	NA	NA	Chlorophyta sp. B	1	0.0105		
Plantae	Chlorophyta	NA	NA	Chlorophyta sp. C	1	0.0105		
Plantae	Chlorophyta	NA	NA	Chlorophyta sp. D	2	0.0105	0.0105	0.0105
Plantae	Chlorophyta	NA	NA	Chlorophyta sp. E	1	0.0105		
Plantae	Chlorophyta	NA	NA	Chlorophyta sp. F	1	0.0105		
Plantae	Chlorophyta	NA	NA	Chlorophyta sp. G	1	0.0105		
Plantae	Chlorophyta	NA	NA	Chlorophyta sp. H	1	0.0105		
Plantae	Chlorophyta	Chlorophyceae	Cladophoraceae	Cladophora sp. A	2	0.0105	0.0105	0.0105
Plantae	Chlorophyta	Chlorophyceae	Cladophoraceae	Pithophora sp. A	9	0.0105	0.0105	0.0105
Animalia	Annelida	Clitellata	NA	Haplotaxida spp.	5413	0.0888	0.0464	0.6887
Animalia	Annelida	Polychaeta	NA	Polychaeta sp. A	2	0.0842	0.0828	0.0857

Animalia	Arthropoda	Arachnida	NA	Acari sp. B	1	0.0248		
Animalia	Arthropoda	Arachnida	NA	Acari sp. C	1	0.0058		
Animalia	Arthropoda	Arachnida	NA	Acari sp. D	4	0.0279	0.0212	0.0315
Animalia	Arthropoda	Arachnida	NA	Acari sp. E	1	0.0099		
Animalia	Arthropoda	Arachnida	NA	Acari sp. G	1	0.0027		
Animalia	Arthropoda	Arachnida	NA	Acari sp. H	1	0.0055		
Animalia	Arthropoda	Arachnida	NA	Acari sp. I	1	0.0068		
Animalia	Arthropoda	Arachnida	NA	Acari sp. K	1	0.0190		
Animalia	Arthropoda	Arachnida	NA	Acari sp. L	1	0.0095		
Animalia	Arthropoda	Arachnida	NA	Mesostigmata sp. F	9	0.0217	0.0042	0.0731
Animalia	Arthropoda	Arachnida	NA	Mesostigmata sp. G	9	0.0131	0.0042	0.0177
Animalia	Arthropoda	Arachnida	NA	Mesostigmata sp. H	1	0.0072		
Animalia	Arthropoda	Arachnida	NA	Mesostigmata sp. I	2	0.0126	0.0114	0.0139
Animalia	Arthropoda	Arachnida	NA	Mesostigmata sp. J	2	0.0119	0.0099	0.0139
Animalia	Arthropoda	Arachnida	NA	*Oribatida* sp. A	1	0.1165		
Animalia	Arthropoda	Arachnida	NA	Prostigmata sp. A	1	0.0093		
Animalia	Arthropoda	Arachnida	Araneidae	*Cyrtophora moluccensis*	2	0.4689	0.4301	0.5077
Animalia	Arthropoda	Arachnida	Ascidae	*Cheiroseius* sp. A	113	0.0096	0.0020	0.0182
Animalia	Arthropoda	Arachnida	Bdellidae	Bdellidae sp. A	7	0.0458	0.0224	0.0941

Continued

Table A2 Abundance and Body Size (Mean, Minimum and Maximum) of All Taxa Recorded—cont'd

Kingdom	Phylum	Class	Family	Species	Abund.	Wet Body Mass (mg)		
						Mean	Min	Max
Animalia	Arthropoda	Arachnida	Cunaxidae	Cunaxidae sp. A	4	0.0193	0.0149	0.0230
Animalia	Arthropoda	Arachnida	Cunaxidae	Cunaxidae sp. B	9	0.0160	0.0052	0.0218
Animalia	Arthropoda	Arachnida	Cymbaeremaeidae	Scapheremaeus sp. A	4	0.0257	0.0217	0.0359
Animalia	Arthropoda	Arachnida	Eupodidae	Eupodidae sp. A	7	0.0173	0.0119	0.0262
Animalia	Arthropoda	Arachnida	Eupodidae	Eupodidae sp. B	1	0.0190		
Animalia	Arthropoda	Arachnida	Ixodidae	Haemaphysalis bancrofti	2	2.3798	0.1018	4.6577
Animalia	Arthropoda	Arachnida	Ixodidae	Haemaphysalis novaeguineae	1	0.0221		
Animalia	Arthropoda	Arachnida	Laelapidae	Cosmolaelaps sp. A	9	0.0242	0.0044	0.0429
Animalia	Arthropoda	Arachnida	Laelapidae	Laelapidae sp. A	37	0.0181	0.0084	0.0329
Animalia	Arthropoda	Arachnida	Laelapidae	Laelapidae sp. B	27	0.0202	0.0060	0.1676
Animalia	Arthropoda	Arachnida	Laelapidae	Laelapidae sp. C	5	0.0108	0.0048	0.0252
Animalia	Arthropoda	Arachnida	Laelapidae	Laelapidae sp. D	3	0.0100	0.0072	0.0114
Animalia	Arthropoda	Arachnida	Linyphiidae	Laperousea sp. A	2	1.5948	1.5375	1.6522
Animalia	Arthropoda	Arachnida	Linyphiidae	Linyphiidae sp. A	2	0.2133	0.0326	0.3940
Animalia	Arthropoda	Arachnida	Linyphiidae	Linyphiidae sp. B	5	0.2847	0.0600	1.0394
Animalia	Arthropoda	Arachnida	Liodidae	Liodidae sp. A	1	0.0194		

Animalia	Arthropoda	Arachnida	Miturgidae	*Cheiracanthium* sp. A	2	0.2909	0.0326	0.5492
Animalia	Arthropoda	Arachnida	Ologamasidae	Gamasiphinae sp. A	14	0.0170	0.0096	0.0344
Animalia	Arthropoda	Arachnida	Ologamasidae	*Gamasiphis* sp. A	6	0.0186	0.0114	0.0344
Animalia	Arthropoda	Arachnida	Ologamasidae	Ologamasidae sp. B	15	0.0057	0.0030	0.0110
Animalia	Arthropoda	Arachnida	Ologamasidae	Ologamasidae sp. C	2	0.0073	0.0055	0.0090
Animalia	Arthropoda	Arachnida	Oonopidae	*Ischnothyreus* sp. A	5	1.5603	1.0394	1.8295
Animalia	Arthropoda	Arachnida	Oonopidae	*Ischnothyreus* spp.	3	0.9425	0.0759	1.4280
Animalia	Arthropoda	Arachnida	Oonopidae	Oonopidae sp. A	4	0.0660	0.0406	0.0759
Animalia	Arthropoda	Arachnida	Oonopidae	Oonopinae sp. A	7	0.4233	0.0989	1.0394
Animalia	Arthropoda	Arachnida	Oonopidae	*Orchestina* sp. A	27	0.3477	0.0198	1.3235
Animalia	Arthropoda	Arachnida	Phytoseiidae	Phytoseiidae sp. A	2	0.0093	0.0072	0.0114
Animalia	Arthropoda	Arachnida	Podocinidae	*Podocinum* sp. A	4	0.0194	0.0135	0.0233
Animalia	Arthropoda	Arachnida	Salticidae	*Lycidas* sp. A	9	1.8530	0.9541	4.6284
Animalia	Arthropoda	Arachnida	Salticidae	*Maratus* sp. A	2	1.6080	1.6080	1.6080
Animalia	Arthropoda	Arachnida	Salticidae	*Opisthoncus* sp. A	2	6.5518	2.2836	10.8201
Animalia	Arthropoda	Arachnida	Scheloribatidae	Scheloribatidae sp. A	133	0.0393	0.0109	1.0605
Animalia	Arthropoda	Arachnida	Scheloribatidae	Scheloribatidae sp. B	2	0.0200	0.0183	0.0217
Animalia	Arthropoda	Arachnida	Scytodidae	*Scytodes thoracica*	5	5.9621	1.5375	13.0337

Continued

Table A2 Abundance and Body Size (Mean, Minimum and Maximum) of All Taxa Recorded—cont'd

Kingdom	Phylum	Class	Family	Species	Abund.	Wet Body Mass (mg)		
						Mean	Min	Max
Animalia	Arthropoda	Arachnida	Tetranychidae	Bryobiinae sp. A	1	0.0030		
Animalia	Arthropoda	Arachnida	Theridiidae	Dipoena sp. A	1	0.0108		
Animalia	Arthropoda	Arachnida	Theridiidae	Euryopis elegans	4	1.3070	0.0600	4.4785
Animalia	Arthropoda	Arachnida	Theridiidae	Theridiidae sp. A	12	0.4906	0.0198	2.1641
Animalia	Arthropoda	Arachnida	Theridiidae	Theridion sp. A	3	1.3177	0.0716	2.4534
Animalia	Arthropoda	Arachnida	Zodariidae	Habronestes sp. A	1	23.9778		
Animalia	Arthropoda	Branchiopoda	Daphniidae	Ceriodaphnia cornuta	7169	0.0041	0.0005	0.0396
Animalia	Arthropoda	Copepoda	NA	Cyclopoida sp. A	28	0.0146	0.0022	0.1325
Animalia	Arthropoda	Insecta	NA	Coleoptera sp. A	11	0.1229	0.0064	0.1638
Animalia	Arthropoda	Insecta	NA	Coleoptera sp. B	1	0.6844		
Animalia	Arthropoda	Insecta	NA	Coleoptera sp. C	4	0.1446	0.0346	0.2514
Animalia	Arthropoda	Insecta	NA	Coleoptera sp. D	5	0.2078	0.1344	0.3962
Animalia	Arthropoda	Insecta	NA	Coleoptera sp. E	1	0.3787		
Animalia	Arthropoda	Insecta	NA	Coleoptera sp. F	10	0.1923	0.0171	0.6755
Animalia	Arthropoda	Insecta	NA	Coleoptera sp. G	2	0.1389	0.0917	0.1861
Animalia	Arthropoda	Insecta	NA	Coleoptera sp. H	3	0.7848	0.0059	2.2919

Animalia	Arthropoda	Insecta	NA	Coleoptera sp. I	1	0.0765		
Animalia	Arthropoda	Insecta	NA	Coleoptera sp. J	1	0.0512		
Animalia	Arthropoda	Insecta	NA	Coleoptera sp. K	1	0.5988		
Animalia	Arthropoda	Insecta	NA	Coleoptera sp. L	2	0.1608	0.1298	0.1918
Animalia	Arthropoda	Insecta	NA	Coleoptera sp. M	3	0.0573	0.0461	0.0772
Animalia	Arthropoda	Insecta	NA	Coleoptera sp. N	3	0.0870	0.0784	0.0957
Animalia	Arthropoda	Insecta	NA	Coleoptera sp. O	38	0.0183	0.0040	0.0390
Animalia	Arthropoda	Insecta	NA	Coleoptera sp. P	1	0.0354		
Animalia	Arthropoda	Insecta	NA	Coleoptera sp. Y	1	0.1240		
Animalia	Arthropoda	Insecta	NA	Collembola sp. A	22	0.0602	0.0197	0.1003
Animalia	Arthropoda	Insecta	NA	Diptera sp. A	12	0.0218	0.0056	0.0435
Animalia	Arthropoda	Insecta	NA	Diptera sp. AA	1	0.0461		
Animalia	Arthropoda	Insecta	NA	Diptera sp. AB	2	0.2619	0.0359	0.4879
Animalia	Arthropoda	Insecta	NA	Diptera sp. AC	5	0.0407	0.0335	0.0473
Animalia	Arthropoda	Insecta	NA	Diptera sp. AD	1	0.1947		
Animalia	Arthropoda	Insecta	NA	Diptera sp. B	2	0.0655	0.0059	0.1252
Animalia	Arthropoda	Insecta	NA	Diptera sp. C	1	0.0583		
Animalia	Arthropoda	Insecta	NA	Diptera sp. D	6	0.1993	0.0512	0.4230

Continued

Table A2 Abundance and Body Size (Mean, Minimum and Maximum) of All Taxa Recorded—cont'd

Kingdom	Phylum	Class	Family	Species	Abund.	Wet Body Mass (mg)		
						Mean	Min	Max
Animalia	Arthropoda	Insecta	NA	Diptera sp. E	1	0.0218		
Animalia	Arthropoda	Insecta	NA	Diptera sp. F	1	0.0773		
Animalia	Arthropoda	Insecta	NA	Diptera sp. G	5	0.0361	0.0078	0.1064
Animalia	Arthropoda	Insecta	NA	Diptera sp. H	2	0.0219	0.0171	0.0266
Animalia	Arthropoda	Insecta	NA	Diptera sp. I	2	0.5224	0.0389	1.0060
Animalia	Arthropoda	Insecta	NA	Diptera sp. J	1	0.0197		
Animalia	Arthropoda	Insecta	NA	Diptera sp. K	1	0.0190		
Animalia	Arthropoda	Insecta	NA	Diptera sp. L	14	1.1037	0.1464	1.4171
Animalia	Arthropoda	Insecta	NA	Diptera sp. M	1	1.9622		
Animalia	Arthropoda	Insecta	NA	Diptera sp. N	13	0.1038	0.0628	0.1521
Animalia	Arthropoda	Insecta	NA	Diptera sp. O	7	0.0958	0.0512	0.1778
Animalia	Arthropoda	Insecta	NA	Diptera sp. Q	1	0.0893		
Animalia	Arthropoda	Insecta	NA	Diptera sp. R	10	0.0239	0.0062	0.0339
Animalia	Arthropoda	Insecta	NA	Diptera sp. S	15	0.0351	0.0054	0.1385
Animalia	Arthropoda	Insecta	NA	Diptera sp. T	13	0.0896	0.0665	0.1098
Animalia	Arthropoda	Insecta	NA	Diptera sp. U	1	0.0207		

Animalia	Arthropoda	Insecta	NA	Diptera sp. V	1	0.0529	
Animalia	Arthropoda	Insecta	NA	Diptera sp. W	1	0.4076	
Animalia	Arthropoda	Insecta	NA	Diptera sp. X	7	0.0681	0.0298	0.1409
Animalia	Arthropoda	Insecta	NA	Diptera sp. Z	1	0.4645	
Animalia	Arthropoda	Insecta	NA	Hymenoptera sp. Q	1	0.2929	
Animalia	Arthropoda	Insecta	NA	Hymenoptera sp. R	1	0.0285	
Animalia	Arthropoda	Insecta	NA	Hymenoptera sp. S	1	0.0433	
Animalia	Arthropoda	Insecta	NA	Insecta sp. A	1	0.0986	
Animalia	Arthropoda	Insecta	NA	Insecta sp. B	2	0.0263	0.0208	0.0319
Animalia	Arthropoda	Insecta	NA	Insecta sp. C	2	0.1915	0.1883	0.1947
Animalia	Arthropoda	Insecta	NA	Insecta sp. D	1	0.0784	
Animalia	Arthropoda	Insecta	NA	Insecta sp. E	1	0.0314	
Animalia	Arthropoda	Insecta	NA	Insecta sp. F	1	0.0285	
Animalia	Arthropoda	Insecta	NA	Insecta sp. G	1	0.0490	
Animalia	Arthropoda	Insecta	NA	Insecta sp. J	1	0.0495	
Animalia	Arthropoda	Insecta	NA	Lepidoptera sp. A	9	0.4649	0.0427	1.2563
Animalia	Arthropoda	Insecta	NA	Lepidoptera sp. B	2	1.3081	0.1697	2.4466
Animalia	Arthropoda	Insecta	NA	Lepidoptera sp. C	5	0.2494	0.0957	0.3454

Continued

Table A2 Abundance and Body Size (Mean, Minimum and Maximum) of All Taxa Recorded—cont'd

Kingdom	Phylum	Class	Family	Species	Abund.	Wet Body Mass (mg) Mean	Min	Max
Animalia	Arthropoda	Insecta	NA	Lepidoptera sp. D	15	0.0492	0.0121	0.1934
Animalia	Arthropoda	Insecta	NA	Lepidoptera sp. E	1	0.0398		
Animalia	Arthropoda	Insecta	NA	Lepidoptera sp. F	1	0.2546		
Animalia	Arthropoda	Insecta	NA	Lepidoptera sp. G	1	0.1947		
Animalia	Arthropoda	Insecta	NA	Lepidoptera sp. H	1	0.0138		
Animalia	Arthropoda	Insecta	NA	Psylloidea sp. A	1	0.0346		
Animalia	Arthropoda	Insecta	NA	Psylloidea sp. B	1	0.0040		
Animalia	Arthropoda	Insecta	Aderidae	Aderidae sp. A	1	0.2904		
Animalia	Arthropoda	Insecta	Aeolothripidae	Aeolothripidae sp. A	2	0.0386	0.0092	0.0679
Animalia	Arthropoda	Insecta	Aeolothripidae	Aeolothripidae sp. B	4	0.0326	0.0184	0.0486
Animalia	Arthropoda	Insecta	Aeolothripidae	*Desmothrips* sp. A	7	0.0260	0.0065	0.0529
Animalia	Arthropoda	Insecta	Archipsocidae	*Archipsocopis* sp. A	69	0.0429	0.0072	0.3368
Animalia	Arthropoda	Insecta	Baetidae	*Cloeon* sp. A	528	1.1888	0.0427	7.0558
Animalia	Arthropoda	Insecta	Bethylidae	Bethylidae sp. A	1	0.0177		
Animalia	Arthropoda	Insecta	Bethylidae	Bethylidae sp. B	2	0.0533	0.0501	0.0565
Animalia	Arthropoda	Insecta	Bethylidae	Bethylidae sp. C	1	0.4473		

Kingdom	Phylum	Class	Family	Species				
Animalia	Arthropoda	Insecta	Blattidae	Blattidae sp. A	1	4.4896		
Animalia	Arthropoda	Insecta	Blattidae	Blattidae sp. B	1	30.1411		
Animalia	Arthropoda	Insecta	Blattidae	Blattidae sp. C	2	0.1888	0.1697	0.2079
Animalia	Arthropoda	Insecta	Braconidae	Microgastrinae sp. A	1	0.7162		
Animalia	Arthropoda	Insecta	Carabidae	Carabidae sp. A	3	1.3497	0.6119	2.6860
Animalia	Arthropoda	Insecta	Carabidae	*Perigona* sp. A	2	1.9552	1.7684	2.1419
Animalia	Arthropoda	Insecta	Carabidae	*Tachys spenceri*	1	1.1865		
Animalia	Arthropoda	Insecta	Cecidomyiidae	Cecidomyiidae sp. A	194	0.0381	0.0056	0.0814
Animalia	Arthropoda	Insecta	Cecidomyiidae	Cecidomyiidae sp. K	1	0.0406		
Animalia	Arthropoda	Insecta	Ceraphronidae	Ceraphronidae sp. A	4	0.0398	0.0172	0.0515
Animalia	Arthropoda	Insecta	Ceraphronidae	Ceraphronidae sp. B	2	0.0384	0.0335	0.0433
Animalia	Arthropoda	Insecta	Ceraphronidae	Ceraphronidae sp. C	1	0.0172		
Animalia	Arthropoda	Insecta	Ceraphronidae	Ceraphronidae sp. D	1	0.0078		
Animalia	Arthropoda	Insecta	Ceratopogonidae	Ceratopogonidae sp. A	343	0.0181	0.0055	0.0493
Animalia	Arthropoda	Insecta	Ceratopogonidae	Ceratopogonidae sp. B	3	0.0363	0.0239	0.0455
Animalia	Arthropoda	Insecta	Ceratopogonidae	Ceratopogonidae sp. C	6	0.0125	0.0063	0.0191
Animalia	Arthropoda	Insecta	Ceratopogonidae	Ceratopogonidae sp. D	17	0.0237	0.0054	0.0775
Animalia	Arthropoda	Insecta	Ceratopogonidae	Ceratopogonidae sp. E	12	0.0180	0.0065	0.0455

Continued

Table A2 Abundance and Body Size (Mean, Minimum and Maximum) of All Taxa Recorded—cont'd

Kingdom	Phylum	Class	Family	Species	Abund.	Wet Body Mass (mg)		
						Mean	Min	Max
Animalia	Arthropoda	Insecta	Ceratopogonidae	Ceratopogonidae sp. F	31	0.0171	0.0059	0.0455
Animalia	Arthropoda	Insecta	Ceratopogonidae	Ceratopogonidae sp. G	25	0.0178	0.0056	0.0455
Animalia	Arthropoda	Insecta	Ceratopogonidae	Ceratopogonidae sp. H	5	0.0178	0.0060	0.0424
Animalia	Arthropoda	Insecta	Ceratopogonidae	Ceratopogonidae sp. I	1	0.0784		
Animalia	Arthropoda	Insecta	Ceratopogonidae	Ceratopogonidae sp. M	4	0.2316	0.0359	0.6907
Animalia	Arthropoda	Insecta	Ceratopogonidae	Ceratopogonidae sp. N	1	0.4879		
Animalia	Arthropoda	Insecta	Ceratopogonidae	Ceratopogonidae sp. O	1	0.0312		
Animalia	Arthropoda	Insecta	Ceratopogonidae	Forcipomyiinae sp. A	966	0.0295	0.0054	0.2426
Animalia	Arthropoda	Insecta	Chironomidae	*Chironomus vitellinus*	4127	0.3828	0.0054	10.1779
Animalia	Arthropoda	Insecta	Chironomidae	*Djalmabatista* sp. A	1	0.1495		
Animalia	Arthropoda	Insecta	Chironomidae	*Paramerina parva*	699	0.1130	0.0054	0.8763
Animalia	Arthropoda	Insecta	Chironomidae	*Polypodium* sp. A	113	0.0633	0.0056	0.2967
Animalia	Arthropoda	Insecta	Chironomidae	Tanypodinae sp. A	7	0.2428	0.1435	0.3540
Animalia	Arthropoda	Insecta	Corylophidae	Corylophidae sp. A	3	0.0352	0.0312	0.0433
Animalia	Arthropoda	Insecta	Cosmopterigidae	Cosmopterigidae sp. A	3	0.3151	0.2800	0.3454
Animalia	Arthropoda	Insecta	Culicidae	*Aedes notoscriptus*	2	0.1168	0.0146	0.2191

Animalia	Arthropoda	Insecta	Culicidae	*Anopheles annulipes*	39	0.0845	0.0056	0.5270
Animalia	Arthropoda	Insecta	Culicidae	*Anopheles farauti*	12	0.0392	0.0105	0.2192
Animalia	Arthropoda	Insecta	Culicidae	*Anopheles* spp.	5	0.0191	0.0088	0.0387
Animalia	Arthropoda	Insecta	Culicidae	*Culex annulirostris*	41	0.1051	0.0062	0.6408
Animalia	Arthropoda	Insecta	Culicidae	*Culex halifaxii*	30	0.1355	0.0056	0.8782
Animalia	Arthropoda	Insecta	Culicidae	*Culex quinquefasciatus*	152	0.1141	0.0056	1.0745
Animalia	Arthropoda	Insecta	Culicidae	*Culex* sp. A	66	0.2399	0.0056	4.3651
Animalia	Arthropoda	Insecta	Curculionidae	*Xyleborus* sp. A	5	0.0730	0.0312	0.2212
Animalia	Arthropoda	Insecta	Diapriidae	Diapriidae sp. A	1	0.0335		
Animalia	Arthropoda	Insecta	Diapriidae	Diapriidae sp. B	1	0.0485		
Animalia	Arthropoda	Insecta	Diapriidae	Diapriidae sp. C	2	0.0477	0.0407	0.0547
Animalia	Arthropoda	Insecta	Dipsocoridae	*Cryptostemma* sp. A	56	0.0489	0.0048	0.7956
Animalia	Arthropoda	Insecta	Dytiscidae	*Chostonectes gigas*	1	5.0231		
Animalia	Arthropoda	Insecta	Dytiscidae	*Copelatus irregularis*	9	11.3991	10.0077	13.5577
Animalia	Arthropoda	Insecta	Dytiscidae	Dytiscidae sp. A	1	0.1722		
Animalia	Arthropoda	Insecta	Dytiscidae	*Hydaticus consanguineus*	2	5.4116	2.8589	7.9642
Animalia	Arthropoda	Insecta	Dytiscidae	*Platynectes* sp. A	1	13.5577		13.5577
Animalia	Arthropoda	Insecta	Ectopsocidae	*Ectopsocus* sp. A	2	0.2652	0.2323	0.2981

Continued

Table A2 Abundance and Body Size (Mean, Minimum and Maximum) of All Taxa Recorded—cont'd

Kingdom	Phylum	Class	Family	Species	Abund.	Wet Body Mass (mg) Mean	Min	Max
Animalia	Arthropoda	Insecta	Ectopsocidae	*Ectopsocus* sp. B	1	0.2223		
Animalia	Arthropoda	Insecta	Encyrtidae	Encyrtidae sp. A	1	0.0529		
Animalia	Arthropoda	Insecta	Entomobryidae	*Acanthocyrtus* sp. A	196	0.0229	0.0059	0.1553
Animalia	Arthropoda	Insecta	Entomobryidae	Entomobryidae sp. A	1	0.1676		
Animalia	Arthropoda	Insecta	Entomobryidae	Entomobryidae sp. B	1	0.0521		
Animalia	Arthropoda	Insecta	Entomobryidae	Entomobryidae sp. C	5	0.0179	0.0093	0.0248
Animalia	Arthropoda	Insecta	Entomobryidae	Entomobryidae sp. D	132	0.0178	0.0059	0.0368
Animalia	Arthropoda	Insecta	Entomobryidae	Entomobryidae sp. E	31	0.0205	0.0078	0.0398
Animalia	Arthropoda	Insecta	Entomobryidae	Entomobryidae spp.	17	0.0155	0.0013	0.0325
Animalia	Arthropoda	Insecta	Entomobryidae	Entomobryinae spp.	1	0.0266		
Animalia	Arthropoda	Insecta	Eulophidae	Entedoninae sp. A	1	0.0383		
Animalia	Arthropoda	Insecta	Eulophidae	Tetrastichinae sp. A	2	0.0366	0.0140	0.0592
Animalia	Arthropoda	Insecta	Eulophidae	Tetrastichinae sp. B	1	0.0105		
Animalia	Arthropoda	Insecta	Flatidae	*Dascalina* sp. A	1	0.0784		
Animalia	Arthropoda	Insecta	Formicidae	*Cardiocondyla nuda*	1	0.3501		
Animalia	Arthropoda	Insecta	Formicidae	*Crematogaster* sp. A	1	1.4659		

Animalia	Arthropoda	Insecta	Formicidae	*Iridomyrmex anceps*	3	0.6876	0.6304	0.7162
Animalia	Arthropoda	Insecta	Formicidae	*Monomorium laeve*	5	0.2665	0.0383	0.3499
Animalia	Arthropoda	Insecta	Formicidae	*Ochetellus glaber clarithorax*	4	0.5158	0.3311	1.0512
Animalia	Arthropoda	Insecta	Formicidae	*Odontomachus* sp. A	2	8.2489	8.2489	8.2489
Animalia	Arthropoda	Insecta	Formicidae	*Oecophylla smaragdina*	18	6.8768	5.7040	9.0423
Animalia	Arthropoda	Insecta	Formicidae	*Opisthopsis haddoni*	2	4.3771	4.0258	4.7285
Animalia	Arthropoda	Insecta	Formicidae	*Paratrechina longicornis*	2	0.4635	0.3501	0.5768
Animalia	Arthropoda	Insecta	Formicidae	*Paratrechina* sp. A	164	0.1917	0.0288	1.3541
Animalia	Arthropoda	Insecta	Formicidae	*Pheidole impressiceps*	34	0.9838	0.6583	1.3763
Animalia	Arthropoda	Insecta	Formicidae	*Pheidole* sp. A	1	3.0960		
Animalia	Arthropoda	Insecta	Formicidae	*Solenopsis* sp. A	39	0.0641	0.0290	0.3311
Animalia	Arthropoda	Insecta	Formicidae	*Spinctomyrmex* sp. A	1	3.6995		
Animalia	Arthropoda	Insecta	Formicidae	*Tapinoma* sp. A	3	0.4634	0.3127	0.7463
Animalia	Arthropoda	Insecta	Formicidae	*Tetramorium simillimum*	60	0.3150	0.0383	0.5768
Animalia	Arthropoda	Insecta	Geometridae	Geometridae sp. A	1	7.2359		
Animalia	Arthropoda	Insecta	Geometridae	Geometridae sp. B	1	3.9274		
Animalia	Arthropoda	Insecta	Hemicorduliidae	*Hemicordulia intermedia*	10	1.3295	0.1580	3.2553
Animalia	Arthropoda	Insecta	Hydraenidae	*Hydraena* sp. A	174	0.1228	0.0069	0.9486

Continued

Table A2 Abundance and Body Size (Mean, Minimum and Maximum) of All Taxa Recorded—cont'd

Kingdom	Phylum	Class	Family	Species	Abund.	Wet Body Mass (mg) Mean	Min	Max
Animalia	Arthropoda	Insecta	Hydraenidae	*Hydraena* sp. B	2	4.4938	4.3717	4.6158
Animalia	Arthropoda	Insecta	Hydraenidae	*Hydraena* sp. C	2	0.5779	0.4462	0.7096
Animalia	Arthropoda	Insecta	Hydraenidae	*Hydraena* sp. D	1	0.0383		
Animalia	Arthropoda	Insecta	Hydraenidae	*Hydraena* sp. E	2	0.2630	0.2630	0.2630
Animalia	Arthropoda	Insecta	Hydrophilidae	*Enochrus deserticola*	168	0.6229	0.0055	6.1085
Animalia	Arthropoda	Insecta	Hydrophilidae	*Enochrus maculiceps*	1	0.9935		
Animalia	Arthropoda	Insecta	Hydrophilidae	*Hydrochus* sp. A	3	0.2942	0.2439	0.3293
Animalia	Arthropoda	Insecta	Hydrophilidae	*Sternolophus marginicollis*	54	6.9787	0.3454	66.8695
Animalia	Arthropoda	Insecta	Isotomidae	*Acanthomurus* sp. A	446	0.0119	0.0014	0.1354
Animalia	Arthropoda	Insecta	Latrididae	*Metophthalmus* sp. A	1	0.0473		
Animalia	Arthropoda	Insecta	Lepidopsocidae	Lepidopsocidae sp. A	1	0.0133		
Animalia	Arthropoda	Insecta	Lepidopsocidae	*Thylacella* sp. A	152	0.0796	0.0048	0.4860
Animalia	Arthropoda	Insecta	Liposcelidae	*Embidopsocus* sp. A	1	0.1239		
Animalia	Arthropoda	Insecta	Liposcelidae	*Liposcelis* sp. A	158	0.0244	0.0059	0.0992
Animalia	Arthropoda	Insecta	Phlaeothripidae	*Baenothrips moundi*	1	0.0340		
Animalia	Arthropoda	Insecta	Phlaeothripidae	*Haplothrips* sp. A	2	0.0211	0.0162	0.0260

Animalia	Arthropoda	Insecta	Phlaeothripidae	*Holothrips* sp. A	1	0.2647		
Animalia	Arthropoda	Insecta	Phlaeothripidae	Phlaeothripidae sp. A	1	0.0312		
Animalia	Arthropoda	Insecta	Phoridae	Phoridae sp. A	1	0.4366		
Animalia	Arthropoda	Insecta	Platygastridae	Platygastridae sp. A	8	0.0397	0.0285	0.0784
Animalia	Arthropoda	Insecta	Platygastridae	Platygastridae sp. B	10	0.0325	0.0140	0.0485
Animalia	Arthropoda	Insecta	Platygastridae	Platygastridae sp. C	1	0.0312		
Animalia	Arthropoda	Insecta	Platygastridae	Platygastridae sp. D	1	0.0140		
Animalia	Arthropoda	Insecta	Platygastridae	Platygastridae sp. E	3	0.0368	0.0368	0.0368
Animalia	Arthropoda	Insecta	Pseudocaeciliidae	Pseudocaeciliidae sp. A	4	0.0380	0.0100	0.0909
Animalia	Arthropoda	Insecta	Pseudococcidae	*Geococcus coffeae*	1	0.0299		
Animalia	Arthropoda	Insecta	Pseudococcidae	Pseudococcidae sp. A	1	0.0325		
Animalia	Arthropoda	Insecta	Pseudococcidae	Pseudococcidae sp. B	1	0.0038		
Animalia	Arthropoda	Insecta	Pseudococcidae	Pseudococcidae sp. C	1	0.0255		
Animalia	Arthropoda	Insecta	Psychodidae	Phlebotominae sp. A	5	0.1484	0.0356	0.4230
Animalia	Arthropoda	Insecta	Psychodidae	*Psychoda* sp. A	3	0.0434	0.0335	0.0583
Animalia	Arthropoda	Insecta	Psychodidae	Psychodidae sp. A	78	0.0592	0.0059	0.3452
Animalia	Arthropoda	Insecta	Psychodidae	Psychodidae sp. B	90	0.0708	0.0059	0.3579
Animalia	Arthropoda	Insecta	Psychodidae	Psychodidae sp. C	1	0.0841		

Continued

Table A2 Abundance and Body Size (Mean, Minimum and Maximum) of All Taxa Recorded—cont'd

Kingdom	Phylum	Class	Family	Species	Abund.	Wet Body Mass (mg) Mean	Min	Max
Animalia	Arthropoda	Insecta	Psychodidae	Psychodidae sp. D	1	0.0643		
Animalia	Arthropoda	Insecta	Psychodidae	Psychodidae sp. E	36	0.0289	0.0060	0.1052
Animalia	Arthropoda	Insecta	Psychodidae	Psychodidae sp. F	5	0.0259	0.0074	0.0402
Animalia	Arthropoda	Insecta	Psychodidae	Psychodidae sp. G	3	0.0400	0.0149	0.0711
Animalia	Arthropoda	Insecta	Ptiliidae	*Nephanes* sp. A	1	0.0123		
Animalia	Arthropoda	Insecta	Scarabaeidae	Melolonthinae sp. A	3	4.9041	3.9076	5.6758
Animalia	Arthropoda	Insecta	Scirtidae	*Pseudomicrocara orientalis*	1	7.8667		
Animalia	Arthropoda	Insecta	Scirtidae	*Scirtes* sp. A	6	1.7882	0.3701	3.9076
Animalia	Arthropoda	Insecta	Scirtidae	Scirtidae sp. A	1	0.9841		
Animalia	Arthropoda	Insecta	Sminthuridae	*Sminthurides pseudassimilis*	689	0.0243	0.0016	0.6056
Animalia	Arthropoda	Insecta	Staphylinidae	Aleocharinae sp. A	1	0.6039		
Animalia	Arthropoda	Insecta	Staphylinidae	*Astenus* sp. A	7	1.2888	0.9009	1.5373
Animalia	Arthropoda	Insecta	Staphylinidae	*Myllaena* sp. A	2	0.6307	0.6039	0.6575
Animalia	Arthropoda	Insecta	Staphylinidae	*Myllaena* sp. B	4	0.3388	0.1606	0.6039
Animalia	Arthropoda	Insecta	Staphylinidae	Oxytelinae sp. A	1	1.3159		
Animalia	Arthropoda	Insecta	Staphylinidae	Oxytelinae sp. B	1	0.2250		

Kingdom	Phylum	Class	Family	Species				
Animalia	Arthropoda	Insecta	Staphylinidae	Pselaphinae sp. A	4	0.2076	0.1385	0.2830
Animalia	Arthropoda	Insecta	Staphylinidae	Staphylinidae sp. A	1	0.0565		
Animalia	Arthropoda	Insecta	Staphylinidae	Staphylinidae sp. B	1	0.0339		
Animalia	Arthropoda	Insecta	Staphylinidae	*Stenus* sp. A	2	1.7951	1.3159	2.2744
Animalia	Arthropoda	Insecta	Syrphidae	*Eristalis* sp. A	11	8.7403	0.2059	15.6122
Animalia	Arthropoda	Insecta	Tabanidae	*Tabanus* sp. A	2	0.0207	0.0192	0.0222
Animalia	Arthropoda	Insecta	Tabanidae	*Tabanus* sp. B	7	2.6829	0.1360	11.2126
Animalia	Arthropoda	Insecta	Thripidae	*Frankliniella schultzei*	7	0.0372	0.0131	0.0515
Animalia	Arthropoda	Insecta	Thripidae	*Pseudodendrothrips* sp. A	19	0.0233	0.0086	0.0939
Animalia	Arthropoda	Insecta	Thripidae	Thripidae sp. A	11	0.0201	0.0068	0.0916
Animalia	Arthropoda	Insecta	Thripidae	Thripidae sp. B	12	0.0222	0.0128	0.0305
Animalia	Arthropoda	Insecta	Thripidae	Thripidae sp. C	8	0.0654	0.0120	0.1322
Animalia	Arthropoda	Insecta	Tipulidae	Tipulidae sp. A	14	0.5891	0.0162	2.1214
Animalia	Arthropoda	Insecta	Triozidae	*Trioza* sp. A	1	0.4377		
Animalia	Arthropoda	Insecta	Veliidae	*Microvelia* spp.	17	0.1007	0.0054	0.4424
Animalia	Arthropoda	Insecta	Veliidae	*Microvelia (Austromicrovelia) torresiana*	1	0.8676		
Animalia	Arthropoda	Insecta	Veliidae	*Microvelia (Picaultia) paranega*	3	0.3824	0.2482	0.4614

Continued

Table A2 Abundance and Body Size (Mean, Minimum and Maximum) of All Taxa Recorded—cont'd

Kingdom	Phylum	Class	Family	Species	Abund.	Wet Body Mass (mg)		
						Mean	Min	Max
Animalia	Arthropoda	Ostracoda	Cyprididae	*Cypretta* sp. A	439	0.0300	0.0046	0.0699
Animalia	Arthropoda	Ostracoda	Cyprididae	*Stenocypris major*	56	0.5313	0.0215	1.1540
Animalia	Cnidaria	Hydrozoa	Hydridae	*Hydra* sp. A	1	0.1053		
Animalia	Mollusca	Gastropoda	Planorbidae	*Segnitila* sp. A	5	12.5835	0.8866	33.2935
Animalia	Nematoda	NA	NA	Nematoda sp. A	2	0.0828	0.0828	0.0828
Animalia	Nematoda	Adenophorea	Dorylaimidae	*Aporcelaimus* sp. A	1	0.0052		
Animalia	Nematoda	Adenophorea	Mermithidae	*Hexamermis* sp. A	3	6.8772	0.9430	10.4997
Animalia	Rotifera	NA	NA	Rotifera sp. S	1	0.0741		

Figure A1 Location of study site (red (grey in the print version) dot) within Goondaloo Creek, in the foothills of the Mount Stuart Range, Townsville, Queensland, Australia.

Figure A2 Photos of the experimental pool bed at various stages of drying.

Figure A3 Method of sampling used in the study, which provided a total core sample of ~2356 cm³. See main text for a detailed description of sampling methods.

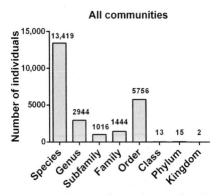

Figure A4 Level of taxonomic resolution of identification for all individuals recorded in the study.

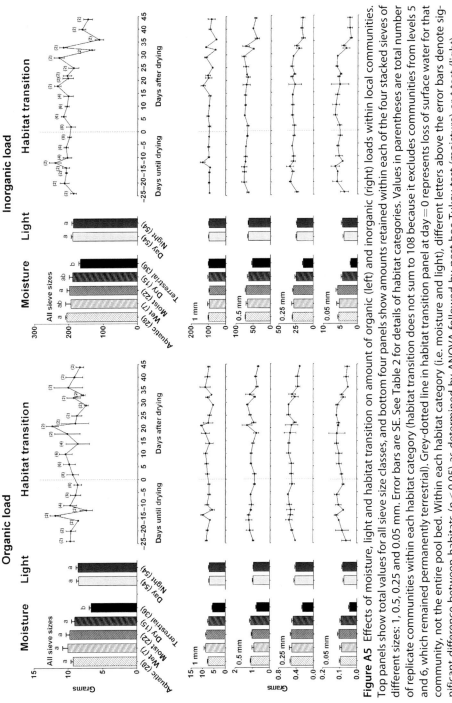

Figure A5 Effects of moisture, light and habitat transition on amount of organic (left) and inorganic (right) loads within local communities. Top panels show total values for all sieve size classes, and bottom four panels show amounts retained within each of the four stacked sieves of different sizes: 1, 0.5, 0.25 and 0.05 mm. Error bars are SE. See Table 2 for details of habitat categories. Values in parentheses are total number of replicate communities within each habitat category (habitat transition does not sum to 108 because it excludes communities from levels 5 and 6, which remained permanently terrestrial). Grey-dotted line in habitat transition panel at day = 0 represents loss of surface water for that community, not the entire pool bed. Within each habitat category (i.e. moisture and light), different letters above the error bars denote significant difference between habitats ($p < 0.05$) as determined by ANOVA followed by post hoc Tukey test (moisture) or t-test (light).

REFERENCES

Achard, F., Eva, H.D., Stibig, H.J., Mayaux, P., Gallego, J., Richards, T., Malingreau, J.P., 2002. Determination of deforestation rates of the world's humid tropical forests. Science 297, 999–1002.

Adis, J., 1992. How to survive six months in a flooded soil: strategies in Chilopoda and Symphyla from Central Amazonian floodplains. Stud. Neotropical Fauna Environ. 27, 117–129.

Adis, J., Junk, W.J., 2002. Terrestrial invertebrates inhabiting lowland river floodplains of Central Amazonia and Central Europe: a review. Freshw. Biol. 47, 711–731.

Allen, D.C., McCluney, K.E., Elser, S.R., Sabo, J.L., 2014. Water as a trophic currency in dryland food webs. Front. Ecol. Environ. 12, 156–160.

Angelsen, A., Kaimowitz, D., 2001. Agricultural Technologies and Tropical Deforestation. CABi, New York.

Brewer, M.C., Dawidowicz, P., Dodson, S.I., 1999. Interactive effects of fish kairomone and light on Daphnia escape behavior. J. Plankton Res. 21, 1317–1335.

Brose, U., Cushing, L., Berlow, E.L., Jonsson, T., Banasek-Richter, C., Bersier, L.-F., Blanchard, J.L., Brey, T., Carpenter, S.R., Blandenier, M.F.C., Cohen, J.E., Dawah, H.A., Dell, T., Edwards, F., Harper-Smith, S., Jacob, U., Knapp, R.A., Ledger, M.E., Memmott, J., Mintenbeck, K., Pinnegar, J.K., Rall, B.C., Rayner, T., Ruess, L., Ulrich, W., Warren, P.H., Williams, R.J., Woodward, G., Yodzis, P., Martinez, N.D., 2005. Body sizes of consumers and their resources. Ecology 86, 2545.

Brose, U., Jonsson, T., Berlow, E.L., Warren, P., Banasek-Richter, C., Bersier, L.F., Blanchard, J.L., Brey, T., Carpenter, S.R., Blandenier, M.F., Cushing, L., Dawah, H.A., Dell, T., Edwards, F., Harper-Smith, S., Jacob, U., Ledger, M.E., Martinez, N.D., Memmott, J., Mintenbeck, K., Pinnegar, J.K., Rall, B.C., Rayner, T.S., Reuman, D.C., Ruess, L., Ulrich, W., Williams, R.J., Woodward, G., Cohen, J.E., 2006. Consumer-resource body-size relationships in natural food webs. Ecology 87, 2411–2417.

Brown, J.H., Gillooly, J.F., Allen, A.P., Savage, V.M., West, G.B., 2004. Toward a metabolic theory of ecology. Ecology 85, 1771–1789.

Chase, J.M., 1999. Food web effects of prey size refugia: variable interactions and alternative stable equilibria. Am. Nat. 154, 559–570.

Chase, J.M., 2000. Are there real differences among aquatic and terrestrial food webs? Trends Ecol. Evol. 15, 408–412.

Davies, J.N., Boulton, A.J., 2009. Great house, poor food: effects of exotic leaf litter on shredder densities and caddisfly growth in 6 subtropical Australian streams. J. N. Am. Benthol. Soc. 28, 491–503.

Dell, A.I., Pawar, S., Savage, V.M., 2011. Systematic variation in the temperature dependence of physiological and ecological traits. Proc. Natl. Acad. Sci. U. S. A. 108, 10591–10596.

Dell, A.I., Pawar, S., Savage, V.M., 2013. The thermal dependence of biological traits. Ecology 94, 1205.

Dell, A.I., Alford, R.A., Pearson, R.G., 2014a. Intermittent pool beds are permanent cyclic habitats with distinct wet, moist and dry faunas. PLoS One 9 (9), e108203.

Dell, A.I., Bender, J., Couzin, I.D., Dunne, J.A., Noldus, L.P.J.J., Perona, P., Straw, A.D., Wikelski, M., Brose, U., 2014b. Automated image-based tracking and its application in ecology. Trends Ecol. Evol. (Amst.) 29, 417–428.

Dell, A.I., Pawar, S., Savage, V.M., 2014c. Temperature dependence of trophic interactions driven by asymmetry of species responses and foraging strategy. J. Anim. Ecol. 83, 70–84.

Denny, M.W., 1990. Terrestrial versus aquatic biology—the medium and its message. Am. Zool. 30, 111–121.

Dieguez, M.C., 2003. Predation by Buenoa macrotibialis (Insecta, Hemiptera) on zooplankton: effect of light on selection and consumption of prey. J. Plankton Res. 25, 759–769.

Dudgeon, D., Wu, K.K.Y., 1999. Leaf litter in a tropical stream: food or substrate for macroinvertebrates? Arch. Hydrobiol. 146, 65–82.

Estes, J.A., Terborgh, J., Brashares, J.S., Power, M.E., Berger, J., Bond, W.J., Carpenter, S.R., Essington, T.E., Holt, R.D., Jackson, J.B., Marquis, R.J., Oksanen, L., Oksanen, T., Paine, R.T., Pikitch, E.K., Ripple, W.J., Sandin, S.A., Scheffer, M., Schoener, T.W., Shurin, J.B., Sinclair, A.R., Soule, M.E., Virtanen, R., Wardle, D.A., 2011. Trophic downgrading of planet Earth. Science 333, 301–306.

Florencio, M., Díaz-Paniagua, C., Gomez-Mestre, I., Serrano, L., 2011. Sampling macroinvertebrates in a temporary pond: comparing the suitability of two techniques to detect richness, spatial segregation and diel activity. Hydrobiologia 689, 121–130.

Fraser, N.H.C., Metcalfe, N.B., 1997. The costs of becoming nocturnal: feeding efficiency in relation to light intensity in juvenile Atlantic Salmon. Funct. Ecol. 11, 385–391.

Gaston, K.J., Chown, S.L., Mercer, R.D., 2001. The animal species-body size distribution of Marion Island. Proc. Natl. Acad. Sci. U. S. A. 98, 14493–14496.

Gergs, A., Hoeltzenbein, N.I., Ratte, H.T., 2010. Diurnal and nocturnal functional response of juvenile Notonecta maculata considered as a consequence of shifting predation behaviour. Behav. Process. 85, 151–156.

Gilbert, J.J., Hampton, S.E., 2001. Diel vertical migrations of zooplankton in a shallow, fishless pond: a possible avoidance-response cascade induced by notonectids. Freshw. Biol. 46, 611–621.

Gouws, J.E., Gaston, K.J., Chown, S.L., 2011. Intraspecific body size frequency distributions of insects. PLoS One 6, e16606.

Hairston, N.G., Hairston, N.G., 1993. Cause-effect relationships in energy-flow, trophic structure, and interspecific interactions. Am. Nat. 142, 379–411.

Hampton, S.E., Duggan, I.C., 2003. Diel habitat shifts of macrofauna in a fishless pond. Mar. Freshw. Res. 54, 797.

Hampton, S.E., Friedenberg, N.A., 2002. Nocturnal increases in the use of near-surface water by pond animals. Hydrobiologia 477, 171–179.

Hawkins, B.A., Field, R., Cornell, H.V., Currie, D.J., 2003. Energy, water, and broad-scale geographic patterns of species richness. Ecology 84, 3105–3117.

Jara, F.G., 2007. Tadpole–odonate larvae interactions: influence of body size and diel rhythm. Aquat. Ecol. 42, 503–509.

Jennings, S., Blanchard, J.L., 2004. Fish abundance with no fishing: predictions based on macroecological theory. J. Anim. Ecol. 73, 632–642.

Jetz, W., Carbone, C., Fulford, J., Brown, J.H., 2004. The scaling of animal space use. Science 306, 266–268.

Kalinkat, G., Brose, U., Rall, B.C., 2013a. Habitat structure alters top-down control in litter communities. Oecologia 172, 877–887.

Kalinkat, G., Schneider, F.D., Digel, C., Guill, C., Rall, B.C., Brose, U., 2013b. Body masses, functional responses and predator–prey stability. Ecol. Lett. 16, 1126–1134.

Kronfeld-Schor, N., Dayan, T., 2003. Partitioning of time as an ecological resource. Annu. Rev. Ecol. Evol. Syst. 34, 153–181.

Lambeets, K., Vandegehuchte, M.L., Maelfait, J.-P., Bonte, D., 2008. Understanding the impact of flooding on trait-displacements and shifts in assemblage structure of predatory arthropods on river banks. J. Anim. Ecol. 77, 1162–1174.

Larned, S.T., Datry, T., Arscott, D.B., Tockner, K., 2010. Emerging concepts in temporary-river ecology. Freshw. Biol. 55, 717–738.

Lensing, J.R., Wise, D.H., 2006. Predicted climate change alters the indirect effect of predators on an ecosystem process. Proc. Natl. Acad. Sci. U. S. A. 103, 15502–15505.

Link, J., 2002. Does food web theory work for marine ecosystems? Mar. Ecol. Prog. Ser. 230, 1–9.

Losos, J.B., 1992. The evolution of convergent structure in Caribbean Anolis communities. Syst. Biol. 41, 403–420.

Losos, J.B., Jackman, T.R., Larson, A., Queiroz, K., Rodriguez-Schettino, L., 1998. Contingency and determinism in replicated adaptive radiations of island lizards. Science 279, 2115–2118.

Mason, N.W.H., Mouillot, D., Lee, W.G., Wilson, J.B., 2005. Functional richness, functional evenness and functional divergence: the primary components of functional diversity. Oikos 111, 112–118.

McCluney, K.E., Sabo, J.L., 2009. Water availability directly determines per capita consumption at two trophic levels. Ecology 90, 1463–1469.

McCluney, K.E., Belnap, J., Collins, S.L., Gonzalez, A.L., Hagen, E.M., Nathaniel Holland, J., Kotler, B.P., Maestre, F.T., Smith, S.D., Wolf, B.O., 2012. Shifting species interactions in terrestrial dryland ecosystems under altered water availability and climate change. Biol. Rev. Camb. Philos. Soc. 87, 563–582.

McGill, B.J., Enquist, B.J., Weiher, E., Westoby, M., 2006. Rebuilding community ecology from functional traits. Trends Ecol. Evol. 21, 178–185.

McLachlan, A.J., Cantrell, M.A., 1980. Survival strategies in tropical rain pools. Oecologia 47, 344–351.

Melguizo-Ruiz, N., Verdeny-Vilalta, O., Arnedo, M.A., Moya-Larano, J., 2012. Potential drivers of spatial structure of leaf-litter food webs in south-western European beech forests. Pedobiologia 55, 311–319.

Melville, J., Harmon, L.J., Losos, J.B., 2006. Intercontinental community convergence of ecology and morphology in desert lizards. Proc. Biol. Sci. 273, 557–563.

Murphy, P.G., Lugo, A.E., 1986. Ecology of tropical dry forest. Annu. Rev. Ecol. Syst. 17, 67–88.

Nathan, R., Getz, W.M., Revilla, E., Holyoak, M., Kadmon, R., Saltz, D., Smouse, P.E., 2008. A movement ecology paradigm for unifying organismal movement research. Proc. Natl. Acad. Sci. U. S. A. 105, 19052–19059.

Pawar, S., Dell, A.I., Savage, V.M., 2012. Dimensionality of consumer search space drives trophic interaction strengths. Nature 486, 485–489.

Pawar, S., Dell, A.I., Van, M.S., 2013. Pawar et al. reply. Nature 493, E2–E3.

Petchey, O.L., Belgrano, A., 2010. Body-size distributions and size-spectra: universal indicators of ecological status? Biol. Lett. 6, 434–437.

Petchey, O.L., Beckerman, A.P., Riede, J.O., Warren, P.H., 2008. Size, foraging, and food web structure. Proc. Natl. Acad. Sci. U. S. A. 105, 4191–4196.

Peters, R.H., 1983. The Ecological Implications of Body Size. Cambridge University Press, Cambridge.

Pielou, E.C.J., 1966. The measurement of diversity in different types of biological collections. J. Theor. Biol. 13, 131–144.

Reddy, M.V., 1995. Soil Organisms and Litter Decomposition in the Tropics. Westview Press, Boulder, CO.

Richardson, J.S., 1992. Food, microhabitat, or both—macroinvertebrate use of leaf accumulations in a montane stream. Freshw. Biol. 27, 169–176.

Riede, J., Brose, U., Ebenman, B., Jacob, U., Thompson, R., Townsend, C., Jonsson, T., 2011. Stepping in Elton's footprints: a general scaling model for body masses and trophic levels across ecosystems. Ecol. Lett. 14, 169–178.

Roman, J., Palumbi, S.R., 2003. Whales before whaling in the North Atlantic. Science 301, 508–510.

Sana, N., Aditya, G., Bal, A., Saha, G.K., 2008. Influence of light and habitat on predation of Culex quinquefasciatus (Diptera: Culicidae) larvae by the waterbugs (Hemiptera: Heteroptera). Insect Sci. 15, 461–469.

Schmidt-Nielsen, K., 1984. Scaling: Why Is Animal Size so Important? Cambridge University Press, Cambridge.

Schramski, J.R., Dell, A.I., Grady, J.M., Sibly, R.M., Brown, J.H., 2015. Metabolic theory predicts whole-ecosystem properties. Proc. Natl. Acad. Sci. U. S. A. 112, 2617–2622.

Shurin, J.B., Gruner, D.S., Hillebrand, H., 2006. All wet or dried up? Real differences between aquatic and terrestrial food webs. Proc. Biol. Sci. 273, 1–9.

Simpson, S.J., Raubenheimer, D., Charleston, M.A., Clissold, F.J., 2010. Modelling nutritional interactions: from individuals to communities. Trends Ecol. Evol. 25, 53–60.

Smith, R.E.W., Pearson, R.G., 1987. The macro-invertebrate communities of temporary pools in an intermittent stream in tropical Queensland. Hydrobiologia 150, 45–61.

Spiller, D.A., Schoener, T.W., 2008. Climatic control of trophic interaction strength: the effect of lizards on spiders. Oecologia 154, 763–771.

Stehr, W.C., Branson, J.W., 1938. An ecological study of an intermittent stream. Ecology 19, 294–310.

Steward, A.L., Marshall, J.C., Sheldon, F., Harch, B., Choy, S., Bunn, S.E., Tockner, K., 2011. Terrestrial invertebrates of dry river beds are not simply subsets of riparian assemblages. Aquat. Sci. 73, 551–566.

Steward, A.L., von Schiller, D., Tockner, K., Marshall, J.C., Bunn, S.E., 2012. When the river runs dry: human and ecological values of dry riverbeds. Front. Ecol. Environ. 10, 202–209.

Tamm, J.C., 1984. Surviving long submergence in the egg stage? A successful strategy of terrestrial arthropods living on flood plains (Collembola, Acari, Diptera). Oecologia 61, 417–419.

Verdeny-Vilalta, O., Moya-Laraño, J., 2014. Seeking water while avoiding predators: moisture gradients can affect predator–prey interactions. Anim. Behav. 90, 101–108.

Webster, J.R., Benfield, E.F., 1986. Vascular plant breakdown in fresh-water ecosystems. Annu. Rev. Ecol. Syst. 17, 567–594.

White, E.P., Ernest, S.K.M., Kerkhoff, A.J., Enquist, B.J., 2007. Relationships between body size and abundance in ecology. Trends Ecol. Evol. 22, 323–330.

Wiens, J.J., Ackerly, D.D., Allen, A.P., Anacker, B.L., Buckley, L.B., Cornell, H.V., Damschen, E.I., Jonathan, Davies T., Grytnes, J.A., Harrison, S.P., Hawkins, B.A., Holt, R.D., McCain, C.M., Stephens, P.R., 2010. Niche conservatism as an emerging principle in ecology and conservation biology. Ecol. Lett. 13, 1310–1324.

Williams, D.D., 1987. The Ecology of Temporary Waters. Croom Helm, London, England.

Woodward, G., Ebenman, B., Emmerson, M., Montoya, J.M., Olesen, J.M., Valido, A., Warren, P.H., 2005a. Body size in ecological networks. Trends Ecol. Evol. 20, 402–409.

Woodward, G., Ebenman, B., Emmerson, M.C., Montoya, J.M., Olesen, J.M., Valido, A., Warren, P.H., de Ruiter, P.C., Wolters, V., Moore, J.C., 2005b. Body size determinants of the structure and dynamics of ecological networks: scaling from the individual to the ecosystem. In: De Ruiter, P., Wolters, V., Moore, J. (Eds.), Dynamic Food Webs: Multispecies Assemblages, Ecosystem Development and Environmental Change. Theoretical Ecology Series, Academic Press, San Diego, CA, pp. 179–197.

Xiong, S., Nilsson, C., 1997. Dynamics of leaf litter accumulation and its effects on riparian vegetation: a review. Bot. Rev. 63, 240–264.

Yvon-Durocher, G., Reiss, J., Blanchard, J., Ebenman, B., Perkins, D.M., Reuman, D.C., Thierry, A., Woodward, G., Petchey, O.L., 2011. Across ecosystem comparisons of size structure: methods, approaches and prospects. Oikos 120, 550–563.

CHAPTER SEVEN

Shifts in the Diversity and Composition of Consumer Traits Constrain the Effects of Land Use on Stream Ecosystem Functioning

André Frainer[*,†], Brendan G. McKie[†,1]

*Department of Ecology and Environmental Science, Umeå University, Umeå, Sweden
†Department of Aquatic Sciences and Assessment, Swedish University of Agricultural Sciences, Uppsala, Sweden
[1]Corresponding author: e-mail address: brendan.mckie@slu.se

Contents

Abstract

Species functional traits provide an important conceptual link between the effects of disturbances on community composition and diversity, and their ultimate outcomes for ecosystem functioning. Across 10 boreal streams covering a gradient of increasing intensity of land-use management, from forested to agricultural sites, we analysed relationships between leaf decomposition, the feeding traits of detritivores and measures of anthropogenic disturbances in two seasons. The direct effect of increasing land-use intensity on decomposition was positive and was associated with increases in nutrient concentrations and current velocities. However, this relationship was countered by negative effects associated with a loss of detritivore functional diversity along the gradient during autumn and shifts in species trait composition during spring, limiting the net change in functioning associated with increasing land-use management overall. Our results highlight the key roles that trait identity and diversity can play in mediating the effects of human disturbance on ecosystem functioning.

1. INTRODUCTION

Species loss can affect ecosystem functioning at similar or even higher levels than other major environmental disturbances (Hooper et al., 2012; Lavorel and Garnier, 2002; Poff et al., 2006; Tilman et al., 2012). However, it is often challenging to disentangle the direct effects of species loss on ecosystem functioning from those of anthropogenic disturbances, which may themselves be driving diversity change, such as alterations in pH (McKie et al., 2009) or nutrient enrichment (Woodward et al., 2012). Species functional traits provide an important conceptual link between the effects of disturbances on community composition and diversity, and their ultimate outcomes for ecosystem functioning (Enquist et al., 2015; Lavorel and Garnier, 2002; Mouillot et al., 2013). However, relationships between functional traits, their responses to disturbances and outcomes for ecosystem functioning remain poorly assessed for most systems. A combined perspective on these linkages should improve prospects for developing a more mechanistic framework that couples disturbances, diversity and functioning in a changing world.

Disturbances affect the distribution and composition of functional traits in predictable sequences, with correlated clusters of traits, termed 'response traits', either favoured or suppressed (Lavorel and Garnier, 2002; McKie et al., 2009; Poff et al., 2006). In general, smaller, fast-growing, short-lived organisms with more generalist feeding traits are more likely to persist over the long term at highly disturbed sites, whereas larger, slow-growing, long-lived and more specialised organisms are often more vulnerable to local extinction (Statzner and Bêche, 2010; Woodward et al., 2012). Such shifts in species trait composition may further impact ecosystem functioning, particularly when traits crucial for key ecosystem processes (known as 'effect traits') are impacted (Lavorel and Garnier, 2002; Pakeman, 2011). However, it is unclear whether the traits typically characterised as response traits are necessarily identical with effect traits (Lavorel and Garnier, 2002; Pakeman, 2011). Indeed, while traits related to mobility or development time are often more closely related to disturbances, traits related to resource consumption and nutrient requirements may be more closely related to ecosystem functioning.

Further complications arise from the potential for disturbances to affect not only the identity of traits but also the dominance hierarchies and inter-action strengths among the species bearing those traits (Thébault and Loreau, 2005). Changes in both species richness and evenness can alter trait relative abundances, and hence the dominance of particular traits, with potentially strong impacts on ecosystem functioning (Hillebrand et al., 2008; McKie et al., 2008; Nilsson et al., 2008). Changes in both the richness and relative abundances of traits can further impact functioning by altering the intensity of antagonistic or facilitative species interactions, which may hinder or enhance key processes, respectively (Jonsson and Malmqvist, 2003; Jousset et al., 2011). Finally, species interactions can also change as a result of sublethal effects of abiotic stressors on the behaviour of particular species (Newcombe and Macdonald, 1991), further highlighting the need to disen-tangle the direct effects of disturbances on functioning from those arising from shifts in species trait composition and diversity.

In this study, we investigated the impacts of an increase in land-use inten-sity on the distribution and diversity of functional traits within a guild of stream-living detritivorous invertebrates, and how these shifts in turn affected a key ecosystem process, leaf litter decomposition. The land-use gradient con-sisted of 10 streams ranging from fully forested to streams flowing through long-established managed meadows, and was associated with changes in stream hydromorphology, nutrient concentrations and riparian vegetation. As relationships between detritivores and leaf litter decomposition may vary greatly across seasons (Frainer et al., 2014), we conducted our study in both autumn and spring. Leaf decomposition is an important ecological process, which can be used in environmental assessment (Gessner and Chauvet, 2002; Pascoal et al., 2003), with rates of leaf decomposition often differing between impacted and control sites (McKie and Malmqvist, 2009). However, classification of ecological status based on ecosystem process rates is not always straightforward (Bergfur et al., 2007; Pascoal et al., 2001), particularly as rela-tionships between disturbance gradients and functional response are often not monotonic (McKie et al., 2006; Woodward et al., 2012).

We expected that decomposition overall would be stimulated by increases in nutrients along the land-use gradient, but that this effect might be countered by a reduction in leaf processing capacity among the detritivores, associated with shifts in trait identity and diversity. Specifically, we hypothesised that (1) increases in the intensity of agricultural land use will affect the identity and evenness of traits present, with a greater dominance of generalist consumers, while pristine streams will have more obligate leaf-consumers, and a more even distribution of functional traits; and (2) the greater dominance of more generalist consumers in the agricultural sites will lower leaf decomposition rates, reflecting their reduced specialisation and dependency on the litter resource.

2. METHODS

2.1 Land-Use Gradient

During autumn 2011 and spring 2012, we investigated multiple parameters associated with the ecosystem-level process of leaf decomposition in 10 streams, which represented a gradient of increasing intensity of land-use management in the catchment.

The major changes along the land-use gradient are detailed below, but included: (i) an increase in the amount of managed (ploughed, sowed and fertilised) meadow adjacent to the study streams (from none to extensive), (ii) a decline in the proportion of forest in the catchment; (iii) a shift in the composition of riparian vegetation (from predominantly coniferous to deciduous) and (iv) an increase in channel modification and other measures (agricultural ditches) to facilitate drainage.

The least managed streams flowed through mature forest (not subject to clear-cutting for >80 years), whereas the most managed streams flowed through agricultural meadows, subject to tillage and fertilisers, and with more modified channel forms and reduced riparian vegetation (Pettersson et al., 2004). Agricultural meadow sites date from at least the mid 1800s and are mostly used for production of hay, but crop farming and limited animal husbandry (mostly cattle) occur upstream in some catchments (Pettersson et al., 2004). All meadows are interspersed with blocks of forest, so that even the most agricultural sites have some forest in the catchment (Table A.1). All streams had predominantly rocky substrates (boulders, cobbles and some gravel), with some coarse sand (Table A.1) and were of comparable width. However, the more agricultural streams were generally more channelised (i.e. hydromorphologically modified to facilitate drainage and transport), being more strongly straightened and deeper than the forested sites and with simpler and coarser stony substrates at the sampling points. All streams had some riparian vegetation, ranging up to 3 m width for the most agricultural streams, to extensive forest around the less impacted sites. The riparian vegetation of the agricultural sites was mostly composed of broadleaf trees, especially birch (*Betula* spp.), alder (*Alnus incana* (L.) Moench) and willow (*Salix* spp.), while birch and Norwegian spruce (*Picea abies* (L.) H. Karst) dominated the forested sites. Across the land-use gradient, dissolved concentrations of total nitrogen ranged from 0.23 to 1.12 mg/L, with nitrate [N–NO$_3$] ranging from 0.01 to 0.56 mg/L. Phosphorus concentrations were mostly below the detection limit (<4 mg/L) (see Table A.1 for mean values of each sampling site).

We measured water velocity and discharge halfway through the experimental period in each season, by taking water velocity measurements at 60% depth every 0.5 m across the stream channel. Discharge was calculated as:

$$Q = \sum (v_i * A_i), \tag{1}$$

where v_i is the water velocity (m/s) at 60% depth at point i and A is the the the cross-sectional area (m^2) of the channel corresponding to point i (Gordon et al., 2004). Stream substratum was estimated as percentage substratum cover of five categories: (1) silt and sand; (2) coarse sand; (3) sand-pebbles; (4) pebbles–boulders and (5) boulders (Giller and Malmqvist, 1998). Riparian vegetation width was also characterised based on estimates of the width of riparian tree cover, from 0 (no trees, only grass) to 5 (riparian tree vegetation extending at least 10 m to each side of the channel).

2.2 Ecosystem Functioning

We studied the decomposition of silver birch (*Betula pendula* Roth) leaves, which were collected freshly abscised from the ground and air-dried for 2 weeks at the start of autumn. An amount of 4 ± 0.05 g dried leaves was placed in litter bags of two different mesh sizes: 0.5 mm (fine mesh bags), which allows microbial colonisation of the leaves while excluding stream invertebrates, and 10 mm (coarse mesh bags), which allows both invertebrates and microbes to colonise the litter. Five litter bag pairs consisting of one coarse and fine mesh bag each were evenly spaced over 50 m of riffle habitat (fast flowing rocky habitat with flow broken by emergent stones) in each stream, and retrieved when leaves at the fastest decomposing site had reached approximately 50% dry mass remaining in the coarse mesh bags

(6 weeks in autumn and 4 weeks in spring). After retrieval, the litter bags were immediately transported to the laboratory for processing, where the leaves were removed from the bags, rinsed under tap water, oven-dried at 60 °C for 48 h and weighed to nearest 0.01 g.

We calculated decomposition as a rate k, using the exponential equation (Petersen and Cummins, 1974):

$$-k = \frac{\ln(M_t) - \ln(M_0)}{dd}, \tag{2}$$

where M_t is the final mass, M_0 is the initial mass and dd is the sum of daily average water temperature in °C.

2.3 Detritivore Richness, Density and Metabolic Capacity

We sorted and identified to species level all leaf-eating detritivores found in the litter bags. After identification, individuals of each species from each sample were pooled and placed in separate pre-weighed aluminium pans, oven-dried for 48 h at 60 °C and weighed to 0.1 mg. We calculated the metabolic capacity (MC) of each species according to the formula:

$$MC = m_i^{0.75} * x_i, \tag{3}$$

where m is the average individual body mass of species i, x is the number of individuals of species I and the exponent 0.75 represents a general relationship between metabolism and body size across a wide range of vertebrate and invertebrate species (Brown et al., 2004). We also quantified detritivore density as the sum of all detritivore individuals found per litter bag.

2.4 Functional Traits

Our analyses focused on the leaf-eating detritivores colonising our litter bags, a functional guild often termed 'shredders' (Cummins and Klug, 1979). However, while all species in this guild consume leaves, they vary both in their feeding mode and in their degree of dependency on litter as a primary nutrient resource. The most obligate leaf-shredders, with diets represented by >70% leaf material (see Table A.2), are best exemplified by the caddisfly genus *Halesus* and the stonefly *Nemoura avicularis* Morton. Less obligate leaf-consumers include those that gain significant proportions of their diet either by scraping biofilm from leaf or rock surfaces (e.g. the stonefly *Amphinemura* spp.) (Lieske and Zwick, 2007) or by gathering deposited and/or drifting fine particles (such as *Taeniopteryx nebulosa* Linnaeus, another stonefly), and these species are more likely to rasp, rather than actively chew, leaf surfaces. Many detritivores also include increasing proportions of prey items in their diets as they complete their larval development, particularly various caddisflies (Wissinger et al., 2004). However, none of our detritivore species fall exclusively into any one of these categories (Göthe et al., 2009; Layer et al., 2013), rather they vary in their association with these different types of feeding traits and behaviours. Accordingly, we used fuzzy coding when scoring species–trait associations, allowing membership of our species in more than one trait grouping simultaneously, with trait scores weighted individually for each species (Table A.2). All trait codings are based on the information compiled in the Freshwater Ecology database (Schmidt-Kloiber and Hering, 2011).

Using these traits to describe the species, we calculated two measures of functional diversity. Firstly, we assessed functional identity ($F_{identity}$) by using community-weighted mean trait values (Garnier et al., 2004) based on the mass ratio hypothesis (Grime, 1998), which captures the identity of the dominant traits within a community. $F_{identity}$ was calculated using the R package *FD* (Laliberté and Shipley, 2011), which yields a matrix where traits are weighted by the abundance of all species sharing it. This matrix was resolved with principal component analysis (R package *vegan*; Oksanen et al., 2011) for each season individually to obtain the axis that explained most of the variation across sites. Secondly, we calculated the functional dispersion ($F_{dispersion}$) of each community using the R package *FD* (Laliberté and Shipley, 2011). $F_{dispersion}$ is a measure of functional trait distribution that accounts for the dissimilarity among traits (Laliberté and Legendre, 2010). To assess $F_{dispersion}$, a centroid was calculated from the trait-based distance matrices for each season, where species were weighted by their abundance. From this, the distances between the centroid and each species, which are weighted again by their abundances, were further calculated. The sum of these distances is $F_{dispersion}$.

2.5 Isotopes

Due to the potential for algal productivity to be stimulated in the agricultural sites by increased nutrient concentrations, we used isotope analyses to investigate the degree to which the potentially omnivorous stream benthic detritivores shifted in their resource assimilation along the disturbance gradient. We sampled invertebrate detritivores from each stream in each of the two seasons (during October and May) by kick-sampling riffles where the litter bags had been previously deployed. From each stream, species were kept separately in clean tap water for 24 h in a climate-controlled room at 6 °C to allow emptying of digestive tracts before processing for isotope analyses. We also sampled resources potentially contributing to detritivore diets. Leaf litter and grass were collected directly from the ground adjacent to the streams (leaf litter and dead grass), while aquatic moss (*Fontinalis* sp.) was sampled from within the streams (water moss) in both autumn and spring. We used algal isotope data (K. Stenroth, unpublished data) sampled in each stream during summer to characterise the isotopic signal of auto-trophic periphyton. Algal samples consisted of biofilm and green algae that were scraped from boulders and separated before isotope analysis. We were unable to obtain sufficient periphyton for isotope analysis during October. Hence, these samples are regarded as representative of the algae consumers that had access to earlier in the season, prior to the autumn dieback, though we interpret the findings with some caution given the potential for the isotopic signature of algae to have shifted from summer to early autumn (Campbell and Fourqurean, 2009). All material was oven-dried at 60 °C for 24 h, weighed to 1 ± 0.5 mg and stored in small tin capsules. Samples were analysed by the Davis Isotope Facilities (Davis, USA) for ^{13}C and ^{15}N.

Isotope results are reported as ∂^{13}C and ∂^{15}N, which is defined as the contribution of heavy isotopes to the samples, relative to an international standard (in ‰), according to the formula

$$\partial X = \left[\left(\frac{R_{\text{sample}}}{R_{\text{standard}}} \right) - 1 \right] \times 1000, \qquad (4)$$

where $X = {}^{13}$C or ^{15}N and $R = {}^{13}$C/^{12}C or ^{15}N/^{14}N (Peterson and Fry, 1987). Positive values denote samples with higher amounts of the heavy isotopes (^{13}C or ^{15}N) than the standard, whereas negative values characterise samples with lower heavy isotopes than the standard. In stream ecosystems, higher concentration of ^{13}C (positive ∂^{13}C) typically refers to carbon orig-inated from terrestrial sources, such as leaf litter or grasses, and depletion of ^{13}C (thus more negative ∂^{13}C) is typically related to carbon originating from aquatic autotrophic sources, such as algal biofilm or aquatic mosses (Leberfinger et al., 2010). ∂^{15}N is indicative of the trophic level, with pri-mary producers having the lowest ∂^{15}N and predators having the highest ∂^{15}N (Fry, 2006). Lipids introduce a bias to isotope signal due to a deple-tion in ^{13}C relative to proteins and carbohydrates (Post et al., 2007).

We corrected the isotope value of consumers characterised by dry mass C:N >3.5% (>5% lipid) following the equations provided by Post et al. (2007).

We conducted our statistical analyses of shifts in isotopic composition by pooling detritivore species into genera. This was necessary for two reasons: (1) all genera but not all species were found at all sites, and (2) we were not able to collect the 1 mg minimum dry weight for each species from each stream required for isotope analysis. While we acknowledge this approach might obscure species-specific differences in isotope signal, we expected larger differences between than within genera, and pooling at the genus level allowed the most extensive analysis of cross-site variability in isotopic composition.

2.6 Fungal Biomass

We quantified fungal ergosterol content as a measure of fungal biomass, an important predictor of decomposition rates (Gessner et al., 2007). Ergosterol was extracted from one of the two sets of five leaf discs cut from each fine mesh bag, with the other set used for mass estimations. Ground leaf discs were shaken on a vortex mixer for 30 min in 1000 μL EtOH 99.5% at 5 °C, followed by centrifugation at 5 °C and 14,000 RPM for 15 min (Dahlman et al., 2002). The liquid extract was then analysed on high-performance liquid chromatography (Mobile phase: 100% MetOH; flow: 1.5 mL; wavelength: 280 nm).

2.7 Data Analyses

The land-use gradient was analysed using a principal components analysis (PCA) based on the environmental data measured for each site in autumn using the package *vegan* (Oksanen et al., 2011) in the software R (R Core Team, 2014). Environmental variables were standardised to zero mean ± 1 SD prior to the PC analysis. When two variables were significantly correlated with a coefficient greater than 0.75 (Pearson's product moment correlation), the variable with most significant correlations across the entire environmental data set was removed from further analyses. The resultant variable set included water temperature, water velocity, total nitrogen (TN), pH, riparian width, stream depth and width and substrate type. The choice of TN as our measure of nutrient variation along the gradient is justified not only on statistical grounds but also because it most accurately reflects the total increase in nutrients associated with changed land use, and the total pool of nutrients available for microbes to utilise. Total nitrogen export from our studied meadows can include a high organic N component (up to 30%, Table A.1), reflecting the use of organic fertilisers (manure) and the high proportion of forests in our catchments. Furthermore, aquatic hyphomycete fungi can utilise, and sometimes are dependent upon, specific forms of organic nitrogen for growth (Bengtsson, 1982; Thornton, 1963).

The PC best describing the land-use gradient (principal component one (PC1)) was used as our measure of land use—the main predictor variable—in subsequent analyses. Two environmental characteristics, discharge and water nutrient concentrations, can differ greatly between autumn and spring, potentially modifying the land-use gradient between the two seasons. To assess the extent to which this natural variability may have affected our land-use gradient, we conducted a similar PCA with the spring data and compared it to the autumn land-use gradient using Pearson product–moment correlation test.

An initial assessment of the relationship between leaf decomposition and land use was obtained using mixed effect model analysis carried out with the R package *nlme* (Pinheiro et al., 2012), which included the land-use gradient, season and mesh size as fixed factors, and all two- and three-way interactions. Stream identity was fitted as a random block factor. Parameters characterising detritivore community structure (density, MC, $F_{identity}$, $F_{dispersion}$ and richness) and fungal biomass (ergosterol) were analysed using the same model. For each model, non-significant interaction terms were removed following stepwise backwards selection. After removing a non-significant interaction term, we compared the competing models using Akaike's Information Criteria (AIC) and selected those models with the lowest AIC. Fixed effect terms were never removed from the models. When only fixed effect terms remained in the analysis, our statistical output was obtained based on Type II ANOVA using the R package *car* (Fox and Weisberg, 2011). This statistical approach prevents the occurrence of equivocal statistical significance in the fixed effect terms, which can originate in an unbalanced design when there are no significant interaction terms (Langsrud, 2003). Error terms of each final model were estimated using Restricted Maximum Likelihood (REML).

Replication for the $F_{identity}$ and $F_{dispersion}$ analyses was lower, due to exclusion of litter bags with only one species present, for which these indices cannot be calculated. Density, richness, MC, $F_{dispersion}$ and ergosterol were log-transformed to normalise variance. Of the 200 litter bags retrieved, data from five coarse mesh bags were excluded from the models (Ängerån stream in autumn ($n = 1$), and Granån (3) and Österån (1) streams in spring), due to extremely low levels of final leaf mass remaining (14% dry mass in one mesh bag and <3% in the other four mesh bags, whereas all other litter bags contained above 40% remaining dry mass). The causes of these isolated cases of extremely high leaf mass loss are not known, but are evidently very localised (e.g. high local turbulence). Such extreme values are typically excluded (Frainer et al., 2015) as they undermine reliable calculation of decomposition rates and related parameters (e.g. detritivore density) using the litter-bag technique and cause the data to irretrievably violate assumptions of normality.

Relationships between $\partial^{13}C$ and $\partial^{15}N$ of consumer body tissues (as response variables) and land use were analysed using multiple linear regressions, which also included detritivore genus and seasons as dummy predictor variables. Consumer $\partial^{15}N$ and $\partial^{13}C$ isotopes may vary across regions simply due to variation in resource $\partial^{15}N$ and $\partial^{13}C$ isotopes, which themselves are influenced by precipitation (Ma et al., 2012) and soil fertilisation (Choi et al., 2005), among other biogeochemical factors (Fry, 2006). Therefore, comparisons of consumer isotopes across regions must be analysed with caution. We addressed this issue by comparing variation in consumer isotopes along the land-use gradient in parallel with a similar comparison of variation in resource isotopes.

Isotope mixing models were generated to characterise relationships between detritivores and their resources using the R package *siar* (Parnell and Jackson, 2013). Such mixing models account for variation in both $\partial^{13}C$ and $\partial^{15}N$ in the resource and consumer to estimate which sources of C and N are most associated with consumer isotopic signal.

To disentangle direct and indirect causal pathways that might explain variability in decomposition rates, we used structural equation modelling (SEM), which allows partitioning causal pathways in complex data (Grace et al., 2010). We tested for the effect of the exogenous variable land use on the detritivore density, MC, functional identity ($F_{identity}$) and functional dispersion ($F_{dispersion}$) and ergosterol, all fitted as endogenous variables (i.e. with potential to be both affected by land use, and to affect other endogenous and response variables), and outcomes for leaf decomposition rates using SEM with the R package *lavaan* (Rosseel, 2012). Due to the role of seasonality in affecting the litter decomposition (Frainer et al., 2014), we had the same model split by seasons, based on a multigroup approach (Rosseel, 2012). Variables were transformed as in the linear regressions described above.

3. RESULTS

3.1 Land-Use Gradient

PCA of the environmental variable set with the smallest degree of autocorrelation among them yielded two axes explaining 36% (axis 1) and 22% (axis 2) of the variation among sites (Fig. 1). Higher water temperature and pH were most associated with the least disturbed sites (Fig. 1, left side of the PC1), whereas coarser substrates and deeper channels, higher total nitrogen concentration and higher water velocity were most characteristic of agriculturally disturbed sites (Fig. 1, right side of the PC1). Autumn and spring PCAs were highly correlated ($r = 0.86$, $p = 0.001$).

3.2 Leaf Decomposition

Temperature-corrected leaf decomposition rates (detailed in Table A.3) were higher in coarse than fine mesh bags ($F_{1,178} = 232.79$, $p < 0.001$) and were affected by a two-way interaction between land use and season

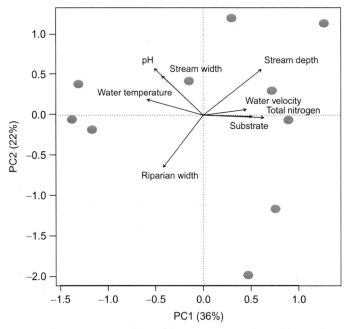

Figure 1 Principal component analysis of the environmental variables, with axis 1 used to represent the land-use gradient.

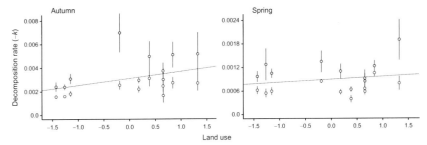

Figure 2 Relationship between land use and leaf decomposition in the fine (open circles) and coarse (grey circles) mesh bags, in both autumn (left panel) and spring (right panel). Each point plots the mean ± 1 SD for a single stream. Trend lines show the relationship between land use and leaf decomposition pooled across the two mesh sizes.

($F_{1,178} = 15.86$, $p < 0.001$). Decomposition rates of both mesh sizes increased moderately along the land-use gradient in autumn (Fig. 2) but were constant across the land-use gradient in spring (Fig. 2).

3.3 Detritivore Biotic Variables

Detritivore density ranged from 1 to 64 individuals per litter bag and was negatively related to the land-use gradient ($F_{1,84} = 7.32$, $p < 0.027$), decreasing as the agricultural impact increased (slope $= -0.67 \pm 0.25$SE). There was no difference in density between seasons ($F_{1,84} = 0.36$, $p = 0.45$). Detritivore metabolic capacity (MC) was higher in spring (mean ± SE $= 0.064 \pm 0.01$) than in autumn (0.031 ± 0.005) ($F_{1,84} = 8.53$, $p = 0.004$), but was not related to the land-use gradient ($F_{1,84} = 4.63$, $p = 0.064$). Detritivore species richness per litter bag ranged from one to six species and was lowered in the more agricultural sites in autumn, but did not vary along the land-use gradient in spring (land use × season interaction: $F_{1,83} = 5.2$, $p = 0.025$).

The principal component analysis of functional identity ($F_{identity}$) strongly differentiated traits associated with feeding and diet. Shredding traits, associated with the most obligate leaf-consumers, occurred on the positive side of PC1, whereas the negative side of PC1 was characterised by biofilm-scraper/grazer traits, associated with less obligate leaf consumer species (Fig. 3). These associations occurred in both autumn and spring, with PC1 overall explaining 89–90% of the variation. The second principal component (PC2) was predominantly associated with variation in predator and FPOM consumer traits, and explained ~10% of the variation in both seasons.

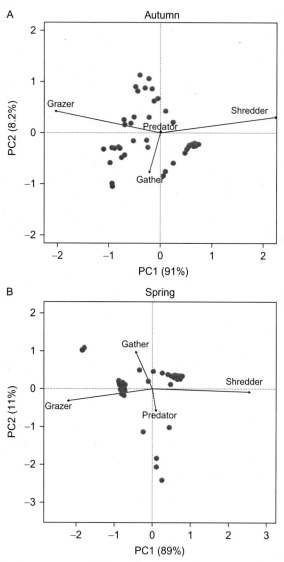

Figure 3 Principal component analysis of the functional traits according to their relative abundances ($F_{identity}$), calculated as the community-weighted mean trait values. Circles represent the sampling sites ($n = 5$ per site, 10 sites in total) in (A) autumn and (B) spring.

$F_{identity}$ did not differ between seasons ($F_{1,75} = 0.03$, $p = 0.60$). There was a trend for increasing $F_{identity}$ along the land-use gradient (slope $= 0.090 \pm 0.04$SE), which was near significance at the 5% level ($F_{1,74} = 5.07$, $p = 0.054$). This result indicates a higher abundance of leaf-shredding traits in the more agricultural sites. Functional dispersion ($F_{\text{dispersion}}$) was affected by an interaction between land use and season ($F_{1,74} = 12.74$, $p < 0.001$), with higher trait dominance associated with the more agricultural sites in autumn, with the opposite relationship observed in spring. There was a weak, negative correlation between $F_{\text{dispersion}}$ and F_{identity} in autumn (Pearson's product moment $= -0.34$, $p = 0.023$), but there was no correlation between $F_{\text{dispersion}}$ and F_{identity} in spring ($p > 0.65$).

3.4 Fungal Biomass

Fungal biomass, measured as ergosterol content, ranged between 161.66 and 2318.16 mg/L. Ergosterol did not differ between the two seasons ($F_{1,83} = 1.41$, $p = 0.24$), and it was not related to the land-use gradient ($F_{1,83} = 0.13$, $p = 0.73$).

3.5 Structural Equation Model

Chi-square, standardised root mean square residual (SRMSR) and root mean square error of approximation (Table A.4) statistics indicated good overall fit for both the autumn and spring structural equations models (SEMs).

Here, we present the significant ($p < 0.05$) standardised correlation coefficients (r values); other results are detailed in Table A.4.

In autumn (Fig. 4A, Table A.4), the direct effect of the land-use gradient on leaf decomposition was positive ($r = 0.50$). Land use was also positively related to F_{identity} ($r = 0.62$), with abundances of shredding traits increasing along the gradient. $F_{\text{dispersion}}$ declined along the land-use gradient ($r = -0.38$), with the more agricultural assemblages characterised by a greater dominance of fewer, similar traits. Detritivore density and MC also decreased along the land-use gradient ($r = -0.45$ and -0.32, respectively). $F_{\text{dispersion}}$ was positively related to leaf decomposition ($r = 0.25$), indicating that higher functional evenness was related to higher decomposition rates. Detritivore density was negatively related to leaf decomposition ($r = -0.40$). Fungal biomass was not related to land use, but was positively related to leaf decomposition ($r = 0.25$). Fungal biomass was also negatively related to F_{identity} ($r = 0.23$), indicating that higher fungal biomass was related to higher abundance of biofilm feeders, rather than leaf eaters. Finally, density and MC were correlated ($r = 0.75$).

In spring (Fig. 4B, Table A.4), F_{identity} increased along the land-use gradient ($r = 0.55$), indicating an increased dominance of more obligate

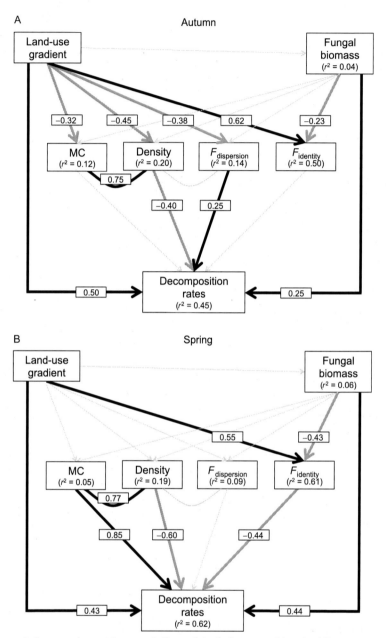

Figure 4 Structural equation modelling of leaf decomposition in (A) autumn and (B) spring. Black and grey lines indicate significant ($p < 0.05$) positive and negative relationships ($p < 0.05$), while dashed lines indicate non-significant relationships. Significant non-standardised coefficients are shown in boxes. r^2 values are shown for each response variable. Model chi-square $= 12.12$, d.f. $= 8$, $p = 0.146$.

leaf-shredder traits at the more agricultural sites. Neither detritivore metabolic capacity, density, $F_{dispersion}$ nor fungal biomass varied with land use, but leaf litter decomposition rates ($r=0.43$) increased along the gradient. Both MC and density affected leaf decomposition. However, while higher MC was positively related to leaf decomposition rates ($r=0.85$), detritivore density negatively affected leaf decomposition ($r=-0.60$). $F_{dispersion}$ had no relationship to leaf decomposition during spring, but increasing $F_{identity}$ had a negative effect ($r=0.44$). Fungal biomass was not affected by the land-use gradient, but was positively related to decomposition ($r=0.44$). Similar to the autumn results, fungal biomass was negatively related to $F_{identity}$ ($r=0.43$). Density was correlated with MC ($r=0.77$).

3.6 Isotopes

The mixing models indicate that all detritivore species feed on multiple resources, but allochthonous carbon constitutes a large portion of their diet, accounting for more than 40% of the isotopic signal across all genera. The detritivore genera analysed in our study differed in $\partial^{13}C$ ($F_{5,38}=7.22$, $p<0.001$), but there was no relationship between $\partial^{13}C$ and land use ($F_{1,38}=1.26$, $p=0.27$, Fig. 5A), and no differences in $\partial^{13}C$ between the two seasons ($F_{1,38}=1.10$, $p=0.30$). $\partial^{15}N$ also differed among the detritivore genera ($F_{5,38}=4.38$, $p=0.003$, Fig. 5B) and had a positive relationship with land use (slope $=3.09\pm2.40$SE, $F_{1,38}=25.51$, $p<0.001$), but did not differ between seasons ($F_{1,38}=0.16$, $p=0.69$). Carbon isotope values differed among resource types (C4 plants, algae and water moss) ($F_{5,58}=38.83$, $p<0.001$), but were not affected by the land-use gradient ($F_{1,58}=0.01$, $p=0.98$, Fig. 5C) and did not differ between seasons ($F_{1,58}=1.58$, $p=0.21$). $\partial^{15}N$ differed between the resource types ($F_{5,58}=10.99$, $p<0.001$) and had a positive relationship with land use (slope $=0.42\pm0.48$SE, $F_{1,58}=21.58$, $p<0.001$, Fig. 5D), but this relationship was more positive for moss than for C4 plants and algae (resource type × season interaction: $F_{2,58}=4.49$, $p=0.015$).

4. DISCUSSION

Shifts in the diversity and composition of consumer functional traits explained a substantial proportion of variation in the key process of leaf decomposition along our gradient of increasing agricultural land-use intensity. Specifically, we found that an overall increase in decomposition

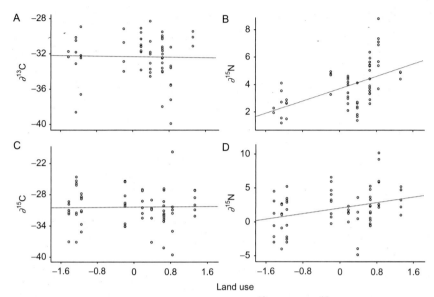

Figure 5 Relationships between land use and (A) $\partial^{13}C$ and (B) $\partial^{15}N$ isotope of inverte-brate detritivores, and between land use and (C) $\partial^{13}C$ and (D) $\partial^{15}N$ isotope of resource types. Detritivore isotopes are pooled across genera. All isotopes are pooled across the two seasons, autumn and spring.

along the land-use gradient was largely offset by negative effects associated with losses of functional diversity and changes in the composition of functional traits within the detritivore consumer guild. These findings highlight the potential for shifts in species traits and functional diversity to mediate the effects of human disturbance on ecosystem functioning.

The direct effect of increasing land use intensity on leaf decomposition was positive in both the autumn and the spring, according to our SEMs. This is in line with the increase in nutrients along the gradient, which stimulate microbial activity and hence decomposition from the bottom up (Ferreira et al., 2006b; Gessner et al., 2007), and also possibly the simultaneous increase in water velocity, which can facilitate greater physical abrasion of leaf litter (Ferreira et al., 2006a; Spänhoff et al., 2007). However, our SEMs further reveal that these positive effects were partly offset by a loss of detritivore functional diversity along the gradient in the autumn, and were completely offset by a shift in functional trait composition in the spring. Consequently, linear regression analysis of the overall effect of land use on leaf mass loss found a net positive effect of land use on decomposition during the autumn only, with no similar relationship apparent during spring.

Thus, while shifts in the functional characteristics of our assemblages modified the effects of land use on functioning in both seasons, the relative importance of functional diversity and functional identity differed.

In autumn, our measure of functional diversity, functional dispersion ($F_{dispersion}$), was positively related to leaf decomposition overall, but declined along the land-use gradient. Disturbances in stream ecosystems often drive decreases in invertebrate richness and evenness, including of detritivores (Dolédec et al., 2011; McKie et al., 2006). This can result in dominance of consumer guilds by a few functional traits (Hillebrand et al., 2008) and altered ecosystem functioning (Frainer et al., 2014; McKie et al., 2008). In our study, higher $F_{dispersion}$ was associated with a more even distribution of dissimilar traits and was positively associated with leaf decomposition during the autumn, similar to findings from a previous study (Frainer et al., 2014). This is indicative of a positive effect of niche differentiation on functioning, whereby a consumer assemblage encompassing contrasting traits and patterns of resource use are able to process the resource more efficiently overall, though further research is required to more explicitly identify the underlying mechanisms (Cardinale, 2011).

Contrary to our expectations, an increase in the intensity of agricultural management did not favour a greater dominance of generalist feeding traits associated with less obligate leaf-consumers in the more disturbed streams (Statzner and Bêche, 2010). Rather, we saw the opposite effect in both autumn and spring. The most obligate leaf-shredding traits were more abundant in the more agricultural streams, whereas traits related to biofilm-grazing/scraping, which indicate the presence of detritivores that can consume substantial amounts of algae, microbes and particulate matter associated with biofilms (Lieske and Zwick, 2007), were more abundant in the forested and less disturbed streams. These findings indicate that dietary flexibility, which normally becomes more prominent in disturbed assemblages, was not strongly associated with other response traits typically conferring tolerance in our assemblages.

In spring, shifts in functional identity helped to offset the generally positive impact of land use on functioning. Unexpectedly, lower processing rates were associated with the shift in trait composition towards the more specialised leaf-eaters in the agricultural sites. The dominant taxa most associated with shredding traits in our agricultural streams were species of the stonefly genus *Nemoura* (e.g. *N. avicularis*), which emerge as winged adults throughout the spring, while the less obligate leaf feeders in the forested sites were dominated by another stonefly genus, *Amphinemura* (e.g. *A. sulcicolis* and *A. borealis*), which emerge later in the summer (Brinck, 1949). *Nemoura* spp., in common with many aquatic insects, reduce or cease feeding and growth several days prior to emergence (Lancaster and Downes, 2013; Svensson, 1966), and thus it is likely these species were not feeding as actively on the litter as the later emerging *Amphinemura* species. Accordingly, life-history traits correlated with feeding traits of our detritivores are likely to explain why less obligate leaf feeders were associated with higher leaf processing rates during the spring.

Further factors with potential to limit the effectiveness of decomposition mediated by detritivores include the availability of alternative food sources and variation in the extent of microbial activity. Algal productivity often increases in agricultural streams (Von Schiller et al., 2008), constituting a potential alternative food resource of high quality, particularly for biofilm-feeding detritivores (Lieske and Zwick, 2007). In our study, both consumer and resource $\partial^{13}C$ isotope values were similar along the land-use gradient, indicating that there was no increase in assimilation of algal carbon along the gradient. This result most likely reflects the low rates of algal productivity attributable to limited light over the winter months preceding our sample date, though we cannot rule out that a shift in the isotope values of algae between the sample dates for periphyton and consumers might have influenced the results. Fungal biomass had a positive effect on decomposition rates in our SEMs, similar to relationships observed previously (reviewed in Gessner et al., 2007). However, there was no relationship between the land-use gradient and fungal biomass, indicating that any effects of elevated nutrients on fungal activity in the agricultural streams did not lead to a net increase in biomass accrual, at least over the time period of our study. Interestingly, the negative relationship between fungal biomass and detritivore functional identity (Fig. 4), associated with a shift from more to less-obligate litter feeding traits, indicates that less-obligate litter consumers were increasingly favoured as fungal biomass increased.

The negative effects of detritivore density on decomposition, apparent in both autumn and spring, are in line with previous observations and are likely to reflect effects of negative density-dependent interactions on per-capita consumption rates (Klemmer et al., 2012; McKie et al., 2008; Reiss et al., 2009). However, disentangling the outcomes of this effect for ecosystem functioning during autumn is complicated by the overall decline in density along the land-use gradient. This decline might partly be a consequence of faster decomposition in the agricultural sites, if the number of individuals who are able to persist in a litter patch declines at a faster rate than the resource. No similar concurrent changes in land-use, decomposition and density occurred in the spring, suggesting other, season-specific, mechanisms might have been important. For example, it is possible that ambient stocks of broad leaf litter were greater in the agricultural streams during the autumn, in line with the higher abundance of broad-leaf tress in the riparian zone, reducing colonisation of our litterbags relative to the more coniferous-forested sites (see McKie and Malmqvist, 2009). During spring, litter standing stocks were very greatly reduced across all sites, possibly obscuring this difference. Regardless of the underlying mechanisms, the lower autumn densities in the agricultural sites might have influenced the overall positive association between land use and functioning, if negative density-dependent effects on leaf processing were alleviated, even if for only part of the study period. These results emphasise the potentially dynamic nature of the relationships between density and leaf mass loss over the whole decomposition process.

The positive effects of metabolic capacity (MC) on decomposition were also in line with previous observations (McKie et al., 2008; Perkins et al., 2010). Higher MC is expected to positively affect leaf decomposition, as increased metabolic demands drive greater resource consumption (Reiss et al., 2009; Vaughn et al., 2007). However, whereas density was important in both seasons, MC was important during spring only. This is opposite to findings from a previous study (Frainer et al., 2014) from the same region, which was, however, conducted solely in minimally disturbed forested streams, with differing communities from those of our agricultural streams. These findings further stress how the relative importance of key biotic drivers of functioning can vary in space and time, influenced by differences in environmental parameters and the taxonomic (and functional) composition of communities (Dell et al., 2015).

In our study, shifts in trait identity and diversity along the land-use gradient appear to have limited the stimulating effect of nutrients and current velocity on net decomposition rates, which might indicate that ecosystem functioning is little impaired in our systems and that management regimes can continue unchanged. However, the detrital food web also represents a key pathway for uptake of nutrients, mediated especially by microbes, and for its subsequent cycling in the production of both biomass and faecal pellets by consumers (Wotton and Malmqvist, 2001). These linkages are thus important in the capacity of a stream ecosystem to retain and utilise additional nutrients, and in an intact ecosystem, energy flows through the detrital food web should be stimulated as nutrient loadings increase (Robinson and Gessner, 2000). It is thus unclear whether the limitation in overall decomposition rates, associated in our study with changes in the functional characteristics of the detritivore assemblages, is necessarily positive for wider ecosystem integrity or for longitudinal patterns of nutrient and energy cycling.

Our study detected important short-term consequences of changes in the diversity and composition of species traits for ecosystem functioning within two seasons, but it is possible that the observed changes are of significance over longer, multi-generational timescales also. Greater biodiversity has been linked with increased system stability and resilience, in part because more diverse assemblages are expected to encompass a greater redundancy of functional traits (Holling, 1973; Yachi and Loreau, 1999). This can assist in buffering the effects of community fluctuations on functioning and may confer a greater capacity to resist or recover from disturbances (Allen et al., 2005; Díaz and Cabido, 2001). Approaches such as those used in our study can assist in identifying which types of species, in terms of their functional attributes, are important in maintaining ecosystem functioning and stability (Cardinale, 2011), at least at small ecological scales (i.e. within single habitats or seasons). However, since resilience depends on the relative distributions of traits both within and across ecological scales (e.g. trophic levels, size classes, habitats and seasons), there is a need for research that is broader in scope to assess the true extent of functional redundancy in ecosystems (Angeler et al., 2014) and its potential to moderate the effects of human disturbances on ecosystem functioning (Allen et al., 2005). Nevertheless, shifts in functional trait indices, such as those observed here, can act as early-warning signals (*sensu* Mouillot et al., 2013) of building stress in an ecosystem, where declines in functional diversity may have longer-term implications for the capacity of an ecosystem to maintain ecosystem functioning if stressor loads increase in the future.

ACKNOWLEDGEMENTS

We are grateful to Emma Fältström, Melina Duarte and Petter Esberg for invaluable lab and field assistance. We also thank Margareta Zetherström and E. Fältström for the ergosterol analyses. Micael Jonsson contributed helpfully to the design of the study. This work was partially supported by a grant from the Kempe Foundation to A. Frainer and from the Swedish Research council (VR 621-2006-375) to Professor Björn Malmqvist.

▷ APPENDIX

Supplementary material on study site description, detritivore trait scores, decomposition rates, and strucure equation modelling results.

Table A.1 Characteristics of the Study Streams

Stream	Discharge (cm³/s)	Width (m)	Depth (cm)	Water Velocity (cm/s)	Temperate (Mean (°C)±SD) Autumn	Temperate (Mean (°C)±SD) Spring	Riparian Vegetation[a]	Substrate Type[b]	Agricultural Area in the Sub-catchment (%)	Nitrogen Runoff from Agricultural Fields (%)[c]	Organic Nitrogen Runoff from Agricultural Fields (%)[d]	N NH4 (mg/L)	N NO3 (mg/L)	Total Nitrogen (mg/L)	PO4 (µg/L)	pH
Bjänsjö	41.3	3.7	19.1	0.54	2.5 ± 1.8	12.8 ± 1.4	4.00	2.00	6	53	26	0.10 ± 0.0	0.03 ± 0.02	0.42 ± 0.0	0.04 ± 0.0	6.93
Tavelån	39.4	5.2	27.7	0.45	2.4 ± 2.2	11.3 ± 1.2	5.00	4.00	11	85	29	0.10 ± 0.0	0.02 ± 0.01	0.24 ± 0.01	0.04 ± n.a.	7.19
Österån	64.4	4.5	22.5	0.64	2.3 ± 1.5	12.2 ± 1.1	5.00	3.00	2	14	28	0.10 ± 0.0	0.05 ± 0.04	0.33 ± 0.04	0.04 ± 0.0	6.71
Ångerån	93.6	4.5	30.0	0.70	n.a.	10.8 ± 1.1	2.00	3.00	9	57	31	0.10 ± 0.0	0.08 ± 0.02	0.54 ± 0.06	0.04 ± 0.0	5.82
Flarkån	134.0	5.0	45.7	0.67	1.5 ± 1.4	9.5 ± 2.0	2.00	3.00	20	103	31	0.24 ± 0.20	0.17 ± 0.15	0.60 ± 0.30	0.04 ± 0.0	6.79
Pell-Febodbäcken	18.8	1.9	20.7	0.62	1.5 ± 1.2	8.4 ± 1.8	5.00	4.00	8	59	26	0.10 ± 0.0	0.01 ± 0.0	0.35 ± 0.02	0.04 ± 0.01	5.12
Vebomarksån	123.7	4.0	32.0	0.57	1.6 ± 1.3	9.9 ± 1.7	5.00	4.00	5	34	31	0.36 ± 0.07	0.15 ± 0.10	0.77 ± 0.15	0.04 ± 0.0	4.81
Kålabodaån	118.9	4.0	38.0	0.76	1.5 ± 1.4	9.9 ± 2.0	2.00	4.00	33	155	32	0.10 ± 0.0	0.12 ± 0.07	0.45 ± 0.05	0.04 ± 0.0	5.94
Bostån	177.7	3.0	35.0	1.33	2.2 ± 2.1	8.3 ± 1.2	4.00	4.00	24	125	27	0.10 ± 0.0	0.37 ± 0.28	0.84 ± 0.40	0.04 ± 0.0	6.46
Granån	123.5	3.0	64.3	0.71	1.9 ± 1.8	11.1 ± 1.4	1.00	5.00	13	72	31	0.10 ± 0.0	0.05 ± n.a.	0.46 ± n.a.	0.04 ± 0.0	6.19

[a]Riparian vegetation was characterised based on estimates of the width of the riparian tree cover: 1 = no trees, only grass; 2 = grass and scattered trees; 3 = tree cover <3 m wide; 4 = tree cover from 5 to 10 m wide; 5 = tree cover extending at least 10 m from each side of the channel.

[b]Substrate type refers to the dominant stream substrate at each sampling site: 1 = silt and fine sand; 2 = mostly coarse sand; 3 = coarse sand and pebbles; 4 = pebbles and boulders; 5 = dominance of boulders.

[c]Modelled nitrogen runoff (kg/year) from agricultural fields relative to the background nitrogen runoff across the entire catchment, summed over all other potential sources, including: forest and logging activities, mires, sewage and lakes. Percentages above 100% denote higher N run-off from agricultural activities than from background concentrations. Data obtained from www.vattenwebb.smhi.se (accessed on 30 January 2015).

[d]Modelled organic nitrogen runoff (kg/year) from agricultural fields calculated as the background nitrogen concentration (assumed to be 100% organic nitrogen) relative to the total N runoff due to agricultural activities alone (assuming a contribution of 100% inorganic nitrogen from the agricultural activities). Data on total N concentration from agricultural meadows and background concentrations were obtained from www.vattenwebb.smhi.se (accessed on 30 January 2015)

Sites are ordered according to their positioning on the land-use gradient (PCA axis one). Values are the average found in autumn (±SD) and n.a. indicates missing data.

Table A.2 Functional Traits Used to Calculate Functional Dispersion and Identity

	Habitat Preferences					Feeding Potential				Emergence Period				Water Velocity		
	Fine Sediment	Coarse Sediment	Detritus	Vegetation	Edge	Grazer	Shredder	Gather	Predator	Winter	Spring	Summer	Autumn	Slow	Median	Fast
Amphinemura spp.	0.50	6.00	1.50	2.00	0.00	4.00	2.50	3.50	0.00	0.00	5.00	5.00	0.00	2.00	2.00	6.00
Capnopsis	3.00	0.00	4.00	3.00	0.00	2.00	4.00	4.00	0.00	1.00	7.00	2.00	0.00	0.00	10.00	0.00
Leuctra spp.	0.00	5.00	3.00	2.00	0.00	3.00	3.00	4.00	0.00	0.00	6.00	4.00	0.00	0.00	10.00	0.00
Nemoura spp.	2.00	2.00	2.00	3.00	1.00	0.00	7.00	3.00	0.00	0.00	5.00	5.00	0.00	2.00	2.00	6.00
N. avicularis	2.00	2.00	4.00	2.00	0.00	0.00	7.00	3.00	0.00	0.00	6.00	4.00	0.00	3.33	3.33	3.33
Protonemura meyeri	0.00	6.00	2.00	2.00	0.00	3.00	5.00	2.00	0.00	1.00	6.00	3.00	0.00	0.00	0.00	10.00
Taeniopteryx nebulosa	3.00	0.00	2.00	5.00	0.00	3.00	2.00	5.00	0.00	0.00	10.00	0.00	0.00	0.00	10.00	0.00
Halesus spp.	0.00	0.00	10.00	0.00	0.00	1.00	7.00	0.00	2.00	0.00	0.00	3.00	7.00	10.00	0.00	0.00
Limnephilidae spp.	0.00	4.50	5.50	0.00	0.00	1.75	6.25	0.00	2.00	0.00	2.00	6.00	2.00	5.00	5.00	0.00
Potamophylax latipennis	0.00	6.00	4.00	0.00	0.00	2.00	6.00	0.00	2.00	0.00	0.00	6.00	4.00	0.00	10.00	0.00
P. rotundipennis	0.00	6.00	4.00	0.00	0.00	2.00	6.00	0.00	2.00	0.00	3.00	6.00	1.00	5.00	5.00	0.00
P. cingulatus	0.00	6.00	4.00	0.00	0.00	2.00	6.00	0.00	2.00	0.00	0.00	8.00	2.00	0.00	10.00	0.00
Aselus aquaticus	2.00	2.00	2.00	2.00	2.00	3.00	3.00	4.00	0.00	0.00	0.00	0.00	0.00	3.33	3.33	3.33

Scores were obtained from the Freshwater Ecology database (Schmidt-Kloiber and Hering, 2011, and references below).

Table A.3 Birch Leaf Litter Decomposition Rates (Mean, SD) of Coarse and Fine-Mesh Bags Across the Land-Use Gradient in Autumn and Spring

Season	Mesh Size	Land-Use Gradient	Decomposition Rate (degree/day)	SD
Autumn	Coarse	−1.42	0.0024	0.0004
		−1.26	0.0024	0.0002
		−1.15	0.0031	0.0004
		−0.19	0.0070	0.0016
		0.18	0.0030	0.0002
		0.38	0.0050	0.0013
		0.65	0.0037	0.0007
		0.65	0.0030	0.0010
		0.84	0.0051	0.0011
		1.32	0.0052	0.0018
	Fine	−1.42	0.0016	0.0001
		−1.26	0.0016	0.0001
		−1.15	0.0018	0.0002
		−0.19	0.0025	0.0004
		0.18	0.0022	0.0003
		0.38	0.0031	0.0007
		0.65	0.0025	0.0005
		0.65	0.0016	0.0006
		0.84	0.0027	0.0005
		1.32	0.0027	0.0006
Spring	Coarse	−1.42	0.0010	0.0001
		−1.26	0.0013	0.0004
		−1.15	0.0011	0.0001
		−0.19	0.0013	0.0003
		0.18	0.0011	0.0002
		0.38	0.0006	0.0001

Continued

Table A.3 Birch Leaf Litter Decomposition Rates (Mean, SD) of Coarse and Fine-Mesh Bags Across the Land-Use Gradient in Autumn and Spring—cont'd

Season	Mesh Size	Land-Use Gradient	Decomposition Rate (degree/day)	SD
		0.65	0.0009	0.0002
		0.65	0.0008	0.0002
		0.84	0.0012	0.0001
		1.32	0.0019	0.0005
	Fine	−1.42	0.0006	0.0001
		−1.26	0.0005	0.0001
		−1.15	0.0006	0.0001
		−0.19	0.0008	0.0000
		0.18	0.0006	0.0001
		0.38	0.0004	0.0001
		0.65	0.0007	0.0001
		0.65	0.0006	0.0001
		0.84	0.0011	0.0001
		1.32	0.0008	0.0002

Table A.4 Standardised Path Coefficients of the Structural Equation Model

	Response	Predictor	Estimate	SE	Z-value	P-value	Standardised Correlation
Autumn	Decomposition rate (degree/day)	Land use	0.28	0.05	5.45	<**0.001**	0.95
		MC	0.03	0.10	−0.28	0.778	−0.05
		CWM	−0.32	0.13	−2.46	**0.014**	−0.42
		$F_{dispersion}$	0.09	0.04	2.37	**0.018**	0.281
		Density	−0.15	0.14	−1.09	0.276	−0.197
		Ergosterol	0.06	0.024	2.47	**0.013**	0.272
	CWM	Land use	0.29	0.04	8.19	<**0.001**	0.74
		Ergosterol	−0.06	0.03	−2.11	**0.03**	−0.19
	$F_{dispersion}$	Land use	−0.37	0.12	−2.95	**0.003**	−0.41
		Ergosterol	−0.06	0.09	−0.68	0.49	−0.09
	Density	Land use	−0.13	0.055	−2.425	**0.015**	−0.34
		Ergosterol	0.04	0.041	0.936	0.349	0.13

Table A.4 Standardised Path Coefficients of the Structural Equation Model—cont'd

	Response	Predictor	Estimate	SE	Z-value	P-value	Standardised Correlation
	MC	Land use	−0.19	0.079	−2.444	**0.015**	−0.35
		Ergosterol	0.03	0.06	0.43	0.667	0.062
	Ergosterol	Land use	−0.27	0.197	−1.373	0.17	−0.203
	Covariances						
	Density − MC		0.02	0.01	4.13	**<0.001**	0.77
	$F_{dispersion}$ − MC		0.01	0.01	2.29	**0.022**	0.23
Spring	Decomposition rate (degree/day)	Land use	0.02	0.02	1.76	0.078	0.06
		MC	0.12	0.02	5.21	**<0.001**	1.02
		CWM	−0.07	0.03	−2.10	**0.036**	−0.33
		$F_{dispersion}$	−0.02	0.01	−1.89	0.058	−0.24
		Density	−0.19	0.04	−4.30	**<0.001**	−0.81
		Ergosterol	0.03	0.01	3.61	**<0.001**	0.45
	CWM	Land use	0.24	0.04	5.47	**<0.001**	0.58
		Ergosterol	−0.10	0.03	−3.14	**0.002**	−0.33
	$F_{dispersion}$	Land use	0.25	0.15	1.7	0.089	0.28
		Ergosterol	0.03	0.10	0.32	0.752	0.05
	Density	Land use	−0.05	0.06	−0.91	0.361	−0.15
		Ergosterol	0.05	0.04	1.29	0.198	0.21
	MC	Land use	−0.10	0.12	−0.87	0.382	−0.14
		Ergosterol	0.05	0.08	0.67	0.503	0.11
	Ergosterol	Land use	−0.56	0.20	−2.75	**0.006**	−0.39
	Covariances						
	Density − MC		0.03	0.01	4.09	**<0.001**	0.77
	$F_{dispersion}$ − Density		0.03	0.01	3.14	**0.002**	0.35

Chi-square = 12.12, d.f. = 8, p = 0.15. SRMR = 0.066. Significant relationships are highlighted in bold.

Trait data-scores: detailed references

Eder, E., Hödl, W., Moog., O., Nesemann, H., Pöckl M., Wittmann, K., 1995. Crustacea (authors depending on taxagroup). In: Moog, O. (Ed.), Fauna Aquatica Austriaca, Lieferungen 1995, 2002. Wasserwirtschaftskataster, Bundesministerium für Land- und Forstwirtschaft, Umwelt und Wasserwirtschaft, Wien.

Graf, W., Murphy, J., Dahl, J., Zamora-Muñoz, C., López-Rodríguez, M.J., Schmidt-Kloiber, A., 2006. Trichoptera Indicator Database. Euro-limpacs Project, Workpackage 7—Indicators of Ecosystem Health, Task 4, www.freshwaterecology.info, Version 5.0 (accessed on 27 September 2011).

Graf, W., Murphy, J., Dahl, J., Zamora-Muñoz, C., López-Rodríguez, M.J., 2008. In: Schmidt-Kloiber, A., Hering, D (Eds.), Distribution and Ecological Preferences of European Freshwater Organisms. Trichoptera, vol. 1. Pensoft Publishers, Sofia-Moscow, 388 p.

Graf, W., Lorenz, A.W., Tierno de Figueroa, J.M., Lücke, S., López-Rodríguez, M.J., Davies, C., 2009. In: Schmidt-Kloiber, A., Hering, D (Eds.), Distribution and Ecological Preferences of European Freshwater Organisms. Plecoptera, vol. 2. Pensoft Publishers, Sofia-Moscow, 262 p.

Graf, W., Lorenz, A.W., Tierno de Figueroa, J.M., Lücke, S., López-Rodríguez, M.J., Murphy, J., Schmidt-Kloiber, A., 2007. Plecoptera Indicator Database. Euro-limpacs project, Workpackage 7—Indicators of Ecosystem Health, Task 4, www.freshwaterecology.info, Version 5.0 (accessed on 27 September 2011).

Graf, W., Schmidt-Kloiber, A., 2011. Additions to and Update of the Trichoptera Indicator Database. www.freshwaterecology.info, Version 5.0 (accessed on 27 September 2011).

Schmedtje, U., Colling, M., 1996. Ökologische Typisierung der aquatischen Makrofauna. Informationsberichte des Bayerischen Land-esamtes für Wasserwirtschaft 4/96, 543 p.

Schmidt-Kloiber, A., Hering D. (Eds.), 2011. www.freshwaterecology. info —The Taxa and Autecology Database for Freshwater Organisms, Version 4.0 (accessed on 27 September 2011).

REFERENCES

Allen, C.R., Gunderson, L., Johnson, A.R., 2005. The use of discontinuities and functional groups to assess relative resilience in complex systems. Ecosystems 8, 958–966.

Angeler, D.G., Allen, C.R., Birgé, H.E., Drakare, S., McKie, B.G., Johnson, R.K., 2014. Assessing and managing freshwater ecosystems vulnerable to environmental change. Ambio 43, 113–125.

Bengtsson, G., 1982. Patterns of amino acid utilization by aquatic hyphomycetes. Oecologia 55, 355–363.

Bergfur, J., Johnson, R.K., Sandin, L., Goedkoop, W., 2007. Assessing the ecological integrity of boreal streams: a comparison of functional and structural responses. Fundam. Appl. Limnol. 168, 113–125.

Brinck, P., 1949. Studies on Swedish stoneflies (Plecoptera). Opuscula Entomol. (Suppl. XI), 250.

Brown, J.H., Gillooly, J.F., Allen, A.P., Savage, V.M., West, G.B., 2004. Toward a metabolic theory of ecology. Ecology 85, 1771–1789.

Campbell, J.E., Fourqurean, J.W., 2009. Interspecific variation in the elemental and stable isotope content of seagrasses in South Florida. Mar. Ecol. Prog. Ser. 387, 109–123.

Cardinale, B.J., 2011. Biodiversity improves water quality through niche partitioning. Nature 472, 86–89.

Choi, W.-J., Chang, S.X., Allen, H.L., Kelting, D.L., Ro, H.-M., 2005. Irrigation and fertilization effects on foliar and soil carbon and nitrogen isotope ratios in a loblolly pine stand. For. Ecol. Manag. 213, 90–101.

Cummins, K.W., Klug, M.J., 1979. Feeding ecology of stream invertebrates. Annu. Rev. Ecol. Syst. 10, 147–172.

Dahlman, L., Näsholm, T., Palmqvist, K., 2002. Growth, nitrogen uptake, and resource allocation in the two tripartite lichens *Nephroma arcticum* and *Peltigera aphthosa* during nitrogen stress. New Phytol. 153, 307–315.

Dell, A.I., Zhao, L., Brose, U., Pearson, R.G., Alford, R.A., 2015. Population and community body size structure across a complex environmental gradient.

Díaz, S., Cabido, M., 2001. Vive la différence: plant functional diversity matters to ecosystem processes. Trends Ecol. Evol. 16, 646–655.

Dolédec, S., Phillips, N., Townsend, C., 2011. Invertebrate community responses to land use at a broad spatial scale: trait and taxonomic measures compared in New Zealand rivers. Freshw. Biol. 56, 1670–1688.

Enquist, B.J., Norberg, J., Bonser, S.P., Violle, C., Webb, C.T., Henderson, A., Sloat, L.L., Savage, V.M., 2015. Scaling from traits to ecosystems: developing a general trait driver theory via integrating trait-based and metabolic scaling theories.

Ferreira, V., Graça, M.A.S., de Lima, J.L.M.P., Gomes, R., 2006a. Role of physical fragmentation and invertebrate activity in the breakdown rate of leaves. Arch. Hydrobiol. 165, 493–513.

Ferreira, V., Gulis, V., Graça, M.A.S., 2006b. Whole-stream nitrate addition affects litter decomposition and associated fungi but not invertebrates. Oecologia 149, 718–729.

Fox, J., Weisberg, S., 2011. An R companion to applied regression, second ed. Sage, Thousand Oaks, CA. Available at: http://socserv.socsci.mcmaster.ca/jfox/Books/Companion.

Frainer, A., McKie, B.G., Malmqvist, B., 2014. When does diversity matter? Species functional diversity and ecosystem functioning across habitats and seasons in a field experiment. J. Anim. Ecol. 83, 460–469.

Frainer, A., Moretti, M.S., Xu, W., Gessner, M.O., 2015. No evidence for leaf trait dissimilarity effects on litter decomposition, fungal decomposers, and nutrient dynamics. Ecology 96, 550–561.

Fry, B., 2006. Stable Isotope Ecology. Springer Verlag, New York.

Garnier, E., Cortez, J., Billès, G., Navas, M.-L., Roumet, C., Debussche, M., Laurent, G., Blanchard, A., Aubry, D., Bellmann, A., Neill, C., Toussaint, J.-P., 2004. Plant functional markers capture ecosystem properties during secondary succession. Ecology 85, 2630–2637.

Gessner, M.O., Chauvet, E., 2002. A case for using litter breakdown to assess functional stream integrity. Ecol. Appl. 12, 498–510.

Gessner, M.O., Gulis, V., Kuehn, K.A., Chauvet, E., Suberkropp, K., 2007. Fungal decomposers of plant litter in aquatic ecosystems. In: Kubicek, C.P., Druzhinina, I.S. (Eds.), Environmental and Microbial Relationships. Springer Verlag, Berlin; Heidelberg, Germany, pp. 301–324.

Giller, P.S., Malmqvist, B., 1998. The Biology of Streams and Rivers. Oxford University Press, Oxford, UK.

Gordon, N.D., McMahon, T.A., Finlayson, B.L., Gippel, C.J., Nathan, R.J., 2004. Stream Hydrology: An Introduction for Ecologists. John Wiley & Sons, Chichester, UK.

Göthe, E., Lepori, F., Malmqvist, B., 2009. Forestry affects food webs in northern Swedish coastal streams. Fundam. Appl. Limnol./Arch. Hydrobiol. 175, 281–294.

Grace, J.B., Anderson, T.M., Olff, H., Scheiner, S.M., 2010. On the specification of structural equation models for ecological systems. Ecol. Monogr. 80, 67–87.

Grime, J.P., 1998. Benefits of plant diversity to ecosystems: immediate, filter and founder effects. J. Ecol. 86, 902–910.

Hillebrand, H., Bennett, D.M., Cadotte, M.W., 2008. Consequences of dominance: a review of evenness effects on local and regional ecosystem processes. Ecology 89, 1510–1520.

Holling, C.S., 1973. Resilience and stability of ecological systems. Annu. Rev. Ecol. Syst. 4, 1–23.

Hooper, D.U., Adair, E.C., Cardinale, B.J., Byrnes, J.E.K., Hungate, B.A., Matulich, K.L., Gonzalez, A., Duffy, J.E., Gamfeldt, L., O'Connor, M.I., 2012. A global synthesis reveals biodiversity loss as a major driver of ecosystem change. Nature 486, 105–108.

Jonsson, M., Malmqvist, B., 2003. Mechanisms behind positive diversity effects on ecosystem functioning: testing the facilitation and interference hypotheses. Oecologia 134, 554–559.

Jousset, A., Schmid, B., Scheu, S., Eisenhauer, N., 2011. Genotypic richness and dissimilarity opposingly affect ecosystem functioning. Ecol. Lett. 14, 537–545.

Klemmer, A.J., Wissinger, S.A., Greig, H.S., Ostrofsky, M.L., 2012. Nonlinear effects of consumer density on multiple ecosystem processes. J. Anim. Ecol. 81, 770–780.

Laliberté, E., Legendre, P., 2010. A distance-based framework for measuring functional diversity from multiple traits. Ecology 91, 299–305.

Laliberté, E., Shipley, B., 2011. FD: measuring functional diversity from multiple traits, and other tools for functional ecology. R Package Version 10–11.

Lancaster, J., Downes, B.J., 2013. Aquatic Entomology. Oxford University Press, Oxford, UK.

Langsrud, Ø., 2003. ANOVA for unbalanced data: use Type II instead of Type III sums of squares. Stat. Comput. 13, 163–167.

Lavorel, S., Garnier, E., 2002. Predicting changes in community composition and ecosystem functioning from plant traits: revisiting the Holy Grail. Funct. Ecol. 16, 545–556.

Layer, K., Hildrew, A.G., Woodward, G., 2013. Grazing and detritivory in 20 stream food webs across a broad pH gradient. Oecologia 171, 459–471.

Leberfinger, K., Bohman, I., Herrmann, J., 2010. The importance of terrestrial resource subsidies for shredders in open-canopy streams revealed by stable isotope analysis. Freshw. Biol. 56, 470–480.

Lieske, R., Zwick, P., 2007. Food preference, growth and maturation of *Nemurella pictetii* (Plecoptera: Nemouridae). Freshw. Biol. 52, 1187–1197.

Ma, J.-Y., Sun, W., Liu, X.-N., Chen, F.-H., 2012. Variation in the stable carbon and nitrogen isotope composition of plants and soil along a precipitation gradient in northern China. PLoS One 7, e51894.

McKie, B.G., Malmqvist, B., 2009. Assessing ecosystem functioning in streams affected by forest management: increased leaf decomposition occurs without changes to the composition of benthic assemblages. Freshw. Biol. 54, 2086–2100.

McKie, B.G., Petrin, Z., Malmqvist, B., 2006. Mitigation or disturbance? Effects of liming on macroinvertebrate assemblage structure and leaf-litter decomposition in the humic streams of northern Sweden. J. Appl. Ecol. 43, 780–791.

McKie, B.G., Woodward, G., Hladyz, S., Nistorescu, M., Preda, E., Popescu, C., Giller, P.S., Malmqvist, B., 2008. Ecosystem functioning in stream assemblages from different regions: contrasting responses to variation in detritivore richness, evenness and density. J. Anim. Ecol. 77, 495–504.

McKie, B.G., Schindler, M., Gessner, M.O., Malmqvist, B., 2009. Placing biodiversity and ecosystem functioning in context: environmental perturbations and the effects of species richness in a stream field experiment. Oecologia 160, 757–770.

Mouillot, D., Graham, N.A.J., Villéger, S., Mason, N.W.H., Bellwood, D.R., 2013. A functional approach reveals community responses to disturbances. Trends Ecol. Evol. 28, 167–177.

Newcombe, C.P., Macdonald, D.D., 1991. Effects of suspended sediments on aquatic ecosystems. N. Am. J. Fish Manag. 11, 72–82.

Nilsson, E., Olsson, K., Persson, A., Nyström, P., Svensson, G., Nilsson, U., 2008. Effects of stream predator richness on the prey community and ecosystem attributes. Oecologia 157, 641–651.

Oksanen, J., Blanchet, F.G., Kindt, R., Legendre, P., Minchin, P.R., O'Hara, R.B., Simpson, G.L., Solymos, P., Stevens, M.H.H., Wagner, H., 2011. vegan: community ecology. R Package Version 2.0-2.

Pakeman, R.J., 2011. Multivariate identification of plant functional response and effect traits in an agricultural landscape. Ecology 92, 1353–1365.

Parnell, A., Jackson, A., 2013. siar: Stable isotope analysis in R. R Package Version 4.2. Available at: http://CRAN.R-project.org/package.siar.

Pascoal, C., Cássio, F., Gomes, P., 2001. Leaf breakdown rates: a measure of water quality? Int. Rev. Hydrobiol. 86, 407–416.

Pascoal, C., Pinho, M., Cássio, F., Gomes, P., 2003. Assessing structural and functional ecosystem condition using leaf breakdown: studies on a polluted river. Freshw. Biol. 48, 2033–2044.

Perkins, D.M., McKie, B.G., Malmqvist, B., Gilmour, S.G., Reiss, J., Woodward, G., 2010. Environmental warming and biodiversity–ecosystem functioning in freshwater microcosms: partitioning the effects of species identity, richness and metabolism. Adv. Ecol. Res. 43, 177–209.

Petersen, R.C., Cummins, K.W., 1974. Leaf processing in a woodland stream. Freshw. Biol. 4, 343–368.

Peterson, B.J., Fry, B., 1987. Stable isotopes in ecosystem studies. Annu. Rev. Ecol. Syst. 18, 293–320.

Pettersson, M., Lundberg, T., Staafjord, J., 2004. Ängs och betesmarker i Västerbottens län. Länsstyrelsen Västerbottens län meddelande.

Pinheiro, J., Bates, D., DebRoy, S., Sarkar, D., R Core Team, 2012. nlme: linear and nonlinear mixed effects models. RPackage Version 31–104.

Poff, N.L., Olden, J.D., Vieira, N.K.M., Finn, D.S., Simmons, M.P., Kondratieff, B.C., 2006. Functional trait niches of North American lotic insects: traits-based ecological applications in light of phylogenetic relationships. J. N. Am. Benthol. Soc. 25, 730–755.

Post, D.M., Layman, C.A., Arrington, D.A., Takimoto, G., Quattrochi, J., Montaña, C.G., 2007. Getting to the fat of the matter: models, methods and assumptions for dealing with lipids in stable isotope analyses. Oecologia 152, 179–189.

R Core Team, 2014. R: A Language and Environment for Statistical Computing. Available at: http://www.R-project.org/.

Reiss, J., Bridle, J.R., Montoya, J.M., Woodward, G., 2009. Emerging horizons in biodiversity and ecosystem functioning research. Trends Ecol. Evol. 24, 505–514.

Robinson, C.T., Gessner, M.O., 2000. Nutrient addition accelerates leaf breakdown in an alpine springbrook. Oecologia 122, 258–263.

Rosseel, Y., 2012. lavaan: an R package for structural equation modeling. J. Stat. Softw. 48, 1–36. Available at: http://www.jstatsoft.org/v48/i02.

Schmidt-Kloiber, A., Hering, D., 2011. The Taxa and Autecology Database for Freshwater Organisms. Available at: http://www.freshwaterecology.info (accessed September 2011).

Spänhoff, B., Augspurger, C., Küsel, K., 2007. Comparing field and laboratory breakdown rates of coarse particulate organic matter: sediment dynamics mask the impacts of dissolved nutrients on CPOM mass loss in streams. Aquat. Sci. 69, 495–502.

Statzner, B., Bêche, L.A., 2010. Can biological invertebrate traits resolve effects of multiple stressors on running water ecosystems? Freshw. Biol. 55, 80–119.

Svensson, P.-O., 1966. Growth of nymphs of stream living stoneflies (Plecoptera) in northern Sweden. Oikos 17, 197–206.

Thébault, E., Loreau, M., 2005. Trophic interactions and the relationship between species diversity and ecosystem stability. Am. Nat. 166, E95–E114.

Thornton, D.R., 1963. The physiology and nutrition of some aquatic hyphomycetes. J. Gen. Microbiol. 33, 23–31.

Tilman, D., Reich, P.B., Isbell, F., 2012. Biodiversity impacts ecosystem productivity as much as resources, disturbance, or herbivory. Proc. Natl. Acad. Sci. U. S. A. 109, 10394–10397.

Vaughn, C.C., Spooner, D.E., Galbraith, H.S., 2007. Context-dependent species identity effects within a functional group of filter-feeding bivalves. Ecology 88, 1654–1662.

Von Schiller, D., Martí, E., Riera, J.L., Ribot, M., Marks, J.C., Sabater, F., 2008. Influence of land use on stream ecosystem function in a Mediterranean catchment. Freshw. Biol. 53, 2600–2612.

Wissinger, S., Steinmetz, J., Alexander, J.S., Brown, W., 2004. Larval cannibalism, time constraints, and adult fitness in caddisflies that inhabit temporary wetlands. Oecologia 138, 39–47.

Woodward, G., Gessner, M.O., Giller, P.S., Gulis, V., Hladyz, S., Lecerf, A., Malmqvist, B., McKie, B.G., Tiegs, S.D., Cariss, H., Dobson, M., Elosegi, A., Ferreira, V., Graça, M.A. S., Fleituch, T., Lacoursière, J.O., Nistorescu, M., Pozo, J., Risnoveanu, G., Schindler, M., Vadineanu, A., Vought, L.B.M., Chauvet, E., 2012. Continental-scale effects of nutrient pollution on stream ecosystem functioning. Science 336, 1438–1440.

Wotton, R.S., Malmqvist, B., 2001. Feces in aquatic ecosystems. Bioscience 51, 537–544.

Yachi, S., Loreau, M., 1999. Biodiversity and ecosystem productivity in a fluctuating environment: the insurance hypothesis. Proc. Natl. Acad. Sci. U. S. A. 96, 1463–1468.

CHAPTER EIGHT

The Role of Body Size Variation in Community Assembly

Samraat Pawar[1]
Grand Challenges in Ecosystems and the Environment, Silwood Park, Department of Life Sciences, Imperial College London, Ascot, Berkshire, United Kingdom
[1]Corresponding author: e-mail address: s.pawar@imperial.ac.uk

Contents

Abstract

Body size determines key behavioral and life history traits across species, as well as interactions between individuals within and between species. Therefore, variation in sizes of immigrants, by exerting variation in trophic interaction strengths, may drive the trajectory and outcomes of community assembly. Here, I study the effects of size variation in the immigration pool on assembly dynamics and equilibrium distributions of sizes and consumer–resource size-ratios using a general mathematical model. I find that because small sizes both, improve the ability to invade and destabilize the community, invasibility and stability pull body size distributions in opposite directions, favoring an increase in both size and size-ratios during assembly, and ultimately yielding a right-skewed size and a symmetric size-ratio distribution. In many scenarios, the result at equilibrium is a systematic increase in body sizes and size-ratios with trophic level. Thus these patterns in size structure are 'signatures' of dynamically constrained, non-neutral

Advances in Ecological Research, Volume 52
ISSN 0065-2504
http://dx.doi.org/10.1016/bs.aecr.2015.02.003

201

community assembly. I also show that for empirically feasible distributions of body sizes in the immigration pool, immigration bias in body sizes cannot counteract dynamical constraints during assembly and thus signatures emerge consistently. I test the theoretical predictions using data from nine terrestrial and aquatic communities and find strong evidence that natural communities do indeed exhibit such signatures of dynamically constrained assembly. Overall, the results provide new measures to detect general, non-neutral patterns in community assembly dynamics, and show that in general, body size is dominant trait that strongly influences assembly and recovery of natural communities and ecosystems.

1. INTRODUCTION

Understanding general patterns or 'rules' in the dynamics of community assembly is of great theoretical and practical importance and remains one of the foremost challenges in ecology (Bascompte and Stouffer, 2009; May, 2009). In nature, local communities assemble through the dynamics of immigration and species interaction-driven extinction, which exert a 'filter' on all possible interactions given the available species pool for immigration (Bastolla et al., 2005; Pawar, 2009; Post and Pimm, 1983). While this general idea seems intuitive enough, we largely lack empirically testable theoretical predictions for general patterns of community assembly rates and trajectories, or for the types of species traits that are likely to facilitate rapid assembly or reassembly following disturbances (an issue of great practical importance). This is coupled with the problem that empirical data on the temporal sequences of community food web assembly remain scarce (May, 2009, but see Fahimipour and Hein, 2014).

In the context of community assembly, body size is a key trait because it strongly determines colonization rates though it effects on locomotion and dispersal (Hein et al., 2012; Schmidt-Nielsen, 1984), the strength of interspecific trophic interactions following colonization (Pawar et al., 2012; Vucic-Pestic et al., 2010), and life history rates that determine population energetics such as of basal metabolism, intrinsic growth and mortality (Brown et al., 2004; Economo et al., 2005; Kleiber, 1961; Peters, 1986; Savage et al., 2004). These wide-ranging effects of body size raise a suite of interesting and potentially important questions about what role body size variation in the global or regional species pool plays in the rate and trajectory of local community assembly. For example, it would be desirable to know if certain properties of the distribution of body sizes in the immigration pool can improve or accelerate the rate of community assembly. Yet, only a handful of previous studies have considered the role of size in community assembly dynamics (Etienne and Olff, 2004; Fukami, 2004; O'Dwyer et al., 2009; Virgo et al., 2006). And of these, the subset that have explicitly considered the filtering effects of interaction-driven extinction in food web dynamics (Fukami, 2004; Virgo et al., 2006) have not studied the effects of size variation in the immigration pool on local community assembly dynamics or the resulting local size or size-ratio distributions.

In this paper, I focus on the fact that body sizes in natural communities can span many orders in magnitude (Brose et al., 2006a; Cohen et al., 2003; Jonsson et al., 2005) and study what role the distribution of body sizes in the immigration pool play in dynamics of interaction-mediated local community food web assembly. In particular, by using a general size-constrained mathematical model of food web assembly, I ask whether assembled food webs at quasi-equilibrium (where species numbers remain relatively constant; Bastolla et al., 2005; Pawar, 2009) are expected to show 'signatures' of non-neutral, size-mediated assembly in (i) distribution of body sizes, (ii) distribution of size-ratios between consumers and resources, and (iii) distributions of sizes and size-ratios across trophic levels, and what role the distribution of sizes in the immigration pool plays in all this. I study not just size but size-ratios as well, because, along with consumer size, size difference between consumer and resource strongly determines tropic interaction strength (Pawar et al., 2012; Vucic-Pestic et al., 2010), a key factor driving individual invasion fitness as well as community stability during and after assembly (Brose et al., 2006b; Emmerson and Raffaelli, 2004; Otto et al., 2007; Pawar, 2009; Tang et al., 2014). I then evaluate the theoretical predictions using food web data from nine terrestrial and aquatic communities.

2. THEORY

I use a Lotka–Volterra (LV) type model of biomass dynamics of an n species community (May, 1974; Pawar, 2009; Tang et al., 2014; Virgo et al., 2006) to model food web assembly dynamics (Fig. 1):

$$\frac{\mathrm{d}x_i}{\mathrm{d}t} = x_i \left(b_i - d_i + \sum_{j=1}^{n} a_{ij} x_j \right), i = 1, 2, \ldots, n \tag{1}$$

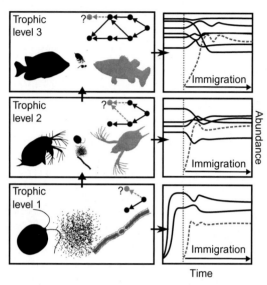

Figure 1 An illustration of the size-constrained community assembly process. At each step of sequential assembly (going bottom to top panel), consumer size and consumer–resource size-ratio determine whether a new immigrant can invade when rare (each dashed line, satisfying the condition in Eq. 20), and then whether the augmented community is stable (the solid lines do not decline to zero, determined by Eq. 24).

Here, x_i is the total biomass of the ith species' population, b_i its intrinsic biomass production rate (0 for consumers) and d_i its intrinsic density-independent biomass loss rate. For the jth consumer and ith resource, coefficient a_{ij} is mass-specific search rate which governs the rate of per-unit biomass loss of the ith species to consumption by the jth species (Pawar et al., 2012). Thus, biomass gain rate of the former (a_{ji}) and loss of the latter (a_{ij}) are related such that,

$$a_{ji} = -ea_{ij}, \text{ (when } j \text{ consumes } i) \qquad (2)$$

where e is the consumer's biomass conversion efficiency of the consumer, which does not scale with body mass within major organismal groups or trophic levels (DeLong et al., 2010; Hechinger et al., 2011). Therefore, henceforth I will assume $e = 0.5$, close to the value observed for carnivores. As long as $e > 0$, this paper's results do not qualitatively depend upon it. Finally, the coefficient a_{ii} (when $i = j$ in the sum in Eq. 1) is a mass-specific intraspecific 'search' rate that governs intraspecific interference between individuals of the ith species (DeLong, 2014; Pawar et al., 2012).

As such, Eq. (1) assumes a Type I (linear) functional response of the consumer. As long as most of the a_{ii}'s are non-zero with magnitudes in the order of the a_{ij}'s (i.e. have some level of intraspecific density dependence), using a Type II functional response instead, where consumer handling times for their resource are taken into account, do not change the results of this paper (Pawar, 2009; Tang et al., 2014).

2.1 Size-Scaling Parameterizations

I use the following body size-based parameterizations of the LV model (Eq. 1). In the following scaling relationships, the normalization constants also include the effects of metabolic temperature (Dell et al., 2011; Gillooly et al., 2001; Savage et al., 2004). For biomass production and loss rates, I use

$$b_i = b_0 m_i^{\beta-1} \qquad (3)$$

and

$$d_i = d_0 m_i^{\beta-1} \qquad (4)$$

where b_0 and d_0 are normalization constants, m is the species' average adult body mass, and $\beta = \frac{3}{4}$. These relationships are empirically well supported (Brown et al., 2004; McCoy and Gillooly, 2008; Peters, 1986; Savage et al., 2004), and have been used previously in similar contexts (Brose et al., 2006b; Pawar et al., 2012; Virgo et al., 2006; Weitz and Levin, 2006; Yodzis and Innes, 1992). For the search rate coefficient between the ith resource and jth consumer species, I use

$$a_{ij} = -a_0 m_j^{-0.25} \varphi_{ij} \qquad (5)$$

where, a_0 is a normalization constant, and $\varphi_{ij} \in [0, 1]$ is a dimensionless function that embodies attack success probability. Thus, from Eq. (2):

$$a_{ji} = e a_0 m_j^{-0.25} \varphi_{ij} \qquad (6)$$

Note that the quarter-power exponent in Eqs. (5) and (6) has been used for the scaling of search rate in numerous previous studies (Brose et al., 2006b; Otto et al., 2007; Virgo et al., 2006; Yodzis and Innes, 1992). However, according to the results of Pawar et al. (2012; also see Giacomini et al., 2013; Pawar et al., 2013), this is approximately the scaling exponent for $2D$ (two spatial dimensions; e.g. benthic) interactions only. For simplicity, I am thus assuming here that all food webs have

2D interactions only (Tang et al., 2014). I will consider the effect of 3D or a mixture of 2D–3D interactions in a subsequent paper.

Next, I assume that attack success probability is unimodal (Gaussian; Fig. 2) with respect to average resource mass (consumer–resource body mass ratio, or size-ratio):

$$\varphi_{ij} = \exp\left(-\left(s\log\left(m_j m_i^{-1}/k\right)\right)^2\right) \tag{7}$$

Here, s determines how rapidly the function reaches its peak k, the size-ratio of maximum consumption rate. This is a simplification of the potentially complex dynamics of consumer–resource encounter and consumption rates in nature, but captures an important feature of empirically observed attack and capture rates—these are unimodal because attack success decreases and handling time increases at extreme size-ratios (Aljetlawi et al., 2004; Brose et al., 2008; Persson et al., 1998). The actual values of k and s are expected to vary with type of consumption (foraging) strategy as well as habitat type. For example, in the case of predator–prey interactions, because smaller consumers have greater mass-specific power relative to larger ones, and can hence handle a larger range of prey sizes, k may be closer 1 (or even <1) for small consumers, and s smaller (a more gradual decline of consumption rate at extreme ratios). Brose et al. (2006a) have shown that invertebrate consumers do indeed have a k closer to 1 than vertebrates across disparate habitat types, suggesting their superior ability to attack and capture prey closer to their own size. In host-parasite and parasitoid interactions on the other hand, k should be <1 because parasites and parasitoids are selected to adopt strategies that increase effective encounter rate as well as exploitation success of resource species much larger than themselves (Cohen et al., 2005; Lafferty et al., 2008; Raffel et al., 2008). Below, I show that my results are largely insensitive over a wide range of choice of the parameters k and s.

Finally, I specify the intraspecific search rate a_{ii}. Because a_{ii} represents biomass loss (including individual mortality) resulting from metabolic stress induced by increasing biomass density of conspecifics, it should scale as

$$a_{ii} = -a_{ii,0} m_i^{-\gamma} \tag{8}$$

where a_{ii} is a normalization constant and γ the scaling exponent (but see DeLong, 2014). Below, I show that if γ ranges from ¼ to ½ (which includes the scaling exponent of mass-specific metabolic rate), empirically tenable scaling between body mass and equilibrium biomass abundance across species is seen in model communities. This is similar to the scaling used in a different set of body size parameterizations of the LV model by

Table 1 Parameters of the Body Size-Based Community Assembly Model and Their Numerical Values

Parameter	Description	Dimensions	Parameter Values
b_i	Mass-specific biomass production rate of the ith basal species	$time^{-1}$	–
d_i	Mass-specific intrinsic death rate	$time^{-1}$	–
a_{ij}	Mass-specific search rate of jth consumer for the ith resource	$area \times time^{-1}$	–
a_0	Scaling constant for a_{ij}	$area \times mass^{0.25} \times time^{-1}$	10^{-3}–1
b_0	Scaling constant for b_i	$mass^{0.25} \times time^{-1}$	10^{-6}–1
d_0	Scaling constant for d_i	$mass^{0.25} \times time^{-1}$	10^{-8}–10^{-2}
k	Location parameter for function φ	–	0.001–1000
s	Scale parameter for function φ	–	0.01–0.1
a_{ii}	Mass-specific intraspecific search rate	$area \times time^{-1}$	–
γ	Scaling exponent for a_{ii}	–	0–0.25
$a_{ii,0}$	Scaling constant for a_{ii}	$area \times mass^{\gamma} \times time^{-1}$	Depending upon target n at IEE

IEE stands for immigration-extinction equilibrium (see main text). Note that these some of these scaling parameters were also used to calculate the community matrix and stability characteristics of real community food webs (Section 3).

Virgo et al. (2006). This completes specification of the size-based assembly model. Table 1 provides a summary of the parameterizations used for the above scaling relationships.

2.2 Assembly Dynamics

I ignore environmental or demographic stochasticity, and focus on purely interaction-driven stable invasions and species sorting events. Previous studies have theoretically shown how food web structure changes in model communities during assembly though a combination of stable invasions (establishment of a new immigrant and its trophic links without any extinctions) and species sorting (unstable invasion followed by one or more extinctions) (Bastolla et al., 2005; Fukami, 2004; Pawar, 2009). Specifically, the ith immigrant arriving at a community already consisting of m residents will invade and grow in biomass density when rare if the condition

$$\sum_{j \in \mathrm{res}_i} e|a_{ij}|\hat{x}_j > d_i + \sum_{k \in \mathrm{con}_i} |a_{ik}|\hat{x}_k, j, k \in 1, 2, \ldots, m \qquad (9)$$

is satisfied (Pawar, 2009; Roughgarden, 1996; Strobeck, 1973). That is, the ith immigrant's biomass gain rate through feeding on its set of resources (res$_i$) must be sufficiently large to offset the loss due to its intrinsic biomass loss rate (d) and consumption exerted upon it by its set of k consumers (con$_i$), measured when all m resident species are at equilibrium. Note that because the immigrant is rare, intraspecific density dependence a_{ii} (Eq. 8) plays a negligible role in invasion success. This successful invasion will also be a _stable_ invasion if subsequently no other species goes extinct, partly determined by the probability of local stability (Pawar, 2009; Tang et al., 2014) of the invaded community. A necessary (but not sufficient) condition for the invasion to be stable is that the new n ($=m+1$) species system is locally asymptotically stable, i.e., the $n \times n$ Jacobian C (the familiar community matrix) with elements,

$$c_{ij} = \left.\frac{\partial(\mathrm{d}x_i/\mathrm{d}t)}{\partial x_j}\right|_{\mathbf{x}=\hat{\mathbf{x}}} = a_{ij}\hat{x}_i, i, j = 1, 2, \ldots, n \qquad (10)$$

has all its n eigenvalues $\lambda_i(C)$ lying in the negative half of the complex plane, i.e., given that $\lambda_{\max}(C) \equiv \max\{\mathrm{Re}(\lambda_i(C))\}, i = 1, 2, \ldots n$,

$$\lambda_{\max}(C) < 0 \qquad (11)$$

must hold (Pawar, 2009; Tang et al., 2014). The element c_{ij} ($i \neq j$) of C represents the population-level effect of a change in the jth species' biomass density on the ith one, or the dependence of the ith species on its own density (if $i = j$), at biomass equilibrium.

Stable invasions (if inequality (9) is satisfied) result in gradual changes in food web structure during assembly, whereas species sorting events (if subsequently, inequality (11) is violated) can cause greater structural upheavals by the extinction of multiple species. Eventually, for a given immigration rate, community assembly reaches quasi-equilibrium where immigrations are approximately balanced by extinctions through failed invasions and species sorting events (immigration-extinction equilibrium, or IEE) (Bastolla et al.,

2005; Fukami, 2004; Pawar, 2009). The condition in Eq. (11) only holds for point equilibria (Allesina and Tang, 2012; May, 1974; Pawar, 2009; Tang et al., 2014). Therefore, I test whether the predictions below (Section 2.3) about effect of size variation on assembly and stability based upon this assumption are more generally valid, using numerical simulations (Section 2.5).

2.2.1 Assembly Without Interactions

To understand the effects of size-mediated interactions on assembly, we can begin by considering the pattern of changes (or lack thereof) in local community body size variation under neutral community assembly—where neither invasion (Eq. 10) nor stability (Eq. 11) matter. This will serve as a null model to test against for the presence of signatures of invasibility and stability constraints during and after assembly.

Assume that the species body mass (size) distribution from which immigrants are drawn can be represented as a probability density function over some body mass range. This function does not merely represent the size distribution of the region to which the community belongs, but is a convolution of the probability density functions of the regional species pool's body size distribution and size-based immigration probability distribution, assuming a relationship between size and dispersal ability (Hein et al., 2012; Jetz et al., 2004). Hence I will refer to this size-distribution as belonging to the 'immigrant pool'. I assume that the probability density function of the sizes in the immigrant pool follows a one-parameter Beta distribution, Beta(1,ω) (Springer, 1979),

$$f_\gamma(\gamma_i) = \omega(1 - \gamma_i)^{\omega-1}, \gamma_i \in [0, 1] \tag{12}$$

rescaled to lie between biologically feasible lower and upper log-size limits (γ'_{min} and γ'_{max}) of the immigrant pool:

$$\gamma' = \gamma'_{min} + \left(\gamma'_{max} - \gamma'_{min}\right)\gamma \tag{13}$$

Thus, here the random variable $\gamma' = \log(m)$, and the distribution function in Eq. (12) gives the probability of immigration of any species with respect to its log-transformed body size. Note that the logarithm of body mass is being used here purely for convenience. Using the Beta probability, density here has two main advantages. First, it has finite bounds, which allows a precise allocation of minimum and maximum feasible body masses in the model. Second, depending upon the choice of ω, different shapes of size distributions in the immigrant pools (e.g. uniform vs. skewed) can be chosen. The latter factor will be considered in greater detail below.

From the immigrant pool, species are assumed to arrive at the local community at a fixed rate. At each immigration event, the ith invader species establishes a trophic link with the jth resident species with some probability, which I assume is independent of the body masses of both species because under neutral assembly, the only limitation on species accumulation is immigration rate (ignoring environmental or demographic stochasticity). In this scenario, the log-size distribution of the local community is expected to reflect that of the immigrant pool. I will concentrate on two properties of the local log-size to measure deviations from this null expectation: its mean and kurtosis. The mean of the local community's log-size distribution under neutral assembly would be approximately,

$$\mu_{\text{log-size}} \cong y'_{\text{min}} + \left(y'_{\text{max}} - y'_{\text{min}} \right) \frac{1}{(1+\omega)} \tag{14}$$

(using properties of the Beta(1,γ) distribution). In addition, because log-size distributions tend to be right skewed to different degrees (Allen et al., 2006a), we are interested in skewness of the community's log-size. Again using the properties of the Beta(1, γ) distribution, this is expected to be

$$sk_{\text{log-size}} \cong \frac{2(\omega-1)\sqrt{\omega+2}}{(\omega+3)\sqrt{\omega}} \tag{15}$$

Equations (14) and (15) are only approximations because there is bound to be finite sample error during stochastic immigration from the immigrant pool. The nature of the log-size-ratio distribution under neutral assembly can also be inferred as follows. Assuming that the log-sizes of consumer and resource species follow the same distribution, if trophic links are assumed to be established independent of species body sizes as well as existing links during assembly, the log-size-ratio of each trophic link is the random variable $y'_c - y'_r$ (because $\log(m_c/m_r) = \log(m_c) - \log(m_r)$), with the subscripts 'c' and 'r' denoting consumer and resource species, respectively. We do not need to determine the actual form of this distribution because its mean and kurtosis can be directly calculated. Firstly, because given two independent random variables X and Y, $E(X-Y) = E(X) - E(Y)$,

$$\mu_{\text{log-size-ratio}} \cong 0 \tag{16}$$

Additionally, because the distribution of differences between two independent and identically Beta-distributed random variables is unimodal and symmetric (with upper and lower bounds equal to $y'_{\text{min}} + y'_{\text{max}}$, in our case), again assuming that the size distributions of consumers and resources are identical, the log-size-ratio distribution must also have approximately zero skewness under neutral assembly, i.e.,

$$sk_{\text{log-size-ratio}} \cong 0 \qquad (17)$$

Any deviation from these characteristics of local community log-size and log-size-ratio distributions (Eqs. 14–17) can be attributed to non-random processes during community assembly (which, following random immigration, bias the success of species towards those with particular body sizes). Next, I will show that one such process is non-random species extinctions driven by the stability constraints of multi-species stability.

2.2.2 Assembly with Size-Mediated Interactions

To obtain predictions about non-neutral assembly with size variation, we need to consider the combined constraints of invasibility (Eq. 9) and stability (Eq. 11). First, because both invasibility and stability depend upon resident species' equilibrium biomass abundances ($\hat{x}$'s) we need to consider whether and how equilibrium biomasses themselves depend upon body size in the LV model, as is typically seen in local communities (Cyr et al., 1997a; Leaper et al., 1999). In Appendix A, I show that while deriving this relationship is analytically intractable for arbitrary community size n, it can be found numerically that the above body mass based LV model yields equilibrium biomass densities that scale across species as

$$\hat{x}_i = x_0 m_i^{\nu} \qquad (18)$$

where x_0 is the intercept and depending upon the parameters ω and k,

$$\nu = z\gamma - c \qquad (19)$$

with z lying between 0.5–0.6 and c between 0.1–015 (see Appendix A). Thus, if γ lies between 0.25 and 0.5, ν lies between 0 and 0.25, which is consistent with data from local communities as well as theory (Cyr et al., 1997a; Leaper et al., 1999; Reuman et al., 2009; Sheldon et al., 1977).

Armed with the scaling of equilibrium abundance, we can now proceed. At the early stages of assembly, because n is small and local stability constraints weak (Pawar, 2009; Tang et al., 2014), the success of each immigrant is dependent mainly on its ability to invade as determined by inequality (9). By substituting Eqs. (4)–(7) and (18) into (9), we get,

$$em_i^{-0.25} \sum_{j \in \text{res}_i} \varphi_{ji} m_j^{\nu} > (d_0/a_0 x_0) m_i^{\beta-1} + \sum_{k \in \text{con}_i} m_k^{-0.25+\nu} \varphi_{ik}, \quad j,k \in 1,2,\ldots,m \quad (20)$$

That is, smaller species are more likely to invade because they tend to have higher mass-specific biomass uptake and production rates (from the negative scaling of these quantities), provided they are able to establish a sufficient number of trophic links with appropriate size-ratios (the function φ) to grow in biomass density when rare. Thereafter, we need to consider how size variation affects local stability during assembly following every successful invasion. For this, I use a previous result for a measure of trophic link strength directly relevant to local stability (Pawar, 2009):

$$\bar{c}_{ij} = |a_{ij}| \sqrt{e\hat{x}_i\hat{x}_j} \qquad (21)$$

(where i consumes j). Each $\bar{c}_{ij}$ is the geometric mean of the pair of coefficients (a_{ij}, a_{ji}) associated with each interspecific interaction, and is proportional to the biomass transfer rate from resource to consumer. Now, substituting Eqs. (5), (6) and (18) into (21), we have the approximate scaling of trophic link strengths:

$$\bar{c}_{ij} \cong a_0 m_i^{-0.25} \varphi_{ij} x_0 \sqrt{e m_i^\nu m_j^\nu} \qquad (22)$$

(where j consumes i). However, because local stability is defined relative to the strengths of the diagonal elements of the community matrix (intraspecific density dependences at biomass equilibrium) (see Pawar, 2009; Tang et al., 2014) which also scale with body size (Eq. 8), the scaling of the $\bar{c}_{ij}$'s by themselves do not provide sufficient information to understand their effects on community stability. So we need a measure that considers effects of the $\bar{c}_{ij}$'s relative to the diagonal elements of C. The destabilizing effect of the $\bar{c}_{ij}$'s is inversely related to the strengths of the diagonal elements (the c_{ii}'s) of the community matrix C, which themselves scale as (combining Eqs. 8, 18 and 19):

$$c_{ii} \cong x_0 a_{ii,0} m_i^{\nu-\gamma}$$

Then, using the result that local stability is inversely related to the quantity $\sum_{i,j=1}^{n} \bar{c}_{ij}$ (also see Tang et al., 2014 for an analogous result), the impact of the jth immigrant's successful invasion on stability during assembly can be measured as the ratio of the sum of the strengths of the new trophic link strengths added to the system, over the new diagonal element (intraspecific interference),

$$\frac{\sum_{i\in res_j}\bar{c}_{ij}}{c_{jj}} \cong \frac{a_0 m_j^{-0.25}\sqrt{em_j^{\nu}}\sum_{i\in res_j}\varphi_{ij}\sqrt{m_i^{\nu}}}{a_{jj,0}m_j^{\nu-\gamma}} \tag{23}$$

where res_j is the set of resource species of the new immigrant. Smaller the ratio in Eq. (23), weaker the impact of the new consumer on community stability. The effects of the immigrant's vulnerability (all the new trophic links in which it is the resource) are not included in Eq. (23), because they can be absorbed into the analogous terms of its consumer species. Simplifying Eq. (23) gives,

$$\frac{\sum_{i\in res_j}\bar{c}_{ij}}{c_{jj}} \cong \frac{a_0\sqrt{e}\sum_{i\in res_j}\left(\varphi_{ij}m_i^{\nu/2}\right)}{a_{jj,0}m_j^{\frac{\nu}{2}-\gamma+\frac{1}{4}}} \tag{24}$$

Equation (24) indicates that if $\gamma \geq 0.25 + \nu/2$, potentially destabilizing effects of the higher mass-specific trophic link strengths associated with smaller species will be counterbalanced by their stronger intraspecific density dependence. Note here that ν itself also increases as a fraction of γ (Eq. 19), and hence it is expected to be >0 for $\gamma \geq 0.25$. If $\gamma = 0.25$, as might be expected from metabolic considerations, the effects of increasing mass-specific biomass acquisition and production rates are canceled out by the biomass loss due to negative intraspecific density dependence. In that case, the denominator of Eq. (24) *increases* with body mass, and larger species tend to be relatively stabilizing. Also, irrespective of its body mass, the stability impact of an immigrant is determined by the terms in the numerator in Eq. (24): the size-ratios (the function φ) associated with its trophic generality (number of res_j's), and provided that $\nu > 0$, the body sizes of these resource species.

Across the range of possible size-ratios over an empirically feasible range of body sizes (see Section 3), Fig. 2 compares the distribution of search rates (a_{ij}'s), trophic link strengths ($\bar{c}_{ij}$'s) and the ratio in Eq. (24) for two different values of γ. Three salient features of the a_{ij}'s are immediately evident from Fig. 2A: (i) they decrease at either extremes of body mass difference, (ii) increase with decrease in body mass of either species (i.e. for a given size-ratio, the peak values—the 'hottest' zones—lie towards the upper left corner of the figure) and (iii) decrease more rapidly at extreme upper values of size-ratios compared to lower extreme values (the hottest zones lie slightly above the size-ratio $= 1$ line). While (i) is a result of the symmetric unimodality of φ_{ij} (Eq. 7), (ii) and (iii) result from

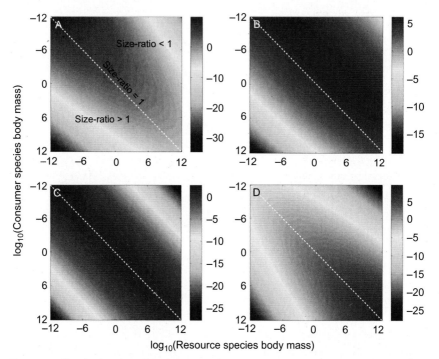

Figure 2 The body size dependence of interspecific trophic interactions across 12 orders of magnitude variation in consumer and resource species' body masses: (A) the variation in search rate (a_{ij}, Eq. 5), (B) trophic link strength ($\bar{c}_{ij}$, Eq. 22, for $\nu = 0.25$), (C) the stability ratio (Eq. 24, for generality $= 1$), and (D) the same stability ratio for $\nu = 0.25$. Note that $\nu = 0$ and 0.25 correspond approximately to $\gamma = 0.25$ and 0, respectively (Appendix A). All these measures are shown in the log scale to facilitate visual interpretation, and for $k = 1$ and $s = 0.1$. Changing k would change the location of the size-ratio $= 1$ line, while increasing (decreasing) s would increase (decrease) spread of the measures around k. The values of the allometric scaling constants a_0, and $a_{ii,0}$ are irrelevant here and were arbitrarily set to 1, and e to 0.5.

the scaling of intrinsic biomass production rates (Eqs. 3 and 6), which decrease with size.

These features of mass-specific interaction strengths are consistent with previous such body size-based models (Brose et al., 2006b, 2008; Yodzis and Innes, 1992). Figure 2B shows that if $\nu > 0$ (the case of $\nu = 0.25$ is shown), when the effects of equilibrium biomasses are added to the a_{ij}'s, the resulting trophic link strengths reduce the bias of the interaction strengths towards smaller sized organisms (the hottest zones extend further along the size-ratio $= 1$ line towards the bottom right corner of the figure), while increasing the bias towards extreme upper values of size-ratios (the hottest zones lie further above the size-ratio $= 1$ line). Finally, Fig. 2C and D show that as expected from the body size-scaling composition of the ratio in Eq. (24), for the two extreme values of γ (0.25 and 0.5), the effects of the mass-specific biomass production scaling are either perfectly negated (Fig. 2C), or overwhelmed (Fig. 2D) (because ν increases linearly with γ; Eq. 19). In other words, in the former case (Fig. 2C), the potentially destabilizing effects of smaller body-sized species are negated (the hottest zone becomes a ridge across the entire size-ratio $= 1$ line), while in the latter (Fig. 2D), larger body-sized species actually become more destabilizing (the hottest zone shifts to the bottom right hand corner along the size-ratio $= 1$ line). Moreover, the more rapid decline in interaction strengths and trophic link strengths at extreme upper values of size-ratios compared to lower extreme ones is no longer seen—cooler areas are larger below the size-ratio $= 1$ line in Fig. 2A–B are no longer larger. These analytical insights can now be used to generate predictions of the effect of body size variation on community assembly.

2.3 Predictions for Size-Driven, Interaction-Mediated Assembly

As assembly proceeds, species' generalities tend to increase as more links are established, and therefore the probability of local stability decreases because the numerator in Eq. (24) overwhelms the denominator. Hence, successful invasions are now more often followed by species sorting events, and so while relatively small sized species continue to invade more successfully, the constraints on trophic link strengths have increased. At such a stage, the numerator of Eq. (24) has to be mitigated for the growing number of species to coexist stably. This is possible if, during assembly, irrespective of the size of the consumer, trophic link strength can be weakened by deviation from peak values of the function φ_{ij} (at size-ratio $= k$) and if $\nu > 0$ (i.e. at least weak increase of equilibrium biomass with body size) by an decrease in the size of the resource species it feeds on (the terms inside the sum of the numerator). These two conditions can simultaneously be satisfied only if the consumer has trophic links with size-ratios > 0. Because their higher invasibility will have favored species in the small end of the spectrum of the immigrant pool's body size at the earlier assembly stages, size-ratios will inherently tend to increase because the later immigrants will be on average larger than the residents. Thus two patterns should emerge during assembly, which as also predicted signatures of size-mediated assembly:

(i) *A negative correlation between a consumer's trophic level and its size-ratios,* and linked to this,

(ii) *A positive correlation between a species' size and its trophic level.*

Predictions (i) and (ii) can be measured by correlation coefficients (specified below), and will be denoted by $r(TL$, size-ratio) and $r(TL$, size), respectively. Note that if community assembly is hierarchical such that later immigrants tend to be consumers rather than resources, occupying higher trophic levels, signatures (i) and (ii) will be stronger because this will predispose the relatively larger later immigrants towards becoming consumers of resident species. Hence the following interdependency between food web structural signatures of stability constraints are predicted (i) $\leftrightarrow$ (ii), i.e., an increase in consumer species' body size and size-ratios with trophic position (measured by $r(TL$, size) and $r(TL$, size-ratio), respectively) will be positively correlated (Fig. 5).

I now consider how size and size-ratio distributions change concurrently with predictions (i) and (ii) above. Firstly, as mentioned above, species with small body sizes are expected to invade with greater success because of their inherently superior rate of mass-specific biomass uptake and production. This is especially true when species richness is low and stability constraints are weak (invasibility constraints dominate community stability constraints). Hence, the local log-size distribution is expected to deviate from the null one in the direction of small body sizes at the early stages of assembly. So $\mu_{\text{log-size}}$ will become increasingly smaller than that expected from neutral assembly (Eq. 14), and skewness of the size distribution $sk_{\text{log-size}}$ higher (more positive, or right-skewed) than that in Eq. 15. At the same time, because trophic link strengths are strongest at size-ratio $= 1$ (Fig. 2C and D), species will be more successful in invading by feeding on those similar to them in size, and $\mu_{\text{log-size-ratio}}$ and $sk_{\text{log-size-ratio}}$ should initially stay around 0 (the null values, Eqs. 16 and 17). As assembly proceeds, because relatively larger species experience increasingly fare better at invading stably (for reasons explained in prediction (ii) above), as the community approaches IEE, $\mu_{\text{log-size}}$ and $sk_{\text{log-size}}$ should decrease at a decreasing rate.

Thus, the invasibility (energetic requirement for initial establishment) of individual species and the stability requirements of multi-species stable coexistence 'pull' the local size distribution in opposite directions during assembly. Thus, the following two additional predictions can be made about the signatures of dynamically constrained assembly on community size and size-ratio distributions:

(iii) *An asymptotic decrease in $\mu_{\text{log-size}}$ and increase in $sk_{\text{log-size}}$ during assembly, culminating in smaller and larger values, respectively, of these measures at IEE relative to those of the immigrant pool's size distribution.*

(iv) *An asymptotic increase in $\mu_{\text{log-size-ratio}}$ during assembly, culminating in a larger value at IEE relative to that of the immigrant pool's size-ratio distribution.*

It is important to note that as in the case of the assembly signatures **(i)** and **(ii)** above, changes in size-ratios are dependent on assembly sequence or pattern. An increase in $\mu_{\text{log-size-ratio}}$ is possible because the local $\mu_{\text{log-size}}$ progressively decreases relative to that of the immigrant pool; hence if interactions are established more or less randomly with respect to body size, $\mu_{\text{log-size-ratio}}$ is bound to increase (because the species arriving from the immigration pool will be on average larger than those in the local community). Obviously, if later immigrants tend to be consumers with greater probability (occupy higher trophic levels), this effect will be magnified. On the other hand, if assembly is completely non-hierarchical such that each immigrant is equally likely to be a consumer or resource, even if an increase in the $\mu_{\text{log-size-ratio}}$ is favored by stability constraints, it may change little from the null expectation of 0 (Eq. 16) irrespective of the local log-size. Also, although the null expectation for $sk_{\text{log-size-ratio}}$ is 0 (Eq. 17), the current theory cannot predict the direction in which a deviation from it may be expected—this will be investigated numerically below.

2.4 Simulations

I evaluated the above predictions about the emergence of non-random food web structural features (signatures) using numerical simulations of community assembly based on the size-based LV model. For this I used the following three-step community assembly algorithm (Pawar, 2009):

- *Immigration.* Beginning with the establishment at least one basal species, at 1000 time step intervals, a species population was introduced at an extinction-threshold biomass abundance x_c. Each immigrant species was generated by sampling a body size from the Beta(1,ω) distribution introduced in Eq. (12). Inter- and intraspecific interaction parameters were determined by body sizes (Eqs. 3–8).

- *Trophic linking.* Upon colonization, the *j*th immigrant established a trophic link to the *i*th pre-existing one with a connectance probability p_c.
  For each assembly simulation, conditional upon p_c, a 'vulnerability probability' p_v ranging between 0.5–1 was set: $p_v = 0.5$ meant that the *j*th
  immigrant was equally likely to be a resource or a consumer of the *i*th
  resident species (provided it was not basal), while $p_v = 1$ meant that
  the *j*th immigrant could only be a consumer.
- *Interaction-driven extinction.* After immigration, the augmented system was
  numerically integrated forward for 1000 time steps, during which most
  populations either reached a non-zero equilibrium size or went extinct.
  A species was considered extinct and deleted from the system if its density dropped below x_e, or decreased during this period.

This algorithm was iterated till the system reached IEE. Simulations were
performed in Matlab using the Runge–Kutta one-step solver ode45. During
each assembly simulation run, changes in key size-related community
characteristics ($\mu_{\text{log-size}}$, $\mu_{\text{log-size-ratio}}$, $sk_{\text{log-size}}$, $sk_{\text{log-size-ratio}}$) were measured
at 1000-time step intervals (coinciding with the interval for numerical
integration). The method used to calculate trophic level is described in
Pawar (2009). Simulation parameters were chosen as follows (again, see
Pawar, 2009):

- *e* was fixed at 0.5 for all species, the approximate midpoint of the range
  reported from empirical data (Brown et al., 2004; Peters, 1986). Values
  ranging from 0.1 to 1 do not change the simulation results qualitatively.
- p_v was set to 0.9 because this is the midpoint of the range of [0.75–1] that
  yields communities with structural and dynamical characteristics similar
  to that of real ones, similar to the niche model (Pawar, 2009). The results
  do not change qualitatively over the full range of [0.5–1].
- x_e was set to 10^{-20}; the results do not change qualitative for values ranging
  from 10^{-32} to 10^{-3}.
- The number of basal species was set as a fixed proportion of the total target community size at IEE. All simulation results shown here are for five
  basal species (~10% of *n* at IEE, which was 43.1 for $\omega = 1$ and 40.1 for
  $\omega = 2$; Table 3). The simulation results do not change qualitatively for
  assembly with fewer or more basal species than this.

In addition, body size-related simulation parameters were chosen as
follows (Table 1):

- ω, which determines the shape of the immigrant pool size distribution
  was varied between 1 (uniform distribution; immigration rate independent of body size) and 2 (power law-like with slope $= -2$; immigration

probability decreases with body mass). This range of ω was chosen because: (a) empirical data show the distributions of sizes at large spatial scales are right-skewed (Allen et al., 2006a; Clauset and Erwin, 2008), probably partly driven by speciation rate, which appears to follow a negative power law relationship with body size (Dial and Marzluff, 1988; Kozlowski and Gawelczyk, 2002; Marzluff and Dial, 1991), possibly linked to metabolic scaling (Allen and Gillooly, 2006; Allen and Savage, 2007; Allen et al., 2006b) and, (b) dispersal ability should increase with body size (Hein et al., 2012; Peters, 1986). Only considering (a) means that ω should be >1; choosing $\omega = 2$ sets a reasonably high upper limit to this size bias in immigration. Considering (b) means that the effect of (a) may be somewhat negated due to dispersal ability. However, because speciation within the local community also adds to the effective bias towards immigration by smaller species, it is unlikely that (b) can overwhelm the effects of (a). Hence, $\omega = 1$, which yields the uniform distribution expected if the effects of (a) and (b) exactly cancel out, is a reasonable lower limit to the immigration rate bias.

- The log-body mass range $[y'_{min}, y'_{max}]$ was chosen to be $[-12, 12]$ because this is approximately the range of species' log-body masses observed across empirical communities (Brose et al., 2006a) (also see Fig. 7).
- k was chosen to vary randomly between 10^{-3} and 10^3 with uniform probability (consumers 1000 times smaller to 1000 times larger than resources), which covers most of the range considered to be 'optimal' (in the sense of viable or evolutionarily stable strategy for the consumer) in previous studies on consumer–resource interactions (Cohen, 2008; Weitz and Levin, 2006), and accommodates potential differences in k across the most common trophic interaction types seen in food webs (i.e. predator–prey, herbivore–plant and parasitoid–host) (Brose et al., 2006a).
- The parameter s was set to 0.1; however, varying it between a wide range (0.05–0.5) does not change the results (simulation results not shown, but see Appendix B).
- The allometric constants b_0 and d_0 were chosen to be 1 and 0.002, respectively, consistent with the observation that baseline birth or production rate is typically two orders of magnitude higher than mortality rate (Brown et al., 2004; Peters, 1986).
- The search rate scaling constant a_0 was varied between 10^{-3} and 1 based upon recent empirical results (Pawar et al., 2012)—the results shown

here are for $a_0 = 1$. Other values in this range do not change the results qualitatively.

- $a_{ii,0}$ was chosen according to the target mean n at IEE; larger values give larger feasible communities (May, 1974; Pawar, 2009; Tang et al., 2014).

2.5 Results

2.5.1 Signatures of Dynamically Constrained Assembly on Size and Size-Ratio Distributions

For both extremes of shapes of the regional log-size distributions ($\omega = 1$ or 2; dashed lines in Fig. 4A and C), Fig. 3A and B shows that as expected, both $\mu_{\text{log-size}}$ and $sk_{\text{log-size}}$ deviate from the null values rapidly, respectively, increasing and decreasing asymptotically, converging on heavily skewed size and symmetric size-ratio distributions (Fig. 4, Table 2). Also, because immigration is strongly dominated by small bodied species for $\omega = 2$, these deviations from the null expectations are proportionally higher. This proportional difference in deviation from the null expectation may be regarded

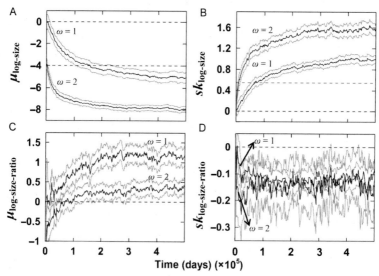

Figure 3 Changes in the log-size and log-size-ratio distributions of model communities during assembly. Mean values with 99% confidence intervals (grey lines) across 150 simulation runs over 400,000 time steps are shown for mean and skewness of the log-size (A and B) and size-ratio distributions (C and D). Each plot compares the trajectories for the two extreme values of ω, with the dashed lines showing the null values of distributional characteristics (Eqs. 14–17). These trends are for the same communities shown in Fig. 4.

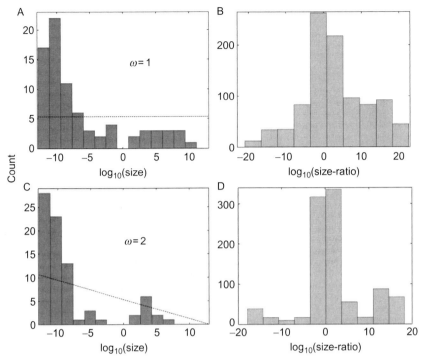

Figure 4 Size distributions (A and C) and the associated size-ratio distributions (B and D) in two different model communities at IEE assembled with $\omega = 1$ (upper panel) and $\omega = 2$ (lower panel). The dashed lines in the size distribution plots show the distribution expected if assembly had been fully neutral (no interactions) and without sampling error.

Table 2 Various Food Web Signatures of Dynamical Constraints on Model Size-Structured Communities at Immigration-Extinction Equilibrium (IEE)

Signatures of Dynamical Constraints	Assembly Type (Distribution of Immigrant Body Sizes)	
	$\omega = 1$ ($n = 43.1 \pm 5.6$)	$\omega = 2$ ($n = 40.1 \pm 5.0$)
$\mu_{\text{log-size}}$	100% (-5.63 ± 1.29)	100% (-7.69 ± 1.11)
$sk_{\text{log-size}}$	100% (1.07 ± 0.36)	100% (1.58 ± 0.47)
$\mu_{\text{log-size-ratio}}$	100% (3.06 ± 1.08)	93.3% (1.56 ± 0.15)
$r(TL, \text{size})$	48% (0.54 ± 0.09)	48% (0.56 ± 0.1)
$r(TL, \text{size-ratio})$	68% (0.31 ± 0.10)	84% (0.31 ± 0.13)

The mean species richness ($\pm$SD in parentheses) at IEE for each assembly type is shown in the header row. For each structural feature, the tabulated values give the percentage of 150 model food webs that showed the expected relationship, along with the mean value ($\pm$SD) of the measure in parentheses. For size and size-ratio distribution characteristics (rows 1–3), the signature of dynamical constraints was considered significant if a deviation in the expected direction from the null value (Eqs. 14–17) was seen. For measures that are correlation coefficients (rows 4–5) (Spearman's rank correlation), the signature was considered significant if two-tailed $p < 0.01$.

as the 'neutral' component of the log-size distribution. That is, under assembly unconstrained by interactions, the relative difference between the two distributions would be the same, even if their absolute $\mu_{\text{log-size}}$ and $sk_{\text{log-size}}$ values were different from those under dynamically-constrained assembly.

In the case of the log-size-ratio distribution as well, the effects of dynamically constrained assembly were as expected from the above theory—on average, $\mu_{\text{log-size-ratio}}$ increases asymptotically (Fig. 3C). For $\omega = 2$, $\mu_{\text{log-size-ratio}}$ starts off well below the null value (dashed line). This is because invasion success of the first few basal species is independent of their body size as they lack consumers (assuming there is no competition between basal species). However, invasibility of the first consumers feeding on these basal species decreases with body size. For $\omega = 2$, there is already a bias towards migration of small species, and hence $\mu_{\text{log-size-ratio}}$ will tend to <0 (consumer smaller than resource) initially. Even after the early stages of assembly, however, $\mu_{\text{log-size-ratio}}$ values for $\omega = 2$ remain below those seen for $\omega = 1$, which, as in the case of the size distribution, indicates that sufficiently strong immigration bias can counteract dynamical constraints on assembly. In the case of $sk_{\text{log-size-ratio}}$ on the other hand, both ω values result in a similarly weak negative deviation from the null value of 0 (note that this null value will be the same irrespective of ω), indicating that skewness of the log-size-ratio distribution may not carry a strong signature of dynamically constrained assembly. Also, the increase in $\mu_{\text{log-size-ratio}}$ during assembly indicates an overall shift of size-ratios away from the hottest zones lying along the size-ratio $= 1$ line towards the cooler zones (weaker interactions) below the line in Fig. 2C.

Table 2 shows that in all predicted signatures (Section 2.3, Fig. 6) of dynamically constrained assembly, a high incidence (high percentage values) of the expected signatures is seen irrespective of ω. Moreover, even in the case where the incidences were low, trends in the opposite direction were not seen. For example, in the case of increase in body size with trophic level $r(TL, \text{size})$, although only 48% of the communities show the expected relationship, none of the remaining ones showed a significant relationship in the opposite direction (decrease in body mass with trophic level). Finally, Fig. 5 shows that the interdependencies between the two signatures are as predicted in Section 2.3. Changes in other, non-body size-related food web structural features were similar to those seen during assembly based on the LV model (see Pawar, 2009), and are not shown here.

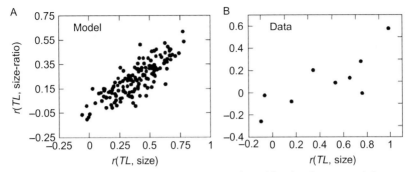

Figure 5 The interdependence between two trait-based food web structural signatures of constrained assembly across (A) 150 model communities at immigration-extinction equilibrium (IEE) and (B) the nine real communities. The relationships are strong and highly significant (A: $p < 0.0001$, $R^2 = 0.68$; B: $p < 0.02$, $R^2 = 0.63$). The model results are for communities assembled at $\omega = 1$. The strengths of these signatures across model communities are summarized in Table 2 and for real communities in Table 6.

3. DATA

I analyzed nine relatively high quality datasets using data–selection criteria that ensured only those communities which have a relatively well-resolved tropic interaction structure were chosen (see Pawar, 2009 for details). In particular, these criteria eliminated community datasets that had been subjected to substantial *a priori* taxonomic aggregation. I did not aggregate taxa into 'trophospecies', wherein taxa with highly similar or identical sets of consumer and resource species are treated as the same functional unit (i.e. two or more network nodes are collapsed into one). Apart from the fact that developing objective criteria for trophospecies is a difficult task (Yodzis, 1988), the use of this approach in the current study is undesirable for two reasons. First, the trophospecies approach has been used mainly in 'static' structural studies of food webs wherein only the presence-absence of trophic links are studied, and not the associated interaction strengths (see Allesina et al., 2008 and references therein). In contrast, this study is focused on effects of invasibility and dynamical stability during food web assembly—hence, even if two species have identical consumers and resources, they are unlikely to have identical strengths of interactions across the links, and hence are not dynamically equivalent. Second, the complete structure of the quantified food web is necessary for analyzing the stability of the community, and collapsing nodes into trophospecies can give a distorted view of the community's dynamical characteristics. Nine food webs were selected (Table 3). In the GC community, only one of the six subcommunities (Clmown1) was chosen because they all had similar food web structural properties (connectance, average trophic degree and food chain length were more similar between these communities than between them and others).

Whenever only data on a species' linear dimension was available, it was converted using allometric relationships of the form: mass $= a$ (length)^b. The parameters a and b were varied according to broad taxonomic groups (typically at the level of class and order) (Enquist and Niklas, 2001; Peters, 1986) (Table 4). In the case of certain groups such as worms and coelenterates, no meta-analyses were available. In these cases, the parameters of groups with similar body form and mass density were used (for example legless herpetofauna for worms). Of the community datasets that had body mass (instead of length) data to begin with, most consisted of at least a few species whose mass information was derived by the original authors using such length-to-mass conversions. The EW community consists of 391 taxa including a number of animal groups for whom length-mass scaling information are unavailable, and whose body forms preclude the use of scaling parameters of other groups (e.g. coelenterates and anthozoans) (Brose et al., 2005). Hence length, instead of mass data, was used for this community.

Clearly, taxonomic class- or order-wide use of the same length-mass scaling parameter is bound to increase the inherent noise in such data; this is exacerbated by the substitution of parameters for groups without information. However, as will be discussed further below, across local communities, >20 orders of magnitude interspecific body mass variation is seen, with each local community having been sampled for at least six orders of magnitude variation (Fig. 7). I expect these errors to be mitigated by this wide size variation.

3.1 Data Analyses

3.1.1 Signatures of Dynamically Constrained Assembly

Four size-related properties of real communities that I was able to compare with the predicted signatures of dynamically constrained assembly (Section 2.3) are illustrated in Fig. 6. Given that data on the regional species pool for each of these communities is unavailable, it was not possible to test for significant different of $\mu_{\text{log-size}}$ or $sk_{\text{log-size}}$ from null expectations. Therefore, I only tested whether $\mu_{\text{sk-size}}$ was positively skewed as predicted by the theory (part of prediction (iii), Section 2.3). A test of significant skewness

Table 3 The Empirical Community Datasets Used to Test Theoretical Predictions About Signatures of Dynamical Constraints on Food Web Structural Characteristics

Community Name	General Habitat	Description	Trophic Link Method[a]	Body Mass Method[b]	Data Sources	Food Web Characteristics				
						n	C_T	G	$\bar{T}_c$	O_{deg}
Broadstone stream (BS)	Aquatic (stream)	Spring-fed acidic headwater stream, Sussex, UK	1, 2	2	Brose et al. (2005) and Woodward et al. (2005)	28	0.37	15.3	4.6	0.81
Caribbean sea (CS)	Aquatic (marine)	Benthic and pelagic communities from surface to 100 m depth	3	2, 3	Bascompte and Melian (2005)	248	0.11	13.5	4.6	0.45
Eastern Weddell Sea (EW)	Aquatic (marine)	Antarctic shelf	1, 2	2, 3	Brose et al. (2005)	391	0.02	9.5	2.9	0.21
Grand Cariçaie marsh (GC)	Terrestrial	Marsh dominated by *Cladietum marisci*, Lake Neuchâtel, Switzerland	1, 3, 4	1, 2, 3	Brose et al. (2005) and Cattin et al. (2004)	163	0.16	24.0	4.8	1.05
Mill Stream (MS)	Aquatic (freshwater)	Lowland chalk stream, Dorset, UK	2	2	Brose et al. (2005)	74	0.14	8.2	1.2	0.03
Scotch Broom (SB)	Terrestrial	Community on *Cytisis scoparius*, Berkshire, UK	1, 2	1, 2, 3	Brose et al. (2005), Cohen et al. (2005) and Memmott et al. (2000)	153	0.03	10.2	3.2	0.14
Skipwith pond (SP)	Aquatic (freshwater)	Large acidic pond, North Yorkshire, UK	1, 2, 3	1, 2	Brose et al. (2005) and Warren (1989)	33	0.61	17.8	4.1	0.58

Continued

Table 3 The Empirical Community Datasets Used to Test Theoretical Predictions About Signatures of Dynamical Constraints on Food Web Structural Characteristics—cont'd

Community Name	General Habitat	Description	Trophic Link Method	Body Mass Method	Data Sources	Food Web Characteristics				
						n	C_T	$\bar{G}$	$\bar{T}_c$	O_{deg}
Tuesday lake (TL)	Aquatic (freshwater)	Small, mildly acidic lake, Michigan, USA	1, 2	1, 2	Brose et al. (2005), Cohen et al. (2003) and Jonsson et al. (2005)	72	0.15	12.3	3.8	0.18
Ythan estuary (YE)	Estuarine	Ythan river mouth, Scotland	1, 3, 4	1, 2, 3	Hall and Raffaelli (1991) and Leaper and Huxham (2002)	79	0.09	5.7	3.2	0.56

The abbreviation of each community's name used in subsequent tables and figures are shown in parentheses. The last four columns show some key food web structural characteristics (Pawar, 2009): species number (n), undirected connectance (C_T), average generality of consumers ($\bar{G}$), average trophic chain length ($\bar{T}_c$), and omnivory degree (O_{deg}). Body mass is expressed in grams throughout this study.

[a]1: Direct observations (lab or field), 2: Gut/stomach content analysis (typically, predators), feeding trials (typically, predators) or rearing (typically, parasitoids), 3: Published account, 4: Unpublished sources (including dissertation theses and Internet sites).

[b]1: Direct measurement, 2: Length-mass regression (see Table 4), 3: Published account, 4: Unpublished sources (including dissertation theses and Internet sites).

Table 4 Parameter Values Used for Converting Length (m) to Wet Mass (g) Using Scaling Models of Species in Different Taxonomic Groups Across Communities Using the Relationship Mass $= a$ Length^b

General Taxonomic Category	a	b	References
Plants	27	3.79	Niklas and Enquist (2001)
Fish	10600	2.57	Peters (1986)
Worms*	720	3.02	Peters (1986)
Mammals	14000	3.23	Peters (1986)
Birds	7390	2.74	Peters (1986)
Legless herpetofauna	720	3.02	Peters (1986)
Legged lizards	28000	2.98	Peters (1986)
Frogs	181000	3.24	Peters (1986)
Arachnids*	8800	2.62	Peters (1986)
Crustaceans*	8800	2.62	Peters (1986)
Insects (terrestrial and aquatic)	8800	2.62	Peters (1986)
Planktonic crustaceans	80	2.1	Peters (1986)
Algae	5.8	1.9	Peters (1986)
Other planktonic invertebrates*	80	2.1	Peters (1986)

For animals, the scaling for snout-vent length was used. Parameters for arachnids, crustaceans and some planktonic vertebrates were substituted from those of other groups with similar body form and density (marked with an asterisk).

without knowledge of what the null expectation should (the real value taken by ω in Eq. (15) be would be pointless in this scenario. Also note that $sk_{\text{log-size-ratio}}$ is not expected to show a strong signature of dynamically constrained (text following prediction **(iv)**, Section 2.3). For all other signatures, i.e., a positive $\mu_{\text{log-size-ratio}}$, $r(TL, \text{size})$ and $r(TL, \text{size-ratio})$, I tested for significance of the measure by calculating the approximate one-tailed p-value as the proportion of 2000 size-randomized communities that had a value of the measure greater than that of the observed, original community. The bivariate relationships $r(TL, \text{size})$ and $r(TL, \text{size-ratio})$ were calculated using the Spearman rank correlation coefficient.

Size randomizations were performed in two ways: full and partial. Under full randomization, species' body masses were randomly permuted while keeping the food web structure intact. This is easily done by permuting body

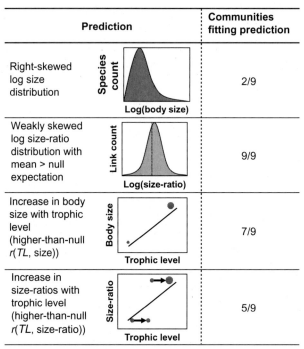

Prediction		Communities fitting prediction
Right-skewed log size distribution	Species count / Log(body size)	2/9
Weakly skewed log size-ratio distribution with mean > null expectation	Link count / Log(size-ratio)	9/9
Increase in body size with trophic level (higher-than-null $r(TL, \text{size})$)	Body size / Trophic level	7/9
Increase in size-ratios with trophic level (higher-than-null $r(TL, \text{size-ratio})$)	Size-ratio / Trophic level	5/9

Figure 6 A graphical overview of the predicted signatures of dynamically constrained assembly on community size structure and observed fits of data to them. The term 'null expectation' refers to the values of characteristics expected under purely neutral trophic linking and assembly. Each filled circle in a figure indicates a single consumer species' node in the food web, with arrows indicating their trophic links with its resources. The size of a circle and the thickness of an arrow represent, respectively, species body mass and interspecific interaction strength. Also see Table 5.

masses across species. Under partial randomization, body masses were randomly permuted only within basal and non-basal species (i.e. trophic level 1 vs. all others). This additional, conservative method of randomization was used because it maintains the body mass ranges of basal and non-basal species, which have been found to show consistent patterns within habitat types, and may be driven by historical, biogeographical and environmental factors rather than the trophic dynamics of the system (Brown and Gillooly, 2003; Hildrew et al., 2007).

3.1.2 Community Stability Properties

I also tested whether the observed communities had indeed self-organized during assembly in a manner that enhanced their stability, by examining whether the observed system was more likely to be locally (Hurwitz) stable

(Eq. 11) than an ensemble of randomized counterparts. For this, I calculated the community matrix C with elements $c_{ij} = a_{ij}\hat{x}_i$ using species' body mass information by substituting interspecific and intraspecific interaction rates calculated according to Eqs. (5)–(8) with parameter values from the ranges shown in Table 1. As in the simulation results, the following empirical results do not change qualitatively over these parameter ranges. I chose to fix the parameter k at 1 (maximum consumption intensity when consumers and resources are equal in size) for simplicity—Appendix B shows that the results are robust to a wide range of variation in this parameter. The value of s was chosen as follows. The function φ_{ij} (Eq. (7)) approaches zero more and more rapidly as s increases. As $s \rightarrow 0.1$, consumption rate falls to zero within the range of the size-ratios observed across the nine real communities Fig. 7, hence values larger than this are unfeasible because if a size-ratio is observed, no matter how extreme, it must be associated with some level of consumption. On the other hand, the consumption rate function becomes flat as $s \rightarrow 0$, and the effects of size-ratios on trophic link strengths are eliminated. Hence, s was set at 0.05, the midpoint of these two extreme values of s. Appendix B shows that the results are remarkably robust to variation in both k and s.

The additional parameter needed to calculate a c_{ij}—equilibrium biomass density—was calculated according to the scaling Eq. (18). For this, the scaling exponent ν was chosen to range between 0 and 0.25, which is consistent with data from local communities (Cyr et al., 1997a,b; Leaper et al., 1999; Tang et al., 2014), as well as the range of exponents predicted by current theories that combine size-metabolic scaling with trophic interactions (Loeuille and Loreau, 2006; Marquet et al., 1995; Rossberg et al., 2008; Woodward et al., 2005). The normalization constant x_0 rescales all the elements of C and was arbitrarily chosen to be 1—a wide range of values, between 0.001 and 100, do not change the results qualitatively.

The exponent γ, which determines the scaling of mass-specific biomass loss rate due to intraspecific interactions, was also chosen to lie between 0 and 0.25, which accommodates density dependence ranging from being independent of species body masses ($\gamma = 0$), to that expected from the scaling of mass-specific metabolic rate ($\gamma = 0.25$) (see Section 2). Finally, for each combination of values of ν and γ, the value of the scaling constant a_{ii} was chosen to be the minimum value across species that would guarantee that the system was locally stable—any nontrivial community matrix can be rendered stable by increasing the magnitudes of the negative diagonal elements (Allesina and Tang, 2012; May, 1974; Pawar, 2009; Tang et al., 2014).

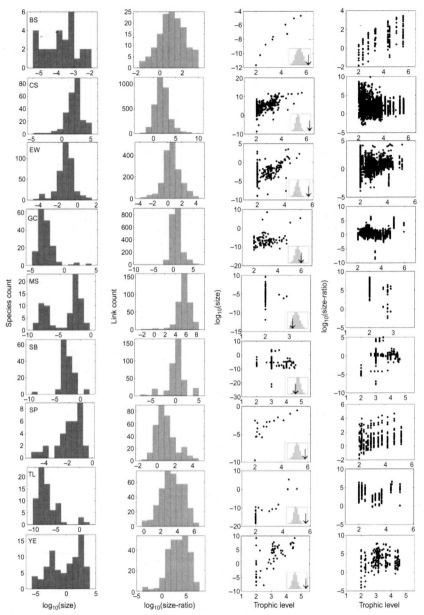

Figure 7 Food web characteristics across nine empirical communities that are expected to show signatures of size-constrained community assembly. Each row of figures represents the four features that should show signatures of size-driven assembly constraints for a particular community (Fig. 6). The histograms embedded in the third column of figures show the location of the correlation coefficient r(TL, size) with respect to the distribution of the coefficients from 2000 size-randomizations of the community. Fits of all these different features to theoretical predictions are summarized in Fig. 6. For the EW community, body size measure is length, not mass.

This was done by first calculating C and assigning a value of c_{ii} that was arbitrarily large (such that the system was locally stable). The value of c_{ii} was then gradually decreased until $\lambda_{\max}(C)$ became positive, thus giving the minimal value needed to make the system Hurwitz stable. This minimal value of c_{ii} was then used across the community matrices of the randomized counterparts of the community.

3.2 Results

3.2.1 Signatures of Dynamically Constrained Assembly

Figure 7 shows the four key community food web characteristics across the nine communities, and Table 5 summarizes the incidence and significance of structural signatures expected from dynamically constrained assembly. Only the results for the more conservative partial randomizations are shown. Full randomizations yielded consistently lower p-values, as expected.

Table 5 shows that the majority of communities indeed show the two signatures that underlie the distributions of size and size-ratios—consistently positive and significant $r(TL, \text{size})$ and $r(TL, \text{size-ratio})$. Only in Mill Stream and Scotch Broom are the opposite trends seen. This is not surprising because as shown in Section 2.3, these features are strongly dependent on hierarchical assembly, wherein later species are more likely to be consumers

Table 5 Signatures of Size-Mediated Dynamical Constraints in Real Communities
Signatures of Size-Based Dynamical Constraints

Community	$sk_{\text{log-size}}$	$\mu_{\text{log-size-ratio}}$	$r(TL, \text{size})$	$r(TL, \text{size-ratio})$
BS	−0.07	1.07*	0.98*	0.58*
CS	−1.17	2.02*	0.53*	0.09*
EW	−0.1	1.61*	0.34*	0.20*
GC	2.32*	0.74*	0.16*	−0.08
MS	−0.49	5.53*	−0.10	−0.26*
SB	−1.03	0.09	−0.07	−0.03
SP	−1.19	0.99*	0.75*	0.28*
TL	1.69*	3.38*	0.76*	−0.01
YE	−0.38	3.36*	0.65*	0.13*

Significant measures are flagged (*). For $sk_{\text{log-size}}$, a positive value of the skew was deemed significant, while for the other three measures, one-tailed significance is at $p < 0.05$, based upon a randomization test (see main text for more details). Note that MS shows a significant correlation in the opposite direction.

than resources. In the case of the MS community, trophic link sampling appears to be poor, with only the average trophic level value being only 1.2 (Table 3). I consider the effects of sampling inadequacy on the results in Appendix C. In the case of the SB community, on the other hand, the observed reversal of the expected pattern suggests a fundamentally different body size organization, as expected for a terrestrial community based on a single primary producer species, and with the largest proportion of parasit-oid–host links among the nine communities. The predicted correlation between $r(TL,$ size$)$ and $r(TL,$ size-ratio$)$ (Section 2.3) seen in model com-munities is also seen in the empirical ones (Fig. 5).

The expected right skew of log-size distributions is seen in only two communities (Table 5). Indeed, Fig. 7 shows that the nine communities have a variety of log-size distributions, including two that are clearly bimodal (YE and MS). Consistent with theoretical predictions (Section 2.3) the mean of the log-size-ratio distributions tends to be >0 (biased towards size-ratios >1) (Table 5, Fig. 7). Moreover, the commu-nities show a $\mu_{\text{log-size-ratio}}$ that is higher than that expected by (partially) ran-dom trophic linking given the observed size distribution (neutral assembly). Interestingly, the mean $\mu_{\text{log-size-ratio}}$ values of most of the communities lie within or close to the range seen for the model communities assembled using the LV model (i.e. 3.06 (± 1.08 SD) and 1.56 (± 0.15 SD), depending upon the parameter ω; Table 2). Also, all the observed log-size-ratio distributions show at most weak left skewness (see Fig. 7), with skewness values ranging from -1.9 (MS community) to 0.81 (SP community). This is consistent with the theoretical prediction (Section 2.3) that skewness of the log-size-ratio distribution is expected to show a weak signature of stability con-straints. In Appendix C, I show that sampling in adequacy partly explains why size and size-ratio distributions show weaker agreement with the pre-dictions than $r(TL,$ size$)$ and $r(TL,$ size-ratio$)$.

3.2.2 Stability of Observed versus Size-Randomized Communities

Table 6 summarizes the local stability properties of the nine communities (proportion of 2000 size-randomized food webs that had $\lambda_{\text{max}}(\mathbf{C})$ smaller than that of the observed one). As might be expected, the $\lambda_{\text{max}}(\mathbf{C})$ of the observed communities deviates more strongly from those of fully random-ized webs compared to the partially randomized ones. Overall, irrespective the scaling of equilibrium biomasses (parameter ν) or the intraspecific density dependence (γ), the observed size and size-ratio structure of the empirical food webs clearly tend to endow some level of stability. Moreover, most

Table 6 The Proportion of 2000 Randomized Community Food Webs that Were More Stable Than the Empirically Observed One (Had Smaller $\lambda_{max}(\mathbf{C})$), Given Various Values of the Scaling Exponents ν (Mass-Biomass Scaling) and γ (Scaling of Mass-Specific Intraspecific Density-Dependent Biomass Loss)

Randomization	γ	ν	Community								
			BS	CS	EW	GC	MS	SB	SP	TL	YE
Part	0	0	0	0.04	0.01	0	0	0.38	0	0	0.05
		0.25	0	0.64	0.13	0	0.008	0.47	0	0	0.67
	0.25	0	0.20	0.001	0.36	0.40	0.20	0.25	0.12	0.32	0.51
		0.25	0.58	0.13	0	0.49	0.53	0	0.15	0.02	0.93
Full	0	0	0	0.03	0.004	0	0	0.29	0	0	0.001
		0.25	0	0.55	0.08	0	0.005	0.37	0	0	0.19
	0.25	0	0.06	0	0.39	0.35	0.03	0.08	0.06	0.33	0.14
		0.25	0.35	0.19	0	0.48	0.60	0	0.10	0.002	0.86

Note that in most cases, $\lambda_{max}(\mathbf{C})$ values are $\ll 1$, that is, the observed webs, with their non-random size structure, are more stable.

of the cases where the stability of observed webs is no better or worse than random (values ≥ 0.5), appear when equilibrium biomasses are assumed to scale strongly with body mass ($\nu = 0.25$). This value of ν is expected under the energetic equivalence rule, but is rarely observed in real communities (Cyr et al., 1997a; Leaper et al., 1999; Marquet et al., 1995; Reuman et al., 2009; Sheldon et al., 1977).

Thus overall, these results provide strong evidence that the observed local food webs show signatures of dynamically constrained assembly in their body size structure.

4. DISCUSSION

In summary, I find that species' body sizes and consumer–resource size-ratios are likely to change systematically with assembly, which, depending upon the pattern or sequence of assembly, may be reflected in **(i)** increase in body size with trophic level, **(ii)** increase in consumer–resource size-ratio with trophic level, and eventually at extinction immigration equilibrium, **(iii)** a strongly right-skewed distribution of body sizes and **(iv)** a symmetric distribution of size-ratios with mean >1 (or >0 in log space). Most importantly, all these signatures of dynamically constrained assembly are likely to emerge irrespective of the distribution of body sizes (trait values) in the immigration pool because invasibility and stability constraints exert a strong filter on interaction strengths and thus body sizes and size-ratios. Because species' body sizes are relatively easy to measure in the field, the quantification of these signatures offers a simple method to gauge the importance of non-neutral processes underlying the assembly and persistence of real communities.

The predicted increase in sizes with trophic level due to stability constraints is a pattern that is commonly observed in natural communities, especially aquatic ones (Hildrew et al., 2007; Riede et al., 2011). This study appears to be the first to show that assembly dynamics can contribute to this aspect of community food web structure. Previous models have mainly invoked species' metabolic constraints and principles of biomass transfer across trophic levels, without the explicit consideration of interaction driven community assembly dynamics (Brown and Gillooly, 2003; Brown et al., 2004; Cohen, 2008).

The distribution of size-ratios along trophic levels is another signature of dynamically constrained assembly, and can explain why the traditional Eltonian paradigm (Elton, 1927) of invariance of size-ratios with trophic level does not always, or even typically, hold (Brose, 2010; Cohen and Fenchel, 1994; Riede et al., 2011). My results also differ from that of Jonsson and Ebenman (1998), who concluded that size-ratio ratios should decrease with trophic level due to stability constraints. Our results differ for two reasons. First, Jonsson and Ebenman only studied predator prey interactions with size-ratios >1. Second, they did not consider the context of community assembly and the tradeoff between invasibility and stability. All these factors act together to result in a gradual increase in size-ratios during assembly, culminating in a positive correlation between tropic level and size-ratio, as predicted (Section 2.3), demonstrated numerically (Section 2.5) and supported empirically (Section 3.2).

From the perspective of trait variation, I have shown that irrespective of the distribution of immigration probability with respect to body size, a right-skewed unimodal log-size distribution emerges through species invasion and extinction dynamics driven by interspecific trophic interactions. The skewness and unimodality of this distribution is a result of the tradeoff between the higher invasibility of smaller body-sized consumer species at early stages of assembly due to their higher mass-specific metabolic rate, and the stabilizing effects (in terms of local stability of the community) of invasion by larger sized species during later stages. That such a unimodal, right-skewed log-size distribution, which is often reported in the empirical literature (Jonsson et al., 2005; Leaper et al., 2001; Siemann et al., 1999; Stead et al., 2005), emerges due to stability constraints is an intriguing result. The origin of the local community's size distribution is an enduring problem in biology, and a variety of explanatory hypotheses have been proposed in the past. Only a small subset of these include interspecific interactions (Allen et al., 2006a). The theory developed in this paper combines the features of three classes of previous size distribution models that exclusively consider, (a) metabolic restrictions of resource use by species' individuals (e.g. Brown et al., 1993), (b) size-based constraints on speciation and dispersal (e.g. Etienne and Olff, 2004) and (c) interspecific interactions (e.g. Hutchinson, 1959).

However, I found little support for a predominance of right-skewness in empirical log-size distributions, which should be the result of the same processes that drive the emergence of the weighted generality-based food web structural signatures. I argue that this lack of right-skewness can be partly attributed to inadequate taxonomic sampling—there is strong evidence that size distributions become more right skewed as the inherent taxonomic bias towards larger organisms is mitigated (Blackburn and Gaston, 1994). Indeed, studies on local communities that have high taxonomic resolution typically find a right-skewed log-size distribution (Allen et al., 2006a) (these could not be included in this study due to lack of trophic interaction data). Also, as shown in Section 2, the local size distribution is dependent upon the size distribution of the immigration pool, and other factors such as the size dependence of speciation and immigration rate. It is currently difficult to determine whether these differ across the different communities studied here. But then why do taxonomic sampling biases also not render $r(TL,$ size) and $r(TL,$ size-ratio) undetectable? This is not surprising because taxonomic sampling *per se* does not affect detection of food web structure as much (Goldwasser and Roughgarden, 1997; Martinez et al., 1999).

The size-ratio distribution emerges concurrently with that of the size distribution. The interest in the effect of stability constraints on the distribution of size-ratios in local communities as a whole appears to have arisen relatively recently (Brose et al., 2006b; Emmerson and Raffaelli, 2004; Jonsson and Ebenman, 1998; Otto et al., 2007), perhaps partly due to increasing availability of food web datasets with body size information (Barnes et al., 2008; Brose et al., 2005). These studies have mainly focused on size-ratios that maximize the fitness or persistence of consumer–resource species pairs (Vasseur and McCann, 2005; Weitz and Levin, 2006; Yodzis and Innes, 1992) without considering assembly dynamics or multi-species coexistence stability. My results support previous conclusions that the size-ratio distribution is constrained by community stability constraints (Brose et al., 2006b; Emmerson and Raffaelli, 2004; Jonsson and Ebenman, 1998). However, it differs from these studies in that it includes a larger spectrum of size-ratios <1, such as those between parasitoids or parasites and their hosts (but see Otto et al., 2007). The fundamental mechanism that interlinks size-ratios to community dynamics identified by those studies and this one are the same (the stabilizing effect of the relatively weaker mass-specific interaction strengths associated with larger size-ratios). However, this study shows that because the size-ratio distribution is tightly linked to that of the community's sizes, it is difficult to separate the feedback between size-ratios *per se* and community stability. For example, assuming that trophic linking probability (p_c) is independent of body sizes, changes in the size-ratio distribution will partly reflect the realized trophic linking due to differences in the sizes of the immigrant pool and the local community.

While I did find support for the effect of stability constraints on the size-ratio distribution, I also found evidence that these apparent fits to the theoretical predictions were *overestimated* by trophic link sampling biases against size-ratios ≤ 1 (Appendix C). Such a bias is understandable because of the practical difficulties associated with detecting interactions wherein consumers are much smaller than their resources (such as parasites and parasitoids). Nevertheless, biases in current food web datasets may seriously hinder our understanding of community stability dynamics because a spectrum of potentially important trophic interactions (mainly with size-ratios <1) remains undersampled.

In general, while the incidences of different signatures (in terms of the percentage of communities showing the expected patterns in each feature) were consistent with those seen in model communities (cf. Table 2), their strengths (e.g. values of the correlation coefficients) lay at or beyond the lower ends of those of model communities. This is not surprising, given the multiple sources of noise inherent in community datasets, combined with the fact that the model communities were assembled without spatial or temporal environmental variation, both of which are expected to have strong effects on real communities. It is also possible that there is a greater influence of biogeographically neutral processes such as immigration and stochastic extinction (which would weaken signatures of interaction driven stability constraints on food web structural features) than what was simulated. I also tested for the possibility that some or all of the observed signatures were artefacts of trophic link sampling bias (Appendix C). I found that this was not the case; in fact, there was some evidence that the strengths and incidences of the observed signatures may actually be underestimated due to sampling bias.

5. CONCLUSIONS

In this paper, I have studied the effect of species' average adult body size, a trait that constrains the magnitudes of mass-specific trophic interactions, on local community assembly. The results show that under biologically feasible assumptions about allometric constraints and community assembly patterns (or assembly sequences), certain body size and consumer–resource size-ratio distributions are likely to emerge due to a combination of which trait values facilitate invasion, and which allow multi-species stable coexistence. I also find strong empirical support for the theoretical results. Body size allows individual level properties to be mapped on to population interactions in assembly dynamics, providing fresh insights into the dynamical organization of natural communities and ecosystems. This is particularly important because the use of species body size data provides a valuable tool to decipher the rather daunting complexity of dynamically assembling natural communities.

ACKNOWLEDGEMENTS

I would like to thank Catalina Estrada, Matthew Leibold, Daniel Bolnick, Lauren Myers, Timothy Keitt, Evan Economo, Richard Law, Sahotra Sarkar and Van Savage for comments, and suggestions that improved this work immensely. I would also like to thank Jennifer Dunne for providing the community food web data.

APPENDIX A. THE SIZE SCALING OF SPECIES' EQUILIBRIUM ABUNDANCES

The vector of equilibrium biomass densities $\hat{x}$ can be found by solving the system

$$-Ax = b - d \tag{A1}$$

where x is the vector of biomasses (x_i), b of the biomass production rates, d of the density-independent biomass loss rates, and A the $n \times n$ matrix of interaction coefficients. Using Cramer's rule to solve equation, the equilibrium density of the ith species' population can be expressed as a ratio of determinants,

$$\hat{x}_i = -\frac{\det(A_i)}{\det(A)} \tag{A2}$$

where A_i is the is the matrix formed by replacing the ith column of A with the vector $b - d$. Now, Leibniz's formula states that for any $n \times n$ matrix Z, $\det(Z) = \sum_{\sigma \in P_n} \text{sgn}(\sigma) \prod_{i=1}^{n} z_{i\sigma(i)}$, where P_n is the set of all possible permutations of the integers 1 to n, and $\text{sgn}(\sigma)$ denotes the signature of the permutation product σ. Thus, while the solution to Eq. (A1) is analytically tractable, deciphering the body size scaling of $\hat{x}$ for an arbitrary community size (n) is not, because both the numerator as well as denominator of Eq. (A2) consist of sums and differences of the $n!$ different permutations of elements (taken n at a time) of the matrices A_i and A, respectively. Hence instead, I determine the relationship numerically by examining the biomass abundances of species in model communities at IEE. The results are shown in Fig. A1. In general, there is a tight linear relationship between ν and γ of the form $\nu = z\gamma + c$ with z lying between 0.5–0.6 and c between 0.1–015, depending upon the choice of the parameter k (log-size-ratio at which consumption intensity peaks). There appears to be no strong dependence of the scaling between ν and γ on ω (distribution of body sizes in the immigrant

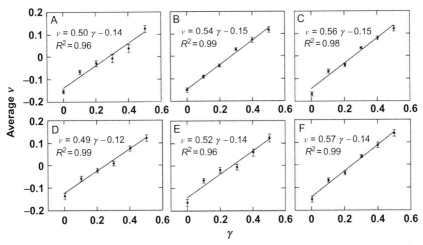

Figure A1 The body size dependence of species' equilibrium biomass abundances ($\hat{x}$'s) in model communities. Each plot shows changes in average ν with increasing values of γ. Each value of average ν ($\pm 99\%$ confidence intervals) for a given γ was calculated from 100 replicated communities at IEE. The least-squares regression line and the associated equation are shown for each plot. All the regressions are highly significant ($p < 0.0001$). Plots (A–C) are for $\omega = 1$ but with orders of magnitude increase in k (10^{-3}, 1 and 10^3, respectively, i.e., size-ratios for peak consumption intensity ranging from 0.001 to 1000), and (D–F) for $\omega = 2$ and the same range of k. Assembly simulation methodology is described in the main text.

pool). Variation in other non-allometric assembly model parameters (see main text) does not change the linearity of this scaling or the range of z and c (results not shown).

APPENDIX B. PARAMETER SENSITIVITY

B.1 Metabolic Scaling Parameters

The fact that I have used a metabolic scaling exponent of ¾ throughout this study might suggest that this scaling exponent relationship is truly universal. Deviations from this exponent can be found within many taxonomic groups (DeLong et al., 2010; Glazier, 2005; Peters, 1986). However, this relationship does hold well across taxonomic groups spanning large orders of magnitude variation in body sizes, which is typically how local communities are composed. Accordingly, the analysis above also encompasses >20 orders of magnitude variation in body size. I have also not included body temperature dependence into the allometry of metabolism for parameterizing the LV model (Dell et al., 2011, 2014; Gilloolly et al., 2001; Pawar et al., 2015). This can be justified to some extent because the vast majority of species in local communities belong to the same metabolic category (ectotherms), so temperature change would similarly change inter- and intraspecific interaction rates across species, leaving the qualitative results unchanged (assuming all species had identical sensitivity to temperature change). Nevertheless, because temperature-dependence can differ significantly between interacting species (Dell et al., 2014), a more realistic body size-based network assembly model should include temperature dependence to account for climatic variation across space, or the effects of climate change.

Another potential problem is the assumption that species' intraspecific density dependence scales with mass according to Eq. (8). I am unaware of any empirical data on the actual scaling of intraspecific density dependence with body mass. The values $\frac{1}{4}$–$\frac{1}{2}$ for γ were chosen because they yield empirically feasible values for the scaling exponent of equilibrium biomasses (Appendix A). Hence, until relevant empirical data become available, the model developed in this study should be viewed for what it is: a possible set of mechanisms for the emergence of biomass scaling in local communities. Note also that as long as equilibrium biomasses in local communities scale with ν between 0 and 0.25, and empirical data strongly support this (Brown and Gillooly, 2003; Leaper et al., 1999), the results of this study remain qualitatively unchanged.

B.2 Scaling of Search Rates

Because variation in encounter rates can effectively change the location of the peak (k) of the consumption intensity function φ_{ij} (Eq. 7), here I examine sensitivity of the results to variation in its two parameters (k and s). In the main text, I explained the reasons for fixing k (the size-ratio at which consumption rate peaks) and s (the speed with which consumption rate declines away from k) at 1 and 0.05, respectively, instead of varying them according to the type of trophic interaction, habitat or the types of organisms involved. To examine the sensitivity of the observed signatures of stability constraints on food web structure to variation in these parameters, here I examine the key signature of dynamically constrained assembly in empirical/communities, $r(TL,$ size) as follows. Using the methods described in the main text, I recalculated $r(TL,$ size) for each of 100 logarithmically spaced values for k between 10^{-3} and 10^3 (consumer 1000 times smaller to 1000 times larger than resource) and 100 linearly placed values for s between 0.01 and 0.1. This range for k covers most of the size-ratios considered to be 'optimal' (in the sense of viable or evolutionarily stable strategies for consumers) in previous studies on body size-based consumer–resource interactions (Cohen, 2008; Weitz and Levin, 2006), and accommodates potential differences in k across the most common trophic interaction types seen in food webs (i.e. predator–prey, herbivore–plant and parasitoid–host) (Brose et al., 2006a). The upper range for s (0.1) was chosen because as $s \rightarrow 0.1$, consumption intensity falls to zero within the range of the size-ratios observed across the nine real communities (Fig. 7). Hence, values larger than this are unfeasible because if a size-ratio is observed, no matter how extreme, it *must* be associated with some level of biomass acquisition by the consumer. The lower limit of s (0.01) was chosen as some value arbitrarily higher than 0 because φ_{ij} becomes flat as $s \rightarrow 0$ and the effects of size-ratios on trophic link strengths are eliminated (Fig. 2). The simulations results are also similarly robust to considerable variation in k and s (Table 1).

Figure B1 shows the resulting variation in the strength of $r(TL, size)$. Clearly, this signature and the others correlated with it (Figs. 5 and 6) are robust to changes in k, with the correlation coefficient remaining similar across its entire range, as long as $s < 0.05$ or so. As $s \rightarrow 0.1$ however, the signatures either become stronger or weaker, depending upon the community and the value of k. This is the effect of elimination of the trophic links associated with extreme size-ratios (because φ_{ij} becomes zero), which strongly biases the interaction data towards trophic links with size-ratios close to k. However, in none of the cases are sign of the relationship reversed, and the value of the correlation coefficient remains within a narrow range across the entire parameter space. Also, the SB community shows a different pattern from the others; whereas the detectability of the correlation signatures becomes weaker as k decreases all the others, it in fact peaks at $k = 10^{-3}$ in the SB community. This is because SB is the only community with a significant number of host-parasitoid interactions. Thus, changing k according to the interaction type makes the signatures of constrained assembly more detectable—k should in fact be $\ll 0$ in host–parasitoid interactions. Similarly, k is typically ≥ 0 in predator–prey interactions, which dominate all communities other than SB, thus explaining why strengths of the signatures increase towards $k = 10^3$ in them. Thus overall, the strengths of the observed signatures, and the main conclusions of this study would be much stronger if trophic link-specific values for k were used. Another community that stands out is EW, where the strengths of both signatures peak at $k = 1$ ($\log_{10}(k) = 0$ in the figures); this is it is the community with the most balanced size-ratio distribution (in terms of the representation of size-ratios < 1; Fig. 7).

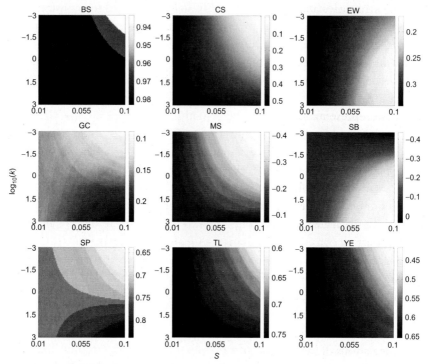

Figure B1 The effect of variation in the parameters *k* and *s* on the predicted measure *r(TL, size)*, a key food web structural signature of dynamically constrained assembly, across nine real communities.

APPENDIX C. DO SAMPLING BIASES AFFECT THE EMPIRICAL RESULTS?

Here, I consider to what extent the above empirical results are affected by sampling inadequacies. This is important because collecting trophic interaction data are an extremely difficult and time labor-intensive task, and at best it is possible to quantify only a subset of the interaction network of any local community (Berlow et al., 2004; Goldwasser and Roughgarden, 1997; Martinez et al., 1999). Errors in community food web datasets arise from two main sources: lack of taxonomic sampling and lack of adequate trophic link sampling. As discussed above, the former probably affects the observed shapes of size distribution and size-ratio distributions, and potentially, food web structural features as well. However, quantifying this source of error is beyond the scope of this study because it would require sampling effort and species accumulation data from each of the local communities. Quantifying sampling inadequacies in trophic link data is a somewhat more tractable proposition, especially because given a sample of species from a local community, the potential distribution of trophic links can be estimated. Furthermore, certain biases in trophic link data are well documented. For example, numerous studies have shown that most food web datasets have an underrepresentation of parasite–host interactions, as well as all trophic interactions where both consumers and resources are small in size (Hechinger et al., 2011; Kuris et al., 2008; Lafferty et al., 2008; Leaper and Huxham, 2002; Memmott et al., 2000; Thompson et al., 2005). Hence here, I consider the effect of trophic link sampling bias on this paper's results.

A number of food web structural features are known to be especially sensitive to inadequate sampling of interactions (Goldwasser and Roughgarden, 1997; Martinez et al., 1999). To obtain a measure of sampling inadequacy across communities, I selected a suite of food web structural characteristics were found to be sensitive to sampling effort by Goldwasser and Roughgarden (Goldwasser and Roughgarden, 1997). These features are connectance (C_T), average generality of consumers ($\bar{G}$), average trophic chain length ($\bar{T}_c$) and omnivory degree (O_{deg}). O_{deg} is the mean of the standard deviations of each consumer species' trophic height (standard deviation of the lengths of all the paths to the species from its basal species; Goldwasser and Roughgarden, 1997; Martinez et al., 1999). Other features such as the number and maximum of trophic chain lengths are also sampling sensitive, but are directly related to one or more of these selected measures, and were not included. Nevertheless, the selected features are still (albeit indirectly) interdependent (for example O_{deg} will typically increase with $\bar{T}_c$); hence, I used a Principal Component Analysis (PCA) to combine them into a single pseudo-variable that can be used as an index of sampling sensitivity. The correlation of strengths of the ostensible signatures of stability constraints across communities with this index would then provide insights into the effects of sampling error. If the signatures become stronger with increasing values of the index, it would suggest that the above results would have been stronger with better sampling. On the other hand, if the signatures become weaker, it would suggest the above results are an artefact of sampling bias.

Data normality is a central assumption of PCA; because the distributions of these food web structural features across communities were somewhat skewed, the data were square-root transformed, which resulted in approximately normal distributions of all four variables (determined using the Lilliefors test). The PCA did indeed result in a single significant component accounting for 86% of variance (with loadings in the following order: $\bar{T}_c >$ $O_{deg} > C_T > \bar{G}$). The scores of this first component were thereafter used as an index of trophic link sampling inadequacy, with which the correlation of observed strengths of the key signature of dynamically constrained assembly $r(TL, \text{size})$ (Section 2.3; $r(TL, \text{size-ratio})$ is correlated with this, see Fig. 5) across communities was calculated using the spearman rank coefficient. The correlation was weakly positive ($R^2 = 0.14$), but insignificant ($p = 0.16$). At the very least, this indicates that the above results are not an artefact of sampling bias.

I also examined the correlation of $\mu_{\text{log-size-ratio}}$ with trophic link sampling inadequacy. Stability constraints are expected to increase the community's $\mu_{\text{log-size-ratio}}$ (Section 2.3), which should therefore be positively correlated with link sampling inadequacy index if sampling bias decreases the observed strength of this signature. However, this correlation was found to be weakly negative ($R^2 = 0.34$) but insignificant ($p = 0.21$). Nevertheless, this suggests that unlike the food web structural signatures, the observed fit of the $\mu_{\text{log-size-ratio}}$ to theoretical predictions may in fact be affected by sampling bias. In other words, while the observed $\mu_{\text{log-size-ratios}}$ are qualitatively consistent with theoretical predictions, their actual values may be overestimated because of an underrepresentation of links with size-ratios <1 (typically observed in the form of parasite–host, parasitoid–host or herbivore–plant interactions). This conclusion is further supported by the fact that that the skewness of the log-size-ratio distribution is also positively and significantly correlated ($R^2 = 0.67$, $p = 0.009$) with the link sampling inadequacy index. Because a higher positive skewness means an inordinately high concentration of values at the right half of the distribution (where size-ratios >1) this indicates that the sampling bias may in fact be against size-ratios ≤ 1, as has been suggested in recent studies highlighting the lack of data on host–parasite links. This is also supported by the fact that an overwhelming majority ($\sim$90%) of links across the nine communities were of predator–prey interactions, which typically have size-ratios >1.

REFERENCES

Aljetlawi, A.A., Sparrevik, E., Leonardsson, K., 2004. Prey-predator size-dependent functional response: derivation and rescaling to the real world. J. Anim. Ecol. 73, 239–252.

Allen, A.P., Gillooly, J.F., 2006. Assessing latitudinal gradients in speciation rates and biodiversity at the global scale. Ecol. Lett. 9, 947–954.

Allen, A.P., Savage, V.M., 2007. Setting the absolute tempo of biodiversity dynamics. Ecol. Lett. 10, 637–646.

Allen, C.R., Garmestani, A.S., Havlicek, T.D., Marquet, P.A., Peterson, G.D., Restrepo, C., Stow, C.A., Weeks, B.E., 2006a. Patterns in body mass distributions: sifting among alternative hypotheses. Ecol. Lett. 9, 630–643.

Allen, A.P., Gillooly, J.F., Savage, V.M., Brown, J.H., 2006b. Kinetic effects of temperature on rates of genetic divergence and speciation. Proc. Natl. Acad. Sci. U.S.A. 103, 9130–9135.

Allesina, S., Tang, S., 2012. Stability criteria for complex ecosystems. Nature 483, 205–208.

Allesina, S., Alonso, D., Pascual, M., 2008. A general model for food web structure. Science 3, 658–661.

Barnes, C., et al., 2008. Predator and body sizes in marine food webs. Ecology 89, 881.

Bascompte, J., Melian, C.J., 2005. Simple trophic modules for complex food webs. Ecology 86, 2868–2873.

Bascompte, J., Stouffer, D.B., 2009. The assembly and disassembly of ecological networks. Philos. Trans. R. Soc. Lond. B Biol. Sci. 364, 1781–1787.

Bastolla, U., Lassig, M., Manrubia, S.C., Valleriani, A., 2005. Biodiversity in model ecosystems, II: species assembly and food web structure. J. Theor. Biol. 235, 531–539.

Berlow, E.L., et al., 2004. Interaction strengths in food webs: issues and opportunities. J. Anim. Ecol. 73, 585–598.

Blackburn, T.M., Gaston, K.J., 1994. Animal body-size distributions change as more species are described. Proc. Biol. Sci. 257, 293–297.

Brose, U., 2010. Body-mass constraints on foraging behaviour determine population and food-web dynamics. Funct. Ecol. 24, 28–34.

Brose, U., et al., 2005. Body sizes of consumers and their resources. Ecology 86, 2545.

Brose, U., et al., 2006a. Consumer-resource body-size relationships in natural food webs. Ecology 87, 2411–2417.

Brose, U., Williams, R.J., Martinez, N.D., 2006b. Allometric scaling enhances stability in complex food webs. Ecol. Lett. 9, 1228–1236.

Brose, U., Ehnes, R.B., Rall, B.C., Vucic-Pestic, O., Berlow, E.L., Scheu, S., 2008. Foraging theory predicts predator-prey energy fluxes. J. Anim. Ecol. 77, 1072–1078.

Brown, J.H., Gillooly, J.F., 2003. Ecological food webs: high-quality data facilitate theoretical unification. Proc. Natl. Acad. Sci. U.S.A. 100, 1467–1468.

Brown, J.H., Marquet, P.A., Taper, M.L., 1993. Evolution of body-size: consequences of an energetic definition of fitness. Am. Nat. 142, 573–584.

Brown, J.H., Gillooly, J.F., Allen, A.P., Savage, V.M., West, G.B., 2004. Toward a metabolic theory of ecology. Ecology 85, 1771–1789.

Cattin, M.F.M., Bersier, L.F., Banasek-Richter, C., Baltensperger, R., Gabriel, J.P., Banas, C., 2004. Phylogenetic constraints and adaptation explain food-web structure. Nature 427, 835–839.

Clauset, A., Erwin, D.H., 2008. The evolution and distribution of species body size. Science 321, 399–401.

Cohen, J.E., 2008. Body sizes in food chains of animal predators and parasites. In: Hildrew, A.G., Raffaelli, D.G., Edmonds-Brown, R. (Eds.), Body Size: The Structure and Function of Aquatic Ecosystems. Cambridge University Press, Cambridge, UK, pp. 306–325.

Cohen, J.E., Fenchel, T., 1994. Marine and continental food webs: three paradoxes? Proc. R. Soc. B Biol. Sci. 343, 57–69.

Cohen, J.E., Jonsson, T., Carpenter, S.R., 2003. Ecological community description using the food web, species abundance, and body size. Proc. Natl. Acad. Sci. U.S.A. 100, 1781–1786.

Cohen, J.E., Jonsson, T., Müller, C.B., Godfray, H.C.J., Savage, V.M., Muller, C.B., 2005. Body sizes of hosts and parasitoids in individual feeding relationships. Proc. Natl. Acad. Sci. U.S.A. 102, 684–689.

Cyr, H., Downing, J., Peters, R., 1997a. Density-body size relationships in local aquatic communities. Oikos 79, 333–346.

Cyr, H., Peters, R.R.H., Downing, J.J.A., 1997b. Population density and community size structure: comparison of aquatic and terrestrial systems. Oikos 80, 139–149.

Dell, A.I., Pawar, S., Savage, V.M., 2011. Systematic variation in the temperature dependence of physiological and ecological traits. Proc. Natl. Acad. Sci. U.S.A. 108, 10591–10596.

Dell, A.I., Pawar, S., Savage, V.M., 2014. Temperature dependence of trophic interactions are driven by asymmetry of species responses and foraging strategy. J. Anim. Ecol. 83, 70–84.

DeLong, J., 2014. The body-size dependence of mutual interference. Biol. Lett. 10.

DeLong, J.P., Okie, J.G., Moses, M.E., Sibly, R.M., Brown, J.H., 2010. Shifts in metabolic scaling, production, and efficiency across major evolutionary transitions of life. Proc. Natl. Acad. Sci. U.S.A. 107, 12941–12945.

Dial, K.P., Marzluff, J.M., 1988. Are the smallest organisms the most diverse? Ecology 69, 1620–1624.

Economo, E.P., Kerkhoff, A.J., Enquist, B.J., 2005. Allometric growth, life-history invariants and population energetics. Ecol. Lett. 8, 353–360.

Elton, C.S., 1927. Animal Ecology. Macmillan, New York.

Emmerson, M.C., Raffaelli, D., 2004. Predator-prey body size, interaction strength and the stability of a real food web. J. Anim. Ecol. 73, 399–409.

Enquist, B.J., Niklas, K.J., 2001. Invariant scaling relations across tree-dominated communities. Nature 410, 655–660.

Etienne, R.S., Olff, H., 2004. How dispersal limitation shapes species-body size distributions in local communities. Am. Nat. 163, 69–83.

Fahimipour, A.K., Hein, A.M., 2014. The dynamics of assembling food webs. Ecol. Lett. 17, 606–613.

Fukami, T., 2004. Community assembly along a species pool gradient: implications for multiple-scale patterns of species diversity. Popul. Ecol. 46, 137–147.

Giacomini, H.C., Shuter, B.J., de Kerckhove, D.T., Abrams, P.A., 2013. Does consumption rate scale superlinearly? Nature 493, E1–E2.

Gillooly, J.F., Brown, J.H., West, G.B., Savage, V.M., Charnov, E.L., 2001. Effects of size and temperature on metabolic rate. Science 293, 2248–2251.

Glazier, D.S., 2005. Beyond the "3/4-power law": variation in the intra- and interspecific scaling of metabolic rate in animals. Biol. Rev. Camb. Philos. Soc. 80, 611–662.

Goldwasser, L., Roughgarden, J., 1997. Sampling effects and the estimation of food-web properties. Ecology 78, 41.

Hall, S.J.J., Raffaelli, D., 1991. Food-web patterns: lessons from a species-rich web. J. Anim. Ecol. 60, 823–842.

Hechinger, R.F., Lafferty, K.D., Dobson, A.P., Brown, J.H., Kuris, A.M., 2011. A common scaling rule for abundance, energetics, and production of parasitic and free-living species. Science 333, 445–448.

Hein, A.M., Hou, C., Gillooly, J.F., 2012. Energetic and biomechanical constraints on animal migration distance. Ecol. Lett. 15, 104–110.

Hildrew, A.G., Raffaelli, D.G., Edmonds-Brown, R., 2007. Body Size: The Structure and Function of Aquatic Ecosystems. Cambridge University Press, Cambridge, New York.

Hutchinson, G.E., 1959. Homage to Santa Rosalia or why are there so many kinds of animals? Am. Nat. XCIII 145–159.

Jetz, W., Carbone, C., Fulford, J., Brown, J.H., 2004. The scaling of animal space use. Science 306, 266–268.

Jonsson, T., Ebenman, B., 1998. Effects of predator-prey body size ratios on the stability of food chains. J. Theor. Biol. 193, 407–417.

Jonsson, T., Cohen, J., Carpenter, S., 2005. Food webs, body size, and species abundance in ecological community description. Adv. Ecol. Res. 36.

Kleiber, M., 1961. The Fire of Life: An Introduction to Animal Energetics. Wiley, New York.

Kozlowski, J., Gawelczyk, A.T., 2002. Why are species' body size distributions usually skewed to the right? Funct. Ecol. 16, 419–432.

Kuris, A.M., et al., 2008. Ecosystem energetic implications of parasite and free-living biomass in three estuaries. Nature 454, 515–518.

Lafferty, K.D., et al., 2008. Parasites in food webs: the ultimate missing links. Ecol. Lett. 11, 533–546.

Leaper, R., Huxham, M., 2002. Size constraints in a real food web: predator, parasite and prey body-size relationships. Oikos 99, 443–456.

Leaper, R., Raffaelli, D., Letters, E., 1999. Defining the abundance body-size constraint space: data from a real food web. Ecol. Lett. 2, 191–199.

Leaper, R., Raffaelli, D., Emes, C., Manly, B., 2001. Constraints on body-size distributions: an experimental test of the habitat architecture hypothesis. J. Anim. Ecol. 70, 248–259.

Loeuille, N., Loreau, M., 2006. Evolution of body size in food webs: does the energetic equivalence rule hold? Ecol. Lett. 9, 171–178.

Marquet, P.A., Navarrete, S.A.S., Castilla, J.J.C., 1995. Body size, population density, and the energetic equivalence rule. J. Anim. Ecol. 64, 325–332.

Martinez, N.D., Hawkins, B.A., Dawah, H.A., Feifarek, B.P., 1999. Effects of sampling effort on characterization of food-web structure. Ecology 80, 1044–1055.

Marzluff, J.M., Dial, K.P., 1991. Life-history correlates of taxonomic diversity. Ecology 72, 428–439.

May, R.M., 1974. Stability and Complexity in Model Ecosystems. Princeton University Press, Princeton, NJ.

May, R.M., 2009. Food-web assembly and collapse: mathematical models and implications for conservation. Philos. Trans. R. Soc. Lond. B Biol. Sci. 364, 1643–1646.

McCoy, M.W., Gillooly, J.F., 2008. Predicting natural mortality rates of plants and animals. Ecol. Lett. 11, 710–716.

Memmott, J., Martinez, N.D., Cohen, J.E., 2000. Predators, parasitoids and pathogens: species richness, trophic generality and body sizes in a natural food web. J. Anim. Ecol. 69, 1–15.

Niklas, K.J., Enquist, B.J., 2001. Invariant scaling relationships for interspecific plant biomass production rates and body size. Proc. Natl. Acad. Sci. U.S.A. 98, 2922–2927.

O'Dwyer, J.P., Lake, J.K., Ostling, A., Savage, V.M., Green, J.L., 2009. An integrative framework for stochastic, size-structured community assembly. Proc. Natl. Acad. Sci. U.S.A. 106, 6170–6175.

Otto, S.B., Rall, B.C., Brose, U., 2007. Allometric degree distributions facilitate food-web stability. Nature 450, 1226–1229.

Pawar, S., 2009. Community assembly, stability and signatures of dynamical constraints on food web structure. J. Theor. Biol. 259, 601–612.

Pawar, S., Dell, A.I., Savage, V.M., 2012. Dimensionality of consumer search space drives trophic interaction strengths. Nature 486, 485–489.

Pawar, S., Dell, A.I., Savage, V.M., 2013. Pawar et al. reply. Nature 493, E2–E3.

Pawar, S., Dell, A.I., Savage, V.M., 2015. From metabolic constraints on individuals to the eco-evolutionary dynamics of ecosystems. In: Belgrano, A., Woodward, G., Jacob, U. (Eds.), Aquatic Functional Biodiversity: An Eco-Evolutionary Approach. Elsevier, Amsterdam, in press.

Persson, L., Leonardsson, K., de Roos, A.M., Gyllenberg, M., Christensen, B., 1998. Ontogenetic scaling of foraging rates and the dynamics of a size-structured consumer-resource model. Theor. Popul. Biol. 54, 270–293.

Peters, R., 1986. The Ecological Implications of Body Size. Cambridge University Press, Cambridge.

Post, W.M., Pimm, S.L., 1983. Community assembly and food web stability. Math. Biosci. 64, 169–182.

Raffel, T.R., Martin, L.B., Rohr, J.R., 2008. Parasites as predators: unifying natural enemy ecology. Trends Ecol. Evol. 23, 610–618.

Reuman, D.C., et al., 2009. Allometry of body size and abundance in 166 food webs. Adv. Ecol. Res. 41, 1–44.

Riede, J.O., Brose, U., Ebenman, B., Jacob, U., Thompson, R., Townsend, C.R., Jonsson, T., 2011. Stepping in Elton's footprints: a general scaling model for body masses and trophic levels across ecosystems. Ecol. Lett. 14, 169–178.

Rossberg, A.G., Ishii, R., Amemiya, T., Itoh, K., 2008. The top-down mechanism for body-mass-abundance scaling. Ecology 89, 567–580.

Roughgarden, J., 1996. Theory of Population Genetics and Evolutionary Ecology: An Introduction. Prentice Hall, Upper Saddle River, NJ.

Savage, V.M., Gilloly, J.F., Brown, J.H., Charnov, E.L., Gillooly, J.F., West, G.B., 2004. Effects of body size and temperature on population growth. Am. Nat. 163, 429–441.

Schmidt-Nielsen, K., 1984. Scaling, Why is Animal Size so Important? Cambridge University Press, Cambridge; New York.

Sheldon, R.W., Sutcliffe, W.H., Paranjape, M.A., 1977. Structure of pelagic food chain and relationship between plankton and fish production. J. Fish. Res. Board Can. 34, 2344–2353.

Siemann, E., Tilman, D., Haarstad, J., 1999. Abundance, diversity and body size: patterns from a grassland arthropod community. J. Anim. Ecol. 68, 824–835.

Springer, M.D., 1979. The Algebra of Random Variables. Wiley, New York.

Stead, T.K., Schmid-Araya, J.M., Schmid, P.E., Hildrew, A.G., 2005. The distribution of body size in a stream community: one system, many patterns. J. Anim. Ecol. 74, 475–487.

Strobeck, C., 1973. N species competition. Ecology 54, 650–654.

Tang, S., Pawar, S., Allesina, S., 2014. Correlation between interaction strengths drives stability in large ecological networks. Ecol. Lett. 17, 1094–1100.

Thompson, R.M., Mouritsen, K.N., Poulin, R., 2005. Importance of parasites and their life cycle characteristics in determining the structure of a large marine food web. J. Anim. Ecol. 74, 77–85.

Vasseur, D.A., McCann, K.S., 2005. A mechanistic approach for modeling temperature-dependent consumer-resource dynamics. Am. Nat. 166, 184–198.

Virgo, N., Law, R., Emmerson, M., 2006. Sequentially assembled food webs and extremum principles in ecosystem ecology. J. Anim. Ecol. 75, 377–386.

Vucic-Pestic, O., Rall, B.C., Kalinkat, G., Brose, U., 2010. Allometric functional response model: body masses constrain interaction strengths. J. Anim. Ecol. 79, 249–256.

Warren, P.H., 1989. Spatial and temporal variation in the structure of a freshwater food web. Oikos 55, 299–311.

Weitz, J.S., Levin, S.A., 2006. Size and scaling of predator-prey dynamics. Ecol. Lett. 9, 548–557.

Woodward, G., Speirs, D.C., Hildrew, A.G., 2005. Quantification and resolution of a complex, size-structured food web. Adv. Ecol. Res. 36, 85–135.

Yodzis, P., 1988. The indeterminacy of ecological interactions as perceived through perturbation experiments. Ecology 69, 508–515.

Yodzis, P., Innes, S., 1992. Body size and consumer-resource dynamics. Am. Nat. 139, 1151–1175.

CHAPTER NINE

Scaling from Traits to Ecosystems: Developing a General Trait Driver Theory via Integrating Trait-Based and Metabolic Scaling Theories

Brian J. Enquist*,†,1,2, Jon Norberg‡, Stephen P. Bonser§,
Cyrille Violle*,¶, Colleen T. Webb‖, Amanda Henderson*,
Lindsey L. Sloat*, Van M. Savage†,#,**,1
*Department of Ecology and Evolutionary Biology, University of Arizona, Bioscience West, Tucson, Arizona, USA
†Santa Fe Institute, Santa Fe, New Mexico, USA
‡Stockholm Resilience Centre, Stockholm University, Stockholm, Sweden
§Evolution and Ecology Research Centre and School of Biological, Earth and Environmental Sciences, University of New South Wales, Sydney, New South Wales, Australia
¶CNRS, Centre d'Ecologie Fonctionnelle et Evolutive, UMR 5175, Montpellier, France
‖Department of Biology, Colorado State University, Fort Collins, Colorado, USA
#Department of Biomathematics, David Geffen School of Medicine at UCLA, Los Angeles, California, USA
**Department of Ecology and Evolutionary Biology, UCLA, Los Angeles, California, USA
2Corresponding author: e-mail address: benquist@email.arizona.edu

Contents

---

[1] Contributed equally to this work.

Advances in Ecological Research, Volume 52
ISSN 0065-2504
http://dx.doi.org/10.1016/bs.aecr.2015.02.001

Abstract

Aim: More powerful tests of biodiversity theories need to move beyond species richness and explicitly focus on mechanisms generating diversity via trait composition. The rise of trait-based ecology has led to an increased focus on the distribution and dynamics of traits across broad geographic and climatic gradients and how these distributions influence ecosystem function. However, a general theory of trait-based ecology, that can apply across different scales (e.g. species that differ in size) and gradients (e.g. temperature), has yet to be formulated. While research focused on metabolic and allometric scaling theory provides the basis for such a theory, it does not explicitly account for differences in traits within and across taxa, such as variation in the optimal temperature for growth. Here we synthesize trait-based and metabolic scaling approaches into a framework that we term 'Trait Driver Theory' or TDT. It shows that the shape and dynamics of trait and size distributions can be linked to fundamental drivers of community assembly and how the community will respond to future drivers. To assess predictions and assumptions of TDT, we review several theoretical studies and recent empirical studies spanning local and biogeographic gradients. Further, we analyze how the shift in trait distributions influences ecosystem processes across an elevational gradient and a 140-year-long ecological experiment. We show that TDT provides a baseline for (i) recasting the predictions of ecological theories based on species richness in terms of the shape of trait distributions and (ii) integrating how specific traits, including body size, and functional diversity then 'scale up' to influence ecosystem functioning and the dynamics of species assemblages across climate gradients. Further, TDT offers a novel framework to integrate trait, metabolic/allometric, and species-richness-based approaches to better predict functional biogeography and how assemblages of species have and may respond to climate change.

1. INTRODUCTION

Understanding and explaining species richness patterns have had far-reaching influence on the development of ecology. Biodiversity science strives to understand the drivers and consequences of variation in the number of species and how species abundances change across spatial and temporal scales (MacArthur, 1972; Rosenzweig, 1995). These changes in species richness have also been linked with changes in ecosystem functioning. The Biodiversity–Ecosystem Functioning (BEF) hypothesis states that ecosystems with greater biodiversity are more productive and stable (Naeem et al., 1994; Tilman, 2001; Tilman et al., 1997). Attempts to answer these questions have led to debates that polarized the field (Wardle, 2002), and a growing consensus that species numbers alone do not inform us about all important aspects of ecosystem functioning and community responses to environmental change (Chapin et al., 2000; Díaz and Cabido, 2001; Diaz et al., 2007; Stevens et al., 2003).

More recently, trait-based approaches have focused on recasting classical questions from the species richness literature (Hillebrand and Matthiessen, 2009; Lamanna et al., 2014; Lavorel and Garnier, 2002; McGill et al., 2006; Violle et al., 2007). Instead of species richness, there is an attempt to focus on functional traits and diversity in trait values (Díaz and Cabido, 2001; Lavorel and Garnier, 2002; Mason et al., 2005; Petchey and Gaston, 2002; Roscher et al., 2012). In addition, metabolic scaling theory or MST has focused on the central role of body size as a critical driver of ecological, ecosystem, and evolutionary patterns and processes (Enquist et al., 1998, 2003; Gillooly et al., 2005; Savage et al., 2004). One could also ask about diversity in the number and/or range of trait or body size values, and to some degree, this depends on how traits are defined. As discussed by Dell et al. (2015) and Pawar (2015) in this issue, the premise is that measures of traits, including body size, can better reveal the mechanisms and forces that ultimately structure biological diversity (Grime, 2006; McGill et al., 2006; Stegen et al., 2009) and increase the generality *and* predictability of ecological models (Díaz et al., 2004; Kattge et al., 2011; Webb et al., 2010). Trait-based approaches have especially received attention for plant life histories and strategies due to a renewed interest in measuring traits across different environments and scales (Craine, 2009). While this has long been part of comparative physiology and ecology (see Arnold, 1983; Grime, 1977), it is now being heralded as its central paradigm (Craine, 2009;

Westoby and Wright, 2006). Similarly, trait-based approaches are being used to disentangle the forces that structure larger scale biodiversity gradients (Belmaker and Jetz, 2013; Han et al., 2005; Reich, 2005; Reich and Oleksyn, 2004; Safi et al., 2011; Swenson and Enquist, 2007) and to predict large-scale ecosystem shifts due to climate change (Elser et al., 2010; Frenne et al., 2013).

1.1 Central Limitations of Trait-Based Ecology

An important limitation to developing a more predictive trait-based ecology is that its focus and implementation have relied almost entirely on empirical correlations and null models (for example, see discussion in Swenson, 2013). There is a need for theory and quantitative arguments to move beyond pattern searching. Further, trait-based ecology has largely developed independently from MST, where the role of body size—arguably a key trait—is central to scaling up organismal processes. Nonetheless, a key focus of trait-based ecology is to identify the general processes underlying trait-based ecology (Enquist, 2010; Shipley, 2010; Suding et al., 2008b; Webb et al., 2010; Weiher et al., 2011). Such an advance would help guide the explosion of trait-based data collection (Dell et al., 2013; Kattge et al., 2011), develop a more predictive ecology, and organize rapidly developing directions in trait-based ecology (Boulangeat et al., 2012; Funk et al., 2008; Lavorel et al., 2011; McGill et al., 2006; Shipley, 2010; Suding et al., 2008b).

Another limitation is the debate about whether biodiversity, trait diversity, or both are important for ecosystem functioning (Hooper et al., 2004; Loreau et al., 2001). We agree with Cardinale et al. (2007) that this debate is largely a false dichotomy. Increasingly, the evidence shows that both the number of species and types of species in an ecosystem impact biomass production. For example, focusing solely on species number has resulted in sometimes positive, negative, or null relationships between species richness and ecosystem functioning (Grace et al., 2007; Roscher et al., 2012).

Lastly, because trait-based ecology measures properties of individuals that are linked to the environment and because it attempts to make predictions for ecosystem functioning, it must be able to scale from individuals to ecosystems. However, achieving this requires an exciting but extremely challenging synthesis of physiology, population biology, evolutionary biology, community ecology, ecosystem ecology, and global ecology (Reich, 2014; Webb et al., 2010). In this chapter, we suggest combining trait-based approaches with MST to make some progress on this problem.

Here, we present a novel theoretical framework to scale from traits to communities to ecosystems and to link measures of diversity. We argue that trait-based ecology can be made more predictive by synthesizing several key areas of research and to focus on the shape and dynamics of trait distributions. Our approach is to develop more of a predictive theory for how environmental changes, including land use and shifts in abiotic factors across geographic and temporal gradients, influence BEF (Naeem et al., 2009). We show how starting with a few simple but general assumptions allows us to build a foundation by which more detailed and complex aspects of ecology and evolution can be added. We show how our approach can reformulate and generalize the arguments of Chapin et al. (2000), McGill et al. (2006) and Violle et al. (2014) by integrating several insights from trait-based ecology (Garnier and Navas, 2012) and MST (Enquist et al., 1998; Gillooly et al., 2001; West et al., 1997). In doing so, we can derive a more synthetic theory that can begin to: (i) assess differing assumptions underlying the assembly of species; (ii) assess the relative importance of hypothesized drivers of trait composition and diversity; and (iii) build a more predictive and dynamical framework for scaling from traits to communities and ecosystems. We call this theory, Trait Driver Theory or TDT, because it links how the dynamics of biotic and abiotic environment then drive the performance of individuals and ecosystems via their traits. Combining MST with trait driver approaches allows TDT to work across scales and also addresses one of MST's key criticisms: it does not incorporate ecological variation—such as trait variation—and cannot be applied to understanding the forces that shape the diversity and dynamics of local communities (Coomes, 2006; Tilman et al., 2004).

2. TRAIT DRIVER THEORY

TDT is based on a synthesis of three influential bodies of work. The first are trait-based approaches that are largely encapsulated in Grime's Mass Ratio Hypothesis or MRH (Grime, 1998). The MRH states that ecosystem functioning is determined by the characteristics or traits of the dominant (largest biomass) species. Implicit in the MRH is the idea that traits of the dominant species are a more relevant measure than species richness. The second component is the generalized and quantitative approach to trait-based ecology through Norberg et al. (2001) who used a mathematical framework to link the distribution dynamics of phenotypic traits with environmental change and ecosystem functioning (Norberg, 2004; Norberg et al., 2001; Savage et al., 2007; Shipley, 2010). The third component is Metabolic

Scaling Theory or MST. MST can be used to predict how variation in organismic size and the traits associated with metabolism will then influence individual performance (growth and resource use), and how these performance measures will then scale up to influence populations, communities, and ecosystems (Enquist et al., 1998, 2003, 2009; Savage et al., 2004; Yvon-Durocher et al., 2012). MST achieves this by showing how variation in individual rates of mass growth, dM/dt, and metabolism, B, can be linked to variation in a few key traits (e.g. body size, M, and traits related to cellular metabolism and allocation; see Enquist et al., 2007b, 2009; von Allmen et al., 2012; West et al., 2002).

2.1 The Central Assumptions of Trait-Based Ecology and the 'Holy Grail' of Trait-Based Ecology

Trait-based ecology assumes that there are traits that are functional, meaning they link the environment to variation in whole-organism performance and ultimately fitness (Schmitz et al., 2015; Violle et al., 2007; see Fig. 1). That is, as shown in Fig. 1 and as described in Schmitz et al. (2015 in this special issue), variation in traits influences organismal performance (e.g. metabolism, growth rate, demographic rates) and ultimately fitness (Ackerly and Monson, 2003; Garnier et al., 2004; Lavorel et al., 2007; Violle et al.,

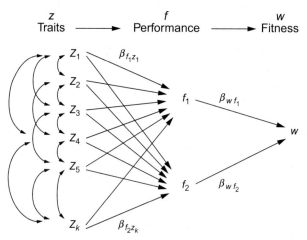

Figure 1 Path diagram representing the linkages between phenotypic traits, z, performance measures, f, and fitness, w. The arrows represent possible mechanistic linkages between traits. The coefficients β represent the correlation coefficients between traits, z, functions, f, and ultimately fitness, w. Note: performance measures include growth rate as well as survivorship and reproductive rates. Metabolic scaling theory explicitly links traits to these performance functions. Trait Driver Theory aims to link the performance function with community assembly and ecosystem functioning. *Figure modified from Kingsolver and Huey (2003) and Arnold (1983).*

2007). This approach has been recently validated with a comparative study linking variation in individual-level traits with variation in life history and demography parameters (Adler et al., 2014). Another key assumption of trait-based ecology is that traits of individuals can be used to predict individual performance that can be effectively summed or scaled up to the functioning of ecosystems (Lavorel and Garnier, 2002; Suding et al., 2008b). The *raison d'être* and the 'Holy Grail' of trait-based ecology is to use functional traits, rather than species identities, to better predict community and ecosystem dynamics (Lavorel and Garnier, 2002; Lavorel et al., 2007; Suding and Goldstein, 2008).

2.2 Linking Traits, Individual Performance, Communities, and Ecosystem Functioning

We start by extending Grime's MRH. Grime argued that dominant traits rather than species number drive ecosystem functioning. As a result, it is crucial to measure the trait frequency distribution defined by biomass for the assemblage. An important question is in order to assess the MRH should one use abundance- or biomass-weighted mean trait values to best estimate the frequency distributions. In Section 5 below we further address this question, however, to start TDT also focuses on the trait frequency distribution $C(z)$—the histogram of biomass across *individuals* characterized by a given trait value, z, summed across all individuals within and across species. Thus, $C(z)$ captures both intra- and interspecific trait differences. However, unlike the MRH and current emphasis on using community-weighted mean traits, we are interested in the overall shape of the distribution of phenotypes of individuals as measured by the central moments—variance, skewness, kurtosis—beyond the mean. We can link individual growth rate and the population *per capita* growth rates via how traits influence organismal performance via the growth function,

$$f(z) = \left[\frac{1}{C(z)}\right]\left[\frac{dC(z)}{dt}\right] \tag{1}$$

where $C(z)/dt$ is the biomass growth rate for *all* individuals with a given trait value z (see Appendix). By integrating the growth equation across all values of the trait across individuals, we can derive dynamic equations for how total community biomass, C_{Tot}, depends on the shape of the biomass-trait distribution, $C(z)$, and how that shape itself changes in time. Consequently, the net production of biomass in the community or Net Primary Productivity (NPP) is

$$\frac{\mathrm{d}C_{\mathrm{Tot}}}{\mathrm{d}t} = \int f(z)C(z)\mathrm{d}z \qquad (2)$$

(see also Lavorel and Garnier, 2002; Norberg et al., 2001; Vile et al., 2006). Equation (2) requires understanding what sets the form of $f(z)$.

2.3 Linking Dynamics of Trait Distributions to Environmental Change and Immigration

Starting with Norberg et al. (2001) and Savage et al. (2007), we focus on how traits that strongly influence organismal growth rate are influenced by the environment, E. In Fig. 2, we show an example of how variation

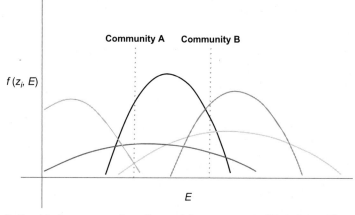

Figure 2 Graphical representation of essential components of Trait Driver Theory. Each curve represents positive performance (here shown as the growth function f) or growth rates of individuals characterized by a unique trait value, z. Each colour then indicates how a given trait value then translates to variation in organismal growth rate across an environmental gradient, E. Because of a trade-off between traits and the environment (see Section 2.3), each trait has an optimal environment where growth is maximized, and each trait exhibits a unimodal response to an environmental gradient, E. At different points along an environmental gradient, different traits are characterized by the highest growth rate, f. Also, species can differ in the width of their performance curves, σ_z^2. Here in two communities, A and B, although several trait values can achieve positive growth, in community A the black phenotype has the highest growth rate in that location and is predicted to then be the most dominant phenotype in a with trait z. in contrast, in community B, the black phenotype is predicted to not be as dominant. This trait value then is the optimal trait value, given the potential species pool, for that community. *Note*: in this example, growth rates are only influenced by E. However, extensions of TDT (see text) can assess how the growth rate of a given trait value then is influenced by the presence and dominance of other trait values. Such trait–trait interactions could then modify the shape and breadth of each growth curve.

in a given trait, z, translates to variation in *per capita* growth rate across an environmental gradient. This example assumes that all individuals in a given community ultimately compete for similar limiting resources, and that there is an optimal environment where growth is fastest (Fig. 2). Although our approach starts with a single trait, trait-based models can straightforwardly incorporate multiple, correlated traits (Savage et al., 2007; see also Appendix). By incorporating temporal environmental forcing into the growth function, TDT predicts how the distribution of traits, $C(z)$, responds to both biotic and abiotic drivers (Norberg et al., 2001; Savage et al., 2007).

The shape and the dynamics of the trait distribution ultimately reflect a balance between two rates—the introduction/immigration of traits, I, into an assemblage and the outcome of variation in the performance, f, of those traits within the assemblage. A general trait-based equation for growth and immigration is given by

$$\frac{dC(z)}{dt} = f[z, E, C(z)]C(z) + I\left[z, E, C(z), C_{\text{landscape}}(z)\right] \qquad (3)$$

Here, we now explicitly include the effects of the environment, E, and the trait distribution $C(z)$ as part of the growth, f, and immigration, I, functions because they can influence both via environmental change, competition, facilitation, sampling effects, or other biological interactions such as density dependence (Savage et al., 2007). The second term, the immigration function, I, reflects the external input of individuals into the community stemming from dispersal as well as the introduction of traits into the assemblage from other factors including evolutionary processes (mutations) and potential seed banks. The immigration function, I, reflects dispersal behaviour of individuals and is a function of other ecological factors, such as organismal traits, and may be influenced by the abundance of that trait already in the local community $C(z)$ as well as the abundance of that trait in the larger landscape, $C_{\text{landscape}}(z)$ (see Gilbert and DeLong, 2015; Laskowski et al., 2015; Schmitz et al., 2015 in this issue).

Next, we use two assumptions to constrain the form of Eqs. (2)–(3). First, a central tenet of TDT and a well-grounded concept in ecology and evolution are that across an environmental gradient, E, organisms will tend to have a unimodal functional response in their performance and fitness functions (Fig. 2). As a result, a shift in the environment, E, will affect the *per capita* population growth rate and thus the traits that are dominant in the community or assemblage (Davis and Shaw, 2001; Whittaker et al., 1973) and the rate of trait evolution (Levins, 1968). Second, there are specific traits that link environmental drivers to individual growth rate, and the trait driven *per capita* biomass growth rate, $dC(z)/dt$ (Fig. 1; Arnold, 1983). The performance or growth function $f(z)$ then is a result from an environment-mediated trade-off between traits, such as investment in growth versus defence or from investment in growth rate versus desiccation resistance. As a result, for a constant environment, E, there are optimal trait values, z_{opt}, that maximize the growth function given in the environment.

In the case of a single-trait optimum, we approximate this as a symmetric function such as a Gaussian or quadratic trade-off $f(z) \propto \left[1 - \left(\frac{z - z_{opt}}{\sigma^2}\right)^2\right]$, where σ^2 is the trait breadth of the trade-off function. If the environment is constant and immigration, I, is zero, individuals with traits that match, z_{opt}, will gradually replace all other individuals, and the trait distribution will collapse on a single point for the optimal trait value z_{opt} (Norberg et al., 2001). Thus, TDT is consistent with a competitive trait hierarchy view of assemblage interactions (Freckleton and Watkinson, 2001; Goldberg and Landa, 1991; Kunstler et al., 2012; Mayfield and Levine, 2010) as well as a population source–sink view of assemblage (Pulliam, 1988) and metapopulation perspective of trait dynamics across environmental gradients (Davis and Shaw, 2001). As we discuss below, additional biotic and abiotic interactions and processes can also be shown to influence the shape of the trait distribution via the growth and immigration functions (see Weiher and Keddy, 1995).

3. PREDICTIONS OF TDT

Next, we emphasize the central predictions of TDT. These predictions are also summarized in Table 1 in terms of how different measures of the trait distribution can provide novel insight and predictions regarding the main drivers of the current composition of the species assemblage as well as the future dynamics of the species assemblage.

Prediction (1): Shifts in the environment will cause shifts in the trait distribution (Fig. 3; Table 1).

Prediction (2): The difference between the optimal trait and the observed mean trait, as well as the trait variance, provides a measure of the capacity of a community to respond to environmental change (Figs. 3 and A1; Table 1).

Shifts in the abiotic or biotic environments, represented by E and $C(z)$, respectively, will lead to corresponding shifts in the community trait distribution. The magnitude of the shift over some time and the rate of change of

Table 1 Summary of the Core Predictions from Trait Driver Theory for How the Different Central Moments of the Trait Distribution Will Respond to Differing Biotic and Abiotic Forces and How They Will Then in Turn Influence Community Dynamics and Ecosystem Functioning

Moment of Community Trait Distribution, $C(z)$	Predictions for Rate of Community Response to a Changing Environment	Predicted Ecosystem Effects
I. Mean	(a) Will shift if environmental change alters value of z_{opt} and time scales are not too rapid and oscillatory (b) Lags z_{opt} by an amount that depends on rate of change in environment, rates of immigration, and the forces that influence the variance	(i) Will shift productivity according to form of growth equation, f
II. Variance	(a) Decreases with strong abiotic filtering (b) Decreases due to competitive exclusion by individuals with trait z_{opt} (c) Can increase with increased immigration, competitive niche displacement, and/or temporal variation in z_{opt} due to a variable environment (d) Under neutral theory, if no immigration or mutation, variance will decrease over time so as to decrease response abilities over time	(i) Increased variance implies lower productivity for fixed or stable environment (ii) Increased variance accelerates community response to environmental changes (iii) Increased variance will lead to increased stability of ecosystem functioning by reducing the lag of $\bar{z}$ and z_{opt} in varying environments
III. Skewness	(a) Skewness values > or < 0 can reflect a lag between $\bar{z}$ and z_{opt} and a rapidly changing community due to an environmental driver or extreme limit to a trait value (b) Increases in skewness can indicate a response to rapid environmental changes or the importance of rare species advantages in local coexistence	(i) Depending upon kurtosis and variance value, productivity should be reduced compared with a community with similar variance but skewness equal to zero

Continued

Table 1 Summary of the Core Predictions from Trait Driver Theory for How the Different Central Moments of the Trait Distribution Will Respond to Differing Biotic and Abiotic Forces and How They Will Then in Turn Influence Community Dynamics and Ecosystem Functioning—cont'd

Moment of Community Trait Distribution, C(z)	Predictions for Rate of Community Response to a Changing Environment	Predicted Ecosystem Effects
IV. Kurtosis	(a) Positive kurtosis – a more peaked distribution – reflects competitive exclusion or other types of biotic exclusion (b) Kurtosis close to -1.2 reflects a uniform distribution consistent with uniform niche partitioning (c) More negative values could reflect the coexistence of contrasting ecological strategies, recent or sudden environmental change	(i) If the trait mean equals z_{opt}, forces that decrease kurtosis will decrease productivity while forces that increase kurtosis will increase productivity (ii) In a temporally varying environment, more negative kurtosis values will lead to increased stability of ecosystem functioning by reducing the lag of $\bar{z}$ and z_{opt}

Trait Driver Theory (TDT) can incorporate each of these forces via the shape of the trait biomass distribution, $C(z)$, to then make specific predictions for how each can drive the dynamics of $C(z)$ and ecosystem functioning (see text). Parameterizing predictions for specific cases depends upon the traits that affect the growth rate, f. Here $\bar{z}$ is the average observed trait value and z_{opt} is the optimal trait value for an given environment, E.

the shift can both be calculated from Eq. (2). According to Eq. (2), the value of the optimal trait will change with the environment (e.g. Ackerly, 2003), while the mean trait of the community will approach the optima but with a lag in time according to how long it takes for either trait plasticity and/or the processes of species sorting and selection to act (Ghalambor et al., 2007). In environments where the optimal trait value is changing quickly relative to generation times or plasticity, there may be little capacity for the mean community trait, $\bar{z}$, to track these changes. In such circumstances, $\bar{z}$, may never, or only rarely, be expressed at an optimal value for the current environment. Nonetheless, we expect that for most communities, the difference between the optimal trait, z_{opt} (E), and the observed mean trait, $\bar{z}$, or $\Delta(E)$, will be a measure of how the community has responded/will respond to environmental change. Norberg et al. (2001) derive the general expression

$$\Delta(E) = z_{opt}(E) - \bar{z} \tag{4}$$

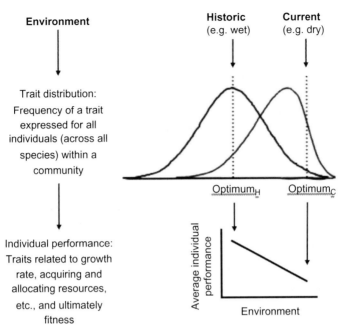

Figure 3 Conceptual diagram linking changes in optimal traits in changing environments with the frequency distributions for that trait in historic versus current environments. In this example, the optimal trait value (dotted line) has shifted to the right. Individual performance (growth, fitness, etc.) is highest at the optimal trait expression in an environment. *Note*: because the response of an assemblage cannot be instantaneous, the trait distribution has an increased skew. Further, since the highest trait frequencies are not yet at the optimal trait expression, the average performance in the current environment is lower than in the historic environment.

where $\Delta(E)$ quantifies the community trait 'lag' in relation to the current environment. This measure is analogous to the 'lag load' in evolutionary theory (Maynard Smith, 1976). We can thus define $d\Delta/dt$ as the response capacity of a community. Equation (4) predicts that capability of the assemblage to respond to directional shifts in the environment will be directly proportional to the trait variance, $d\Delta/dt \propto V$. Importantly, within TDT, directional selection for a given optimal trait value need not always lead to an *increase* in *per capita* growth rate, f. Because of trade-offs between traits and frequency and density-dependent effects on performance and fitness, the performance (and fitness) associated with the new optimum value likely differs from the fitness and growth rate in the previous environment (Antonovics, 1976; Dieckmann and Ferrière, 2004; Ferriere and Legendre, 2013). For example in Fig. 3, we highlight a hypothetical example of a shift in the community trait distribution from wet to dry that comes with a decrease in optimal performance. Extensions of TDT can in principle include these effects (Savage et al., 2007).

In sum, predictions 1 and 2 formalize Chapin et al.'s conceptual framework (Chapin et al., 2000) for the development of a predictive trait-based ecology. In the case of multiple traits underlying growth, f, differing trait combinations could lead to similar growth rates in differing environments (see also Marks and Lechowicz, 2006). We note that these predictions implicitly ignore the effects of frequency dependence but elaborations of TDT can include these effects (see equation 5 in Savage et al., 2007).

Prediction (3): The skewness of the trait distribution can be an indicator of past or ongoing immigration and/or environmental change due to lags between growth, reproduction, and mortality (Table 1).

Because of time lags between environmental change and the time scale of organismal responses (growth, demography, etc.), the trait distribution of an assemblage will not be able to instantaneously track environmental change, and skewness in the trait distribution will develop (see also Figs. 3 and A1). Alternatively, skewness may reflect differential immigration of traits from one side of a habitat or a community that contains 'sink' populations (Pulliam, 1988) supported via immigration (see also Pawar, 2015 in this issue). Neutral theory (Hubbell, 2001), in which traits have no demographic effects, could also lead to skewed distributions due to neutral trait evolution. The implication is that trait-based ecology can infer the dynamics of trait assemblages via assessing the shape of contemporary trait distributions. Combining information on the shape trait distributions with additional information such as dispersal history and/or size distributions would help separate lag effects from drift and differential immigration.

Prediction (4): The rate of change of net ecosystem productivity in response to environmental change can be predicted via the growth function, f, and the shape of the community biomass-trait distribution $C(z)$ at some initial time (Table 1).

In the simple case of a single trait with a single environmental driver, Norberg et al. (2001) derived a general expression linking the dynamics of the trait distribution by noting Eq. (2), Eq. (3) can be approximated as

$$\frac{dC_{\text{Tot}}}{dt} \approx \left[f(z, E, C(z))_{z=\bar{z}} + \frac{\partial^2 f(z, E, C(z))}{\partial z^2} \bigg|_{z=\bar{z}} V \right] C_{\text{Tot}} + I \qquad (5)$$

Equation (5) follows from a Taylor expansion that effectively linearizes the equations. If the terms in brackets depend on total biomass, dC_{Tot}/dt would scale non-linearly with total biomass, but in the simplest case, these terms are independent of total biomass implying that production scales linearly with total biomass. In Eq. (5), the net primary production, C_{Tot}/dt, is equal to the growth rate of the mean community trait, $\bar{z}$, plus the second term that accounts for how much variation there is in the community trait distribution, V. Because the growth function, $f(z,E)$, has a maximum at z_{opt}, we expect the second derivative term to be negative, as long as the average observed trait value, $\bar{z}$ is in the neighborhood of the optimal trait value, z_{opt} (Norberg et al., 2001; see also discussion in Appendix) reflecting the increasing reduction in growth rate as trait values increasingly differ from z_{opt} (see Fig. 2 and Eq. 4). The unimodal shape is the simplest assumption requiring only the mean and variance. There is reason to expect that f can be approximated as unimodal. For example, growth rates typically exhibit unimodal response with measures of temperature, pH, etc. (McGill et al., 2006). Again, the term I gives the addition of biomass through immigration/dispersal.

Prediction (5): Within a community whose growth rate depends on a single trait, an increase in the variance of that trait will lead to a decrease in net primary production (Table 1).

An additional prediction is that for communities whose growth is driven by a single key trait, larger trait variance, V, will decrease the net primary production for the whole community because a higher proportion of individuals differ from z_{opt} (Norberg et al., 2001). This idea of a trade-off between short-term productivity and long-term response to environment is reflected by agricultural imperatives with agricultural issues, where short-term productivity is emphasized and variance in traits is minimized in trait values and short-term productivity is maximized. Elaborations of TDT have shown that incorporating multiple limiting resources, multiple traits, and trait covariation (Savage et al., 2007) can weaken, nullify, or even reverse the predicted negative relationship between dC_{Tot}/dt and V.

From Eqs. (3) and (4), the rate at which a community can track environmental change will be greater when there is greater trait variance. Intuitively, greater variance leads to more extreme traits being immediately available to respond to environmental change. Thus, the rate of response will also depend upon the specific form of the growth function f (e.g. for a given value of z, how does f vary across an environmental gradient? see Savage et al., 2007 and discussion in Appendix). For example, if f is a simple Gaussian or polynomial function with E (Fig. 1), then the value of $d\Delta/dt$ can be approximately proportional to the community trait variance, V. Building on the work of Norberg et al. (2001) and Savage et al. (2007), these equations can be extended to include higher-order moments such as skewness and kurtosis.

3.1 Extending TDT via Recasting and Assessing Different Ecological Hypotheses About Diversity and Trophic Interactions

TDT predicts that over time only one phenotype should dominate a given environment. An important question is what maintains diversify (trait variation) within an assemblage? According to TDT trait, variance can be increased by many different ways (also see Dell et al., 2015; Gilbert and DeLong, 2015; Pawar, 2015 in this issue; Pettorelli et al., 2015). Immigration, I, from outside the assemblage, as well as from a directionally shifting or a temporally variable environment (Norberg et al., 2001; Savage et al., 2007), can increase and/or maintain trait variation. Further, theoretical elaborations of TDT have shown that the diversity of phenotypes (traits) present in a given assemblage can be influenced by trade-offs between traits that influence growth. For example, trade-offs between allocation to predator defence and growth rate (Norberg et al., 2001; Savage et al., 2007) can increase the variance of a given trait. In a variable environment, correlations between traits that underlie the growth function, f, lead to the survival of organisms with trait values that are less favourable in the current environment but may be well suited for new environments that arise. Thus, phenotypic trait correlations among traits can ramify to have quantitative effects on ecosystem dynamics (lowering NPP) and enable assemblages to better track environmental change (Savage et al., 2007).

Additionally, trait variation can also stem from additional ecological hypotheses for biological diversity. An exciting aspect of TDT is that differing ecological hypotheses based on species richness can be recast in terms of traits. In Table 1 and Fig. A1, we overview the predictions of the different theories as recast in the light of TDT. As a result, TDT can then be used to 'scale up' the implications of many differing classic and current hypotheses for species richness via the assumptions of trait distributions implicit in these theories. Another exciting potential of TDT is to show (i) how different trophic interactions and variation in traits controlling 'predation risk' and 'prey selection' would then influence community trait and size distributions (see Pettorelli et al., 2015 in this issue) and (ii) how individual trait variation could potentially have opposing effects on predator–prey dynamics (see Gilbert and DeLong, 2015 in this issue). TDT would then offer a framework to show how these differing ecological processes then scale up to influence ecosystem functioning.

As we show in Appendix A.1 we can use TDT to recast several different ecological theories in terms of the distribution of fucntional traits. Different ecological theories based on species richness (neutral theory, abiotic filtering, competitive exclusion, Chesson's storage effect, or rare species advantages; see discussion in Appendix) as well as the effects of abiotic processes (shifts due to environmental change, disturbance) will uniquely influence the shape of the community trait distribution and the potential of the assemblage to maintain diversity, as reflected in changes in the trait variance, dV/dt. The relative strengths of abiotic, biotic, and neutral processes will lead to different shapes of trait distributions that will have different implications for responses of community to directional shifts in the environment as well as ecosystem functioning.

4. EXTENDING AND PARAMETERIZING TDT

4.1 Scaling from Individuals to Ecosystems Using MST

So far, TDT assumes that there is no variation in organismal size. Instead, the total biomass associated with a trait, z, is denoted by $C(z)$. This notation avoids ever needing to account for *individual organismal* mass, M, or even the number of individuals with mass. However, body size can vary greatly—it is also an important trait that influences variation in organismal metabolism (Peters, 1983), population growth rate (Savage et al., 2004), and abundance (Damuth, 1981; Enquist et al., 1998). The scaling equations in MST differ from TDT so far as they are phrased in terms of individual mass. In order to integrate these theories, we use three insights from MST (Enquist et al., 2007b, 2009; Savage et al., 2004; West et al., 1997, 2009) to explicitly formulate TDT to work across scales in organismal size, M, and environmental changes or gradients in temperature, T. As we show, MST provides the basis to formally link traits, organismal growth rate, and ecosystem fluxes (Enquist et al., 2007a,b).

First, MST is explicit about how the organismal growth is dependent upon the size of the organism. In the case of MST, we start with how organismal biomass growth rate, dM/dt, is related to whole-organism metabolic rate, B, and organismal mass, M, as

$$\frac{dM}{dt} = b_0(z)M^\theta \tag{6}$$

where $b_0(z)$ is a metabolic coefficient that depends on a single or set (meaning z is a vector) of traits. The allometric scaling exponent θ is hypothesized to reflect the branching geometry of vascular networks (Enquist et al., 2007b). Theory and empirical data point to $\theta \approx 3/4$ for large size ranges (Enquist et al., 2007c; Savage et al., 2008). Equation (6) has recently been shown to be a good characterization of tree growth (Stephenson et al., 2014) and is a specific case of a more generic growth function (Moses et al., 2008; West et al., 2001) that can be applied to both plants and animals. While we focus here on a specific plant growth model, we note that other trait-based models have recently been developed for animals and phytoplankton (Litchman and Klausmeier, 2008; Muller et al., 2001; Ricker, 1979; West et al., 2001), and they could also be used to parameterize TDT. Below, we elaborate Eq. (6) to explicitly include the traits for plants that underlie b_0 and how we can use this equation as the basis for a general trait-based growth function.

To integrate MST into TDT, we first recognize that, $C(z)$, the biomass associated with trait z can be expressed as $C(z) = \int dM C(z, M) = \int dM N(z, M) M$, where $C(z,M)$ is the mass density of individuals with both trait value z and individual mass M, while $N(z,M)$ is the number density of individuals that have both trait value z and individual mass M. In this expression, we have integrated over all possible values of mass, M, so that we have the total biomass of *all* individuals with trait z. Furthermore, note that integrating this over all traits, z, gives the total biomass, $C_{Tot} = \int dz C(z) = \int dz \int dM N(z, M) M$.

To integrate MST into TDT, we solve for the conditions of steady state where $N(z,M)$ is not changing in time. It can be shown (see Appendix) that the equation for the scaling of NPP with the total biomass of the assemblage is,

$$\frac{dC_{Tot}}{dt} = \left\langle b_0(z) M^{-1/4} \right\rangle_C C_{Tot} \tag{7}$$

where dC_{Tot}/dt scales isometrically with C_{Tot} and the C subscript denotes that the average, denoted by brackets $\langle \cdot \rangle$, is taken with respect to the biomass. This equation is in the most generic form of a general TDT equation. Note that the TDT growth function, f, is now

$$f(z) = b_0(z) M^{-1/4} \tag{8}$$

and can be expanded and expressed in terms of the biomass-weighted central moments of the trait z, such as the variance, skewness, and kurtosis (see below). Again, the exponent, $-1/4$ is the idealized case and empirical values that may deviate from $\theta \approx 3/4$ can be used.

4.1.1 Incorporating Environmental Trade-Offs in the Growth Function, f

Equation (8) alone would predict that the *per capita* growth rate will increase forever as the trait, b_0, and mass-specific metabolic rate increase. In reality, though, there is some range of trait values at which the organism can grow. This is because there are trade-offs in performance and fitness: decrease in growth when, for a given value of E, the trait value gets either too small or too large. In the case of a given leaf trait such as leaf size or leaf investment (closely associated with variation in photosynthetic rates and the specific leaf area (SLA)), at some point, continued increases in leaf nitrogen may ultimately limit resource uptake as high N would result in individuals more prone to herbivores, pathogens, etc. and/or will result in water transport demands that would increasingly be maladaptive for a given local environment. Ultimately, one cannot have an infinitely large leaf, an infinitely thin leaf, or a plant that is all leaf area. Thus, deviation away from $b_{0,opt}$ would be associated with a trade-off between specific trait values and plant performance (such as growth rate, survivorship, and/or reproduction; see also Ghalambor et al., 2007).

Incorporating trade-offs between trait values, the environment, and performance is central to TDT. We can incorporate these trade-offs in a general form by multiplying the scaling relationship by a quadratic function. As a result, $f(b_0)$ is maximal at the optimal trait, $b_{0,opt}$, and the niche width defined by $\sigma_{b_0}^2$ where

$$f(z) = b_0(z)M^{-1/4}\left(1 - \frac{\left(b_0 - b_{0,opt}\right)^2}{\sigma_{b_0}^2}\right) \tag{9}$$

Here, the second term is the trade-off function, and c_0 is an overall constant coefficient. Expressing $f(b_0)$ across an environmental gradient, $f(b_0, E)$, would then reveal a unimodal growth function (Fig. 2).

The second insight from MST shows that the metabolic normalization, b_0, can be linked to specific traits. For the plant growth function, building on the insights from the relative growth rate literature (Evans, 1972; Lambers et al., 1989; Poorter, 1989), Enquist et al. (2007b) derived an extension to Eq. (6) that explicitly details the traits that together define b_0 and hence f so that

$$b_0 \propto \frac{c}{\omega}\left(\frac{a_L}{m_L}\right)\dot{A}_L\beta_L \tag{10}$$

Equation (11) shows that, in addition to plant size, M, the rate of growth is governed by the scaling exponent, θ, and five traits: (i) $\dot{A}_L$, the net leaf photosynthetic rate (grams of carbon per area per unit time); (ii) a_L/m_L, the SLA, the quotient of area of the leaf, a_L, and the mass of a leaf, m_L;

(iii) ω, the carbon fraction of plant tissue; (iv) c, the carbon use efficiency of whole-plant metabolism, and (v) β_L, the leaf mass fraction (the ratio of total leaf mass to total plant mass) which is a measure of allocation to leaves. As a result, we can parameterize TDT with specific traits that underlie b_0.

5. ADDITIONAL PREDICTIONS OF TDT

In the second column of Table 1, we summarize additional TDT predictions for scaling up community or assemblage trait distributions to predict several ecosystem-level effects. Specifically, the shape of the trait distribution as measured via the central moments of the distribution.

Prediction (6): Ecosystem net primary productivity, dC_{Tot}/dt, will scale with the total biomass but will be influenced differently by the mean and variance of the community trait distribution.

A third insight from MST allows us to more formally link TDT with MST by including organismal mass dependence into TDT. In particular, most assemblages of organisms will be characterized by a distribution of sizes. For plants, following the arguments in Enquist et al. (2009), we can substitute the distribution of the number of individuals as a function of their size, M, or the size spectra, $N(M)$. For the idealized case of $\theta \approx 3/4$, they show that $N(M) \propto M^{-11/8}$ and link the total biomass, C_{Tot}, with the size of the largest individual, M_b, where $M_b \propto C_{Tot}^{8/5}$. This allows us to consider a few special cases of Eq. (7) that relate TDT and scaling equations already in the literature.

In the case of a given assemblage where there is no size distribution and only a single mass value, M^*, or a very small range of mass values, the scaling of NPP becomes

$$\text{NPP} = \frac{dC_{Tot}}{dt} = (M^*)^{-1/4} \langle b_0(z) \rangle_C C_{Tot} \qquad (11)$$

The term $(M^*)^{-1/4}$ can be thought of as an overall normalization to the growth function $f(z)$ from TDT. As such, this result reveals that TDT, as originally formulated (see Eq. 3), ignores variation in individual mass. Thus, based on Eq. (6), growth functions within TDT should have a roughly $(M^*)^{-1/4}$ hidden with the normalization constant for their growth function. In the case where (i) organisms within the community or assemblage can differ greatly in their sizes; (ii) z and M are uncorrelated; and (iii) the number density is a separable function, such that $N(z, M) = N(M)N(z)$; it can be shown that the growth equation can be expressed in two different ways. Each way depends on how one averages the trait distribution. In the first case, we have

$$\frac{dC_{\text{Tot}}}{dt} = k\langle b_0(z)\rangle C_{\text{Tot}}^{3/5} \qquad (12)$$

and in the second case, we have

$$\frac{dC_{\text{Tot}}}{dt} = k\langle b_0(z)\rangle \left\langle M^{-1/4}\right\rangle_C C_{\text{Tot}} \qquad (13)$$

where k is a proportionality constant. Equation (13) is equivalent to Eq. (12) but expresses the growth function more in terms of the TDT framework such that the right side appears to have an overall *linear* dependence in C_{Tot}. As a result, in Eq. (12) we have a mixture of types of averages, with $\langle b_0(z)\rangle$ being the abundance average of the function $b_0(z)$, while $\langle M^{-1/4}\rangle_C$ is the biomass average of $M^{-1/4}$. Both equations are equivalent ways to express the scaling of NPP function. Equation (12) is a more simple expression and only involves using the abundance average of the trait distribution. Equation (12) consolidates the organism mass average with the 3/5 scaling dependence of C_{Tot}. These derivations help to clarify when trait-based studies should use biomass- or abundance-weighted values.

Both equations assume a community steady-state approximation where $N(z,M)$ is not changing in time. If this is violated (e.g. the community trait abundance or number distribution $N(z,M)$ is changing), then deviations from Eqs. (12) and (13) are expected. Nonetheless, these equations provide a basis for linking the scaling of organismal growth rate and trait variation of individuals with ecosystem-level processes. For all of these equations and cases, the functions inside the averages can be expanded in terms of moments as done for TDT for biomass-weighted averages or as done in Savage (2004) for abundance-weighted averages.

Putting all of this together with Eq. (5) yields the prediction that Eqs. (12) and (13) are then modified by the shape of the trait distribution, where for a given E, growth is reduced with departure from $b_{0,\text{opt}}$,

$$\frac{dC_{\text{Tot}}}{dt} \approx k\langle b_0(z)\rangle + \left[\left(1 - \frac{(b_0 - b_{0,\text{opt}})^2}{\sigma_{b_0}^2}\right)\frac{d^2 f}{dz^2}\bigg|_{z=\bar{z}} V(b_0(z))\right] C_{\text{Tot}}^{3/5} \qquad (14)$$

The second term captures the reduction of production due to deviation from b_{opt}. Equation (14) represents a formal integration of foundations of TDT from Norberg et al. (2001) with MST from West et al. (1997) and Enquist et al. (2009). Because the growth function, $f(z,E)$, has a maximum at z_{opt} (Fig. 2), we expect the second derivative term, $d^2 f/dz^2$, to be negative, as long as $\bar{z}$ is in the neighbourhood of z_{opt}. Importantly, Eq. (14) enables one to parameterize TDT with a specific trait-based growth function. Further, it enables the integration of physiological performance curves for how the key integrative trait, b_0, varies across a given environmental gradient, E. Equation (8) predicts that dC_{Tot}/dt will increase with increasing community biomass, C_{Tot}. Note, here, the role of the relative breadth of species performance curves (see Fig. 1) is represented by $\sigma_{b_0}^2$.

> Prediction (7): Equation (14) generates specific and testable relationships for the scaling of trait means, dispersion, and ecosystem production (see Table 3).

Equation (14) predicts that there is a range of mean trait values for which increases in the variance of traits, $V(b_0)$, will *decrease* Net Primary Productivity, dC_{Tot}/dt, and shifts in $\langle b_0 \rangle$ will lead to corresponding shifts in dC_{Tot}/dt. This will occur whenever the mean trait value is near the maximum of the growth function, which should occur frequently because evolution is driving the mean trait to match the optimal trait with some lag time. However, there are also mean trait values for which increases in the variance of traits, $V(b_0)$, will *increase* Net Primary Productivity (NPP), dC_{Tot}/dt, and shifts in $\langle b_0 \rangle$ will lead to corresponding shifts in dC_{Tot}/dt. This will happen when the mean trait value is further from the optimal trait value and below an inflection point in the growth function that occurs for small trait values (see the example of a shift from historically wet climate regime to a dry regime in Fig. 3). Intriguingly, this scenario suggests that trait variance can potentially act to either increase *or* decrease NPP depending on if the current trait distribution is close to the local trait optimum or not. So, if the assemblage is close to the optimal value, increases in trait variance will typically decrease NPP. This contrasts with biodiversity theories in which increasing variance (increased trait diversity) tends to increase NPP. Importantly, Eq. 14 shows the influence of variation, V, of the traits that underlie b_0 observed within the community.

Integrating MST into a more generalized TDT lists several key traits for TDT. We explore predictions of TDT in the special case of a single-trait driver such as $SLA = (a_L/m_L)$. First, a change in the environment will likely be associated with a shift in the mean value of SLA. Second, using Eqs. (12) and (14), we expect that

$$\frac{\mathrm{d}C_{\mathrm{Tot}}}{\mathrm{d}t} \propto (\langle \mathrm{SLA} \rangle - \langle V_{\mathrm{SLA}} \rangle) C_{\mathrm{Tot}}^{3/5} \tag{15}$$

Thus, according to Eq. (15), a shift that increases the abundance-weighted mean trait value of $\langle \mathrm{SLA} \rangle$ will lead[1] to an increase in NPP. Preliminary support of this prediction comes from empirical studies that have noted that increases in the mean community SLA are closely linked with increases in ecosystem productivity (Garnier et al., 2004; Violle et al., 2007) and follows from TDT. Third, due to the productivity–variance trade-off predicted by TDT, holding the other variables constant, an increase in the community variance in SLA or V_{SLA} will lead to a *decrease* in productivity so long as the mean community SLA is close to the optimum value.

As we discuss below, TDT provides a foundation that can be modified by additional factors. When there are multiple trait shifts that may covary, measuring all of the traits listed in Eq. (11) would allow more detailed predictions. Of all the traits specified by Eq. (11), there appears to be evidence that SLA may vary more across environmental gradients than other traits and be more important for linking changes in a trait driver or environment, E, with variation in local plant growth (see Appendix). It varies across taxa (up to 3 orders of magnitude or approximately 1000-fold) and directionally varies across environmental gradients in soil moisture, irradiance, and temperature (see Garnier et al., 2004; Poorter et al., 2009; Wright et al., 2005). SLA has also been noted to vary considerably *within* species in response to local changes in climate and abiotic conditions (Cornwell and Ackerly, 2009; Jung et al., 2010; Shipley, 2000; Sides et al., 2014).

6. METHODS: ASSESSING TDT ASSUMPTIONS AND PREDICTIONS

6.1 Quantifying the Shape of Trait Distributions

In order to assess predictions of TDT, it is necessary to quantify the biomass distribution of traits, $C(z)$, in a species assemblage. This involves enough measurements of the trait values and body masses to obtain accurate

---

[1] *Note*: Because the growth equation f is for more instantaneous measures of growth, this prediction is based on rates of more instantaneous NPP and not necessarily annual net primary production. So, accurate testing of this prediction with annual productivity data should make sure to standardize for growing season lengths.

estimates of the underlying distributions, as guided by sampling theory and statistics (Baraloto et al., 2010; Paine et al., 2011). The sampling must occur across all individuals within our group and thus incorporates both inter- and intraspecific trait variability (see Appendix and Violle et al., 2012). The sampling protocols often make choices that limit accurate measurements of within-species variability more than across-species variability. Indeed, simultaneous measurements of intra- and interspecific trait measures are rarely collected (Ackerly, 2003; Baraloto et al., 2010). However, intraspecific variation in traits is important to determine the breadth of the distribution (Sides et al., 2014; Violle and Jiang, 2009). Trait abundance or biomass distributions, $C(z)$, can be approximated through sampling (see Appendix), so that predictions of TDT can be tested without explicitly measuring the traits of all individuals.

There are two reasonable approximations for community trait distributions. The first approximation method calculates the weighted trait distribution by taking the mean species trait value and multiplying by a measure of dominance (cover, biomass, abundance; Grime, 1998). This method can be implemented by calculating the central moments of the joint distribution. In the Appendix, we show the equations used to approximate the community trait moments in particular the community-weighted mean, variance, skewness, and kurtosis (CWM, CWV, CWS, and CWK). Community-weighted metrics, however, ignore intraspecific variation. Increasingly, it is becoming clear that intraspecific variation can contribute to a considerable amount of trait variation (Messier et al., 2010) and that relying on species mean trait values may not provide a robust measure of the shape of a trait distributions (Violle et al., 2012).

A second method utilizes sampling theories to help avoid the time-consuming work of sampling the traits of all individuals. Subsampling individuals can be used to better approximate how intraspecific variation influences the community distribution. In the Appendix, we develop this new method (see discussion in Appendix). The method utilizes subsampling individuals to obtain a better approximation of how intraspecific variation influences the community distribution. By subsampling individuals to estimate intraspecific trait variation within each species, one can begin to incorporate intraspecific variation around mean trait values for each species. We expect that utilizing this method in addition to the incorporation of how MST influences the scaling of the number of individuals will improve estimates for the shape of trait distributions. In short, by subsampling individuals for each species within a given assemblage, one can begin to incorporate intraspecific variation around mean trait values for each species.

6.2 Testing Predictions of TDT

We tested several of the specific TDT predictions (Table 1) and assumptions using several examples that allow us to assess temporal and spatial variations of trait distributions. First, we searched the literature to determine if trait distributions measured from individuals actually do shift across local environmental gradients. Second, we assessed the dynamics of trait distributions and ecosystem carbon flux measures using data from an elevational gradient in Colorado. Third, we assessed the temporal dynamics of trait distributions and ecosystem net primary productivity using the Park Grass Experiment (PGE) from Rothamsted, UK. Lastly, to assess potential linkages with larger scale biogeographic gradients, we review recent studies that assess shifts in trait distributions across large-scale biogeographic environmental gradients.

We primarily focus on assemblage variation in one trait, SLA, because it appears to vary more than other traits in Eq. (11). Thus, we begin to assess predictions from TDT (Table 3) by substituting the mean SLA value for $<b_0>$. If other traits in Eq. (11) also vary or covary with each other across gradients, TDT would allow us to explore this as well. For example, a shift in the mean community carbon use efficiency (or c) will lead to a decrease in NPP as observed (DeLucia et al., 2007). Utilizing Eq. (11), we can now codify several additional TDT predictions based on SLA (see Table 3).

6.3 Shifts in Trait Distributions and Ecosystem Measures Across Local Abiotic Gradients

We tested several predictions generated by TDT (Table 1) with data collected along an elevational gradient in Colorado. We (Henderson, Sloat, and Enquist) have measured community composition, ecosystem fluxes, and traits of *all* individuals in several communities across an elevational gradient within subalpine communities near the Rocky Mountain Biological Laboratory (RMBL, Gunnison Co., CO, USA). Sites ranged from 2460 to 3380 m and had similar slope, aspect, and vegetation. The lowest elevation site is characterized as a semi-arid sagebrush scrub, whereas subalpine meadow communities dominate at the higher elevations. To estimate $C(z)$ leaf traits and biomass was measured from every individual within a 1.2×1.2 m plot. Measures of total ecosystem carbon production, community-weighted SLA, and variances were obtained by harvesting biomass and measuring total ecosystem carbon fluxes or net ecosystem production (NEP; $\mu mol\ CO_2\ m^{-2}\ s^{-1}$). A more detailed listing of the methods used in our analyses is given in the Appendix.

6.4 Local Tests of TDT Predictions

We next tested several predictions generated by TDT (Table 1) with data from the PGE from Rothamsted, UK. The PGE follows grassland plant community composition over a 140-year period. Started in 1956, it is the oldest ecological experiment in the world (Silvertown et al., 2006). The dataset is unique as it allows us to assess community responses to an environmental driver—the experimental altering of soil nutrient availability. We use this dataset to assess how a change in the environment, E, in this case soil nutrients, differentially influences community composition and ecosystem function via the trait distribution, $C(z)$. Within this experimental setup, the main environmental driver is a nutrient addition in the fertilized plot. We first focused on quantifying community SLA frequency distributions. However, as a more direct test of TDT, we also assessed two other key traits, plant height and seed size (see Appendix). To approximate the trait distribution, $C(z)$, we assigned species mean traits to species found within the PGE from the LEDA database (Kleyer et al., 2008). A detailed listing of the methods used in our analyses is given in the Appendix including background of the PGE.

7. RESULTS

7.1 Community Trait Shifts Across Local Gradients

Numerous studies have documented shifts in the traits of communities and assemblages across environmental gradients (Ackerly, 2003; Choler, 2005; Fonseca et al., 2000; Swenson and Enquist, 2007). However, many studies generally calculate a species mean trait as part of a species list, thus ignoring intraspecific variation. In contrast, several recent studies have measured traits within communities to assess community-level trait shifts (Albert et al., 2010; Gaucherand and Lavorel, 2007; Hulshof et al., 2013; Lavorel et al., 2008). For example, Cornwell and Ackerly (2009) show that across a gradient of water availability, the community mean and intraspecific mean SLA significantly shifted such that drier environments have lower mean SLA (Fig. 4A).

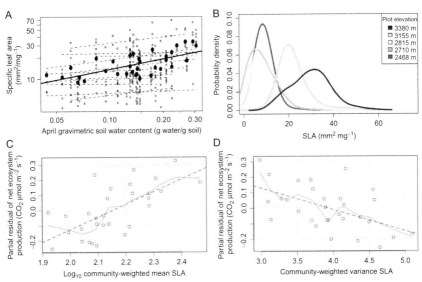

Figure 4 Two examples of shifts in community trait distributions across gradients. (A) Data from Cornwell and Ackerly (2009) showing shifts in the community mean and intraspecific mean value of the trait distribution for specific leaf area or SLA. Dashed lines represent least-squares fits for a given species and show the change in the population mean intraspecific variation across the gradient in soil water. Solid points and the solid line show the least-squares regression for the arithmetic mean community or interspecific value for SLA. *Note*: consistent with the assumption of a shift in an 'optimal phenotype' across environmental gradients, intraspecific trait shifts are in the same direction as the community or interspecific shift. (B) Data from our Colorado elevational gradient showing the probability density distributions of SLA based on all individuals in a 1.3 × 1.3 m plot at five sites along an elevational gradient. The number of individuals at each site is 2468 m = 234, 2710 m = 639, 2815 m = 938, 3155 m = 282, and 3380 m = 160. Across the elevational gradient, the community trait distribution of all individuals significantly shifts with changes in the mean community trait and variance. In (C) and (D), for the Colorado plots, we assess how changes in the community-weighted mean and variance of five plots within each site contributes to variation in net ecosystem production of CO_2 (NEP). These plots are partial residual plots showing linearization of NEP relationships with the community-weighted values of mean SLA and variance in SLA. As predicted by TDT, ecosystem carbon flux is positively related to a shift in mean SLA but negatively related to an increase in community trait variance. In contrast, variation in species diversity explains none of the variation NEP in this system.

7.2 Shifts of Community Trait Distributions Across Environmental Gradients

Data from our elevational gradient at the Rocky Mountain Biological Lab, Gothic, CO (see Bryant et al., 2008; Sides et al., 2014; Sloat et al., 2014) provide one of the first studies to measure the functional traits of every single individual within a given community (Fig. 4B). Few studies have fully documented the community trait distribution by measuring trait values from every individual within the community. In this system, increasing elevation is associated with a decrease in temperature and increase in precipitation, and these changes drive the observed increase in SLA with elevation (Sides et al., 2014). With increasing elevation, leaves have less structural durability, and lower life spans due to a shorter growing season but have higher photosynthetic rates (B.J. Enquist et al., unpublished data). According to TDT, the elevational trend in the SLA distribution is due to a shift in the optimum trait value based on temperature and water availability. This shift should also correspond with a shift in ecosystem functioning. Further, any change in the variance of the distribution will also have impacts on ecosystem functioning and the ability of the site to respond to future shifts in the environment.

Assessing shifts in inter- and intraspecific trait variation allows us to assess two central assumptions of TDT. First, trait distributions show directional shifts across gradients. According to Eq. (10), for a given E, those individuals with phenotypes that are closer to the mean community value should have, on average, the highest growth rates. Previous studies along this same gradient and study site have documented a rapid turnover of species with elevation (Bryant et al., 2008). Figure 4B indicates that the strong species diversity gradient is also reflected by a shift in traits. Note the range of trait variation is approximately 2–3 orders of magnitude. This range of trait values observed within communities in our gradient is approximately half of the fraction of the variation observed in SLA across the globe in all plants (Reich et al., 1997). So, across the span of ~25 km distance between their study sites, we observe a significant fraction of the trait variation that is observed within species across the globe. As more studies document shifts in SLA across strong environmental gradients (such as elevation, flooding, soil water availability, disturbance), it is becoming clear that the magnitude of change can be nearly as large the global variation in the trait (Elser et al., 2010; Violle et al., 2012). These results suggest that more local studies of community trait distributions are reasonable proxies or natural laboratories for scaling up trait-based ecology across large global climate gradients as well as to predict future climate change scenarios.

Second, analyses from Sides et al. (2014) and Cornwell and Ackerly (2009) also provide a key assessment to a core assumption of TDT. According to TDT, for a given environment, E, if there is a mean optimal phenotype that maximizes growth rate given an environmental trade-off

(Fig. 2), then external filters and/or selection/plasticity will then promote convergence of traits around this local optimal phenotype (Norberg et al., 2001; see also Violle et al., 2012). Both of these studies show that patterns of intraspecific mean trait shifts across an environmental gradient are in the *same* direction as the interspecific community shift across the gradient (see Fig. 4A). This is consistent with the expectation that either selection and/or phenotypic plasticity has resulted in individuals that adjust their phenotypes to better match an optimal phenotype within each community.

7.3 Shifts of Assemblage Trait Distributions Across Broad Environmental Gradients

Across broad-scale geographic gradients, recent geographic trait mapping analyses from Swenson et al. (2012) and Šímová et al. (2014) show that the mean assemblage trait value of many plant functional traits varies directionally across biogeographic scales. Geographic variation in the mean tree assemblage SLA as well as tree size (height, a proxy for plant mass, m) shows significant shifts in both traits across gradients at the biogeographic scale (Fig. 5). As is assumed in TDT, across a given environmental gradient, E, the mean community value, $C(z)$ or $C(b_0)$, will shift. Indeed, across North America, Šímová et al. find that the mean assemblage SLA is positively correlated with annual precipitation ($r^2 = 0.539$) but negatively related to annual temperature seasonality ($r^2 = -0.440$) (see Table 2 in Šímová et al., 2014).

7.4 Building Better Models for Variation in Ecosystem Function via the Shape of Trait Distributions and MST

At the local scale, measures of trait distributions associated with our theory in principle can be used to scale up to ecosystem function as well as to predict potential future community responses to climate change. Focusing on a key trait, SLA, and the total community biomass, C_{Tot}, we fit a simplified version of the TDT scaling model $NEP \propto \langle CWM_{SLA} \rangle \cdot \langle CWV_{SLA} \rangle \cdot C_{Tot}^b$, where NEP is the Net Ecosystem Production (μmol CO_2 m^{-2} s^{-1}). To focus on just the effects of SLA and C_{Tot} on NEP, we allowed each sample site along the gradient to be a random factor in the model. Here, b is the fitted exponent. Here, the values $\langle CWM_{SLA} \rangle$ and $\langle CWV_{SLA} \rangle$ are the community abundance-weighted mean and variance in SLA, respectively, based on log transformed trait values. The fitted model explains $\sim 71\%$ of the variation in NEP (df $= 22$, $F = 11.04$, $p < 0.0001$, AIC $= -24.39$). In support of TDT,

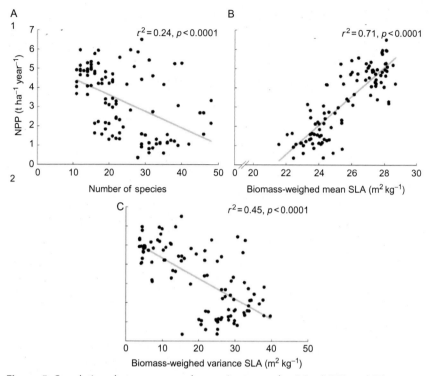

Figure 5 Correlations between annual net primary productivity (NPP) and (A) species diversity per plot; (B) the plot community biomass-weighted mean, CWM, of the SLA distribution; and (C) the plot community biomass-weighted variance, CWV, of the SLA distribution for the four selected plots from the Park Grass experiment. Each dot represents the annual aboveground biomass production of a given plot in a given year. Compare these relationships to the predictions with Trait Driver Theory and Biodiversity and Ecosystem Functioning Theory (Table 3). Note that there is a negative correlation between NPP and diversity, while the opposite is predicted by classical Biodiversity–Ecosystem Functioning theory. However, in accordance with TDT, there is a *negative* relationship between the variance of the abundance-weighted community trait distribution and a *positive* relationship between mean-weighted trait and NPP. These results show a direct linkage between a critical trait that influences plant growth, forces that influence the shape of the trait distribution, and how both changes in trait mean and variance then shape ecosystem functioning.

both increases in $\langle \mathrm{CWV_{SLA}} \rangle$ and C_{Tot} each increase community NEP ($p=0.023$, $t=0.337$; and $p=0.058$, $t=2.332$ respectively) while increases in $\langle \mathrm{CWV_{SLA}} \rangle$ tends to *decrease* NEP ($p=0.068$) (see Fig. 4). The fitted exponent for b is different than the predicted value of 3/5 ($b=0.13 +/- 0.14$) but the range of variation in plot biomass, C_{Tot} in this study is generally less than an order of magnitude. In contrast, variation in species diversity explains none of the variation in NEP in this system either as a single predictor ($r^2 = 0.029$, $p = 0.3617$). Allowing each sample site along the gradient to be a random factor in the model shows that species richness has a negative effect on NEP in the model ($t = 0.01$, $p = 0.03$). Additional analyses underscore the importance of both $\langle CWM_{SLA} \rangle$ and $\langle CWV_{SLA} \rangle$ on influencing variation NEP across this elevational gradient (see Appendix). Together, these results support several key predictions of TDT—the shift in the mean of $C(b_0)$ is closely tied to environmental drivers and that shifts in the mean and variance of $C(b_0)$ are a primary driver of variation in community carbon flux (Fig. 4C and D).

Across broad climatic gradients, recently Michaletz et al. (2014) utilized MST to predict variation in annual net primary productivity (NPP, grams of biomass per area per year). In support of MST, rates of growing season NPP scaled with total autotrophic biomass indistinguishable from the allometrically ideal value of 3/5 predicted value in Eq. (12). Their analysis support another prediction of TDT that controlling for scaled effects of total biomass, C_{Tot}, and stand age on NPP shows that shifts in $<b_0>$, primarily because of shifts in $\langle SLA \rangle$, will also shift variation in NPP (Fig. 7).

At larger biogeographic scales, assemblage trait maps such as Fig. 5 could then be used to predict ecosystem functioning. At these larger geographic scales, if the distribution of SLA still reasonably approximates rates of biomass production, then, according to TDT, regions with high mean SLA and low variance should have the highest rates of instantaneous net primary production. In general, recent compilation of geographic variation in instantaneous rates of terrestrial ecosystem NPP from remotely sensed data indicates that areas with the highest instantaneous rates of NPP do generally correspond[2] to assemblages with high mean and low variance in SLA. However, according to Eq. (16), one should also control for total system biomass (which correlates with variation in tree height) as well as variation in the other traits that also can influence NPP. Nonetheless, the correspondence between biogeographic variation traits and predictions from TDT is a promising future direction.

7.5 Temporal Trait Shifts Across Fertilization Gradients

The results from the long-term dynamics and fertilization experiment from Rothamsted are given in Tables 2 and 3 as well as in Figs. 6 and A3–A5. Within the Rothamsted dataset, all of the traits studied, the biomass-

---

[2] http://daac.ornl.gov/NPP/npp_home.shtml#.

Table 2 Observed Temporal Changes in the Central Moments of the Community Trait Distribution C(z) in the Park Grass Experiment (see Figures A3–5 in Appendix)

	Moment of Community Trait Biomass Distribution, C(z)	Control (Unfertilized)	Fertilized	Corresponding TDT Predictions in Table 1
SLA	Mean (CWM)	0.65*** (−)	0.40** (+)	See I(a) and I(b)
	Variance (CWV)	0.48** (+)	0.26* (−)	See II(a)–(e)
	Skewness (CWS)	0.13$^{ns.}$	0.29* (−)	See III(a)–(b)
	Kurtosis (CWK)	0.69*** (−)	0.10$^{ns.}$	See IV(a)–(b)
Height	Mean (CWM)	0.61** (−)	0.04$^{ns.}$	See I(a) and I(b)
	Variance (CWV)	0.27* (−)	0.51** (+)	See II(a)–(e)
	Skewness (CWS)	0.17$^{ns.}$	0.02$^{ns.}$	See III(a)–(b)
	Kurtosis (CWK)	0.38** (+)	0.20$^{ns.}$	See IV(a)–(b)
Seed mass	Mean (CWM)	0.21$^{ns.}$	0.01$^{ns.}$	See I(a) and I(b)
	Variance (CWV)	0.38** (−)	0.01$^{ns.}$	See II(a)–(e)
	Skewness (CWS)	0.56** (−)	0.00$^{ns.}$	See III(a)–(b)
	Kurtosis (CWK)	0.31* (−)	0.01$^{ns.}$	See IV(a)–(b)

Predictions from Trait Driver Theory correspond to the cells indicated in Table 1. As an estimate of the central moments of C(z), we estimated the community-weighted values for the mean, variance, skewness, and kurtosis (CWM, CWV, CWS, and CWK, respectively). Values are the Pearson product–moment correlations, r, between time since the start of the experiment and the corresponding trait distribution moment for the three main traits investigated, specific leaf area or SLA, adult height, and seed size. In accordance with TDT, for the trait SLA, the distributions of the fertilized and unfertilized plots have diverged for all trait moments. Further, in accordance with TDT, increasing fertilization leads to an increasing skew of the SLA distribution, but not for seed size and reproductive height, traits that do not directly underlie the growth equation, indicating that fertilization is a strong environmental driver that influences plant growth. Correlations are for both unfertilized: plot 2 and fertilized: plot 16. *$p < 0.05$; **$p < 0.01$; ***$p < 0.001$; ns., not significant; (−), negative relationship; (+), positive relationship. Observed shifts in the central moments of SLA are generally in accordance to predictions from TDT (see Table 1 and text for details).

weighted distribution of the trait SLA was associated with the most prominent shifts in the central moments of communities in response to fertilization (Fig. A1; Table 2; see also Appendix). For both fertilized and control plots, all four moments of the community SLA distribution changed significantly. The overall effect of fertilization on the community trait

Table 3 Correlations Between Trait Moments and Annual Net Primary Productivity for the Park Grass experiment

	Central Moment of Community Trait Distribution	Observed Correlations	Predicted Response–Trait Driver Theory	Predicted Response–Biodiversity Theory
SLA	Mean (CWM)	0.71*** (+)	+	NP
	Variance (CWV)	0.45*** (−)	−	+
	Skewness (CWS)	0.28*** (+)	NP	NP
	Kurtosis (CWK)	0.19*** (+)	+	−

Also listed are the predicted signs of the correlations made from Trait Driver Theory (see text and Table 1) and from Biodiversity–Ecosystem Functioning theory (assuming the theory of Tilman et al., 1997 where variability in trait SLA is a good proxy for variation in species richness via niche or trait space). *Note:* both theories make opposite predictions for the signs of the correlations. NP, no specific prediction is made as such predictions would depend upon specifics of system. As an estimate of the central moments of $C(z)$, we estimated the community biomass-weighted values for the mean, variance, skewness, and kurtosis (CWM, CWV, CWS, and CWK, respectively).
***$p < 0.001$.
(−), Negative relationship; (+), positive relationship.

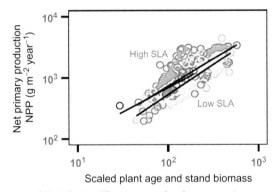

Figure 6 In support of Trait Driver Theory, woody plant net primary production scales with total assemblage biomass (see Eqn. 12). Further, for a given amount of biomass, NPP will be modified by shifts in the mean community leaf trait, specific leaf area or SLA (see Eqn. 15). *Figure modified from Michaletz et al. (2014).*

distribution is consistent with fertilization differentially favouring certain phenotypes (Chapin and Shaver, 1985; Suding et al., 2005; Tilman, 1982) and a replacement of slower growing species with faster growing species (Chapin, 1980; Grime and Hunt, 1975; Knops and Reinhart, 2000). Similar to past findings, across all plots, the community mean SLA increased (see Craine et al., 2001; Knops and Reinhart, 2000) and the variance decreased. The directional community trait shift is reflected in increased skewness values (differing from zero). Fertilization also led to a shift in the kurtosis but only for SLA values. Specifically, SLA shifted from negative kurtosis values to zero or positive kurtosis values, suggesting that fertilization increased rates of competitive exclusion of suboptimal trait values leading to a more peaked trait distribution. Intriguingly, the direction and rate of change for fertilized versus control plots differed in sign and magnitude indicating that the trait distributions of control and experimental plots steadily diverged over time.

The PGE supports several predictions from TDT. First, TDT predicts that a shift in an environmental driver (fertilization in this case) should primarily be seen as a shift in traits associated with growth rate. Of all of the traits assessed, SLA showed the strongest shifts over time (Table 1). The other traits showed relatively little to no change over time. The disproportionate shift in SLA is consistent to expectations from TDT as SLA is the only trait directly linked to the growth function, $f(b_0)$. Second, consistent with a shifting in an optimal phenotype, the skewness of the fertilized plot increased over time. Third, consistent with a productivity–trait variance trade-off, the annual net primary productivity (NPP) was positively correlated with community mean and kurtosis of the SLA distribution but negatively correlated with variance in SLA (Fig. 6; Table 3). Lastly, shifts in the trait distribution are more closely tied to NPP than species richness (see Table 3). We observe a weak to negative correlation between species richness and NPP (Fig. 6A; Table A1). Together, our approach provides: (i) a more mechanistic understanding of the long-term response of the species and communities found within the PGE experiment and (ii) an alternative trait-based approach that can in principle be integrated with past 'species richness based' theories invoked to explain the decrease species richness with fertilization (Tilman, 1982).

8. DISCUSSION

We have shown that TDT can formalize numerous assumptions and approaches in trait-based ecology. We provide examples of how this can be done for several different biodiversity hypotheses in terms of the dispersion of traits (Tables 1 and 3). We further argue that ecological theories need to move beyond species richness and be recast in terms of organismal performance via functional traits. As a result, TDT offers an alternative framework to the standard taxonomic approach for linking biodiversity and ecosystem functioning, where primacy has been placed on the importance of species richness. TDT instead focuses on the importance of 'trait diversity' via the shape of the trait distribution of individuals and shared performance currencies (e.g. growth). Because TDT incorporates intraspecific variation, it necessarily includes natural selection as a process that shapes the trait distribution. By incorporating interspecific trait variation, it also includes 'selective' processes at higher levels of organization within the community, such as species sorting (see also Shipley, 2010). We show that using TDT to analyze these processes leads to several predictions opposite to predictions made by Biodiversity–Ecosystem Function theory (Naeem and Wright, 2003; Tilman et al., 1997; see also Table 3); it also offers a useful alternative hypothesis by which to assess the linkage between 'diversity' (whether measured by species richness or via the trait distribution) and ecosystem functioning.

TDT also offers a predictive framework for management. Increasingly, trait-based approaches to management have shown that a focus on trait shifts due to land use, as well as management and agricultural practice, can yield deeper insight into the processes of concern to managers (Garnier and Navas, 2012). For example, biomass production, the timings of peak production and plant digestibility, and response time to disturbance can be predicted from the shape of the community trait distribution as well as many of the plant traits underlying our general growth equation (see studies and references listed in Garnier and Navas, 2012). It is intriguing to note that TDT predicts a trade-off between short-term productivity and long-term response to the environment. Agriculture focused on maximizing short-term productivity reduces trait variance by planting genetically homogeneous monocultures.

Our analyses find that none of the central moments of the trait distribution in the Park Grass dataset are correlated with species richness (see Table A1 in the Appendix). Indeed, species richness does not appear to be a reliable proxy for how the diversity of phenotypes and trait distributions respond to environmental change. Further, across large biogeographic gradients, recent studies have found the trait variance and the total functional trait space is often unrelated to species richness (Lamanna et al., 2014; Safi et al., 2011; Šímová et al., 2014; Fig. 7). The potential for improved predictions using trait and size distributions that can be linked to metabolic scaling fundamentally comes down to the increase in information contained in traits that is not necessarily present in a species-richness-based approach (Tilman et al., 1997) or even a phylogenetic approach to community ecology

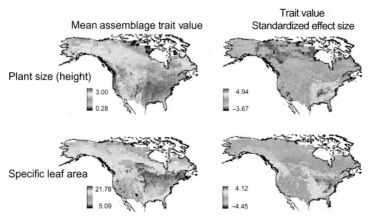

Figure 7 Biogeographic variation in the (A) mean species assemblage functional trait specific leaf area (SLA, $m^2\ g^{-1}$) and plant height (m) and (B) the standardized effect size (SES) of the mean trait values across woody plant communities across North America. The SES measures is the observed value is significantly higher or lower than expected given the species richness. Thus, high SES values indicate greater mean trait values than expected from the observed number of species and low SES values indicate lower mean values than expected. The mean SLA is lowest in the low elevation-latitude tropical forests and tends to be highest in temperate forest and grassland regions. Across broad geographic gradients, both the mean and the variance significantly vary. According to Trait Driver Theory, areas that correspond to high mean SLA and larger standing stocks of biomass (corresponding to taller mean heights) should have the highest rates of instantaneous net primary production, NPP. However, NPP will also be modified by the assemblage variance in SLA. *Figure modified from Šímová et al. (2014).*

(Cavender-Bares et al., 2009). Although TDT is based on simplifying assumptions, it helps to better connect and scale trait-based ecology and MST with large-scale ecology and biogeography. It integrates and builds upon prior work that developed highly mechanistic trait-based models (Norberg et al., 2001; Zhang et al., 2013) and other work that predicted how individual growth rates change across scale with size and temperature (Enquist et al., 2007a). TDT thus enables ecological theories to be 'scaled up' to predict and test the consequences of how organismal response to climate change will ramify at the community and ecosystem levels across both local- and large-scale gradients in geography (space) and fluctuations in time (climate change).

It is becoming clear that multiple assembly processes—abiotic filtering, biological enemies, competition, facilitation, below-ground competition—likely operate simultaneously and at differing scales to structure communities and larger scale species assemblages (Cavender-Bares et al., 2009; Grime,

2006; Mayfield and Levine, 2010; Swenson and Enquist, 2009). The result is that the distribution of some traits may be more over- or under-dispersed than others. Future work should better link empirical data with theory to test the community and ecosystem responses when multiple trait drivers influence trait variance and when trait optima are strongly influenced by differing levels of ecological interactions (competition, predation, mutualism, etc.).

Our work has primarily focused on the traits and environmental drivers that underlie growth rate. Indeed, refining and extending TDT will also require better identification of the above- and below-ground traits that influence growth rate. Lastly, analysis of the shift in the mean and variance of assemblage values of many plant functional traits as well as stand biomass across biogeographic scales may provide the necessary basis to predict ecosystem functioning across large scales. Focusing on variation in SLA, TDT predicts that the distribution of SLA will influence rates of biomass production. Correcting for the effects of stand biomass (see Eq. 8), assemblages with high mean SLA and low variance should have the highest rates of instantaneous net primary production. In general, recent compilation of geographic variation in instantaneous rates of NPP from remotely sensed data indicates that areas with the highest instantaneous rates of NPP generally do correspond (Zhao and Running, 2010; Zhao et al., 2005) to assemblages with high mean and low variance in SLA regions identified by Swenson et al. (2012) and Šímová et al. (2014). TDT also predicts that while regions with relatively lower trait variance will be more sensitive to rapid directional climate change, assemblages with greater variance, however, would be expected to more closely track climate change. Future tests of TDT at the scale of global ecology should more formally assess the predictions of TDT at this scale by assessing the specific relationships between the trait distribution, vegetation biomass, and possible covariation of other traits.

The TDT prediction of an inverse relationship between trait variance and production is not necessarily in conflict with either 'positive species complementarity'—niche partitioning allows species to capture more resources in ways that are complementary in both space and time (Tilman, 1999)—or 'transgressive overyielding' where species use resources in ways that are complementary in space or time to stably coexist with one another so that more diverse communities capture a greater fraction of available resources and produce more biomass than even their most productive species (see Tilman et al., 1997). TDT needs to be reconciled with these ideas because recent studies confirmed that within biodiversity experiments, 'positive species complementarity' does enhance ecosystem productivity (Cardinale et al., 2007). Given TDT, a natural question is how are trait distributions modified when complementarity effects are strong? Effects such as complementarity can be incorporated into TDT growth functions as explained in Savage et al. (2007).

In sum, we have argued that more powerful tests of biodiversity theories need to move beyond species richness and explicitly focus on mechanisms generating diversity via trait composition and diversity via the shape of trait distributions. The rise of trait-based ecology has led to an increased focus on the distribution and dynamics of traits across broad geographic and climatic gradients and how these distributions influence ecosystem function. However, a trait-based ecology that is explicitly formulated to apply across different scales (e.g. species that differ in size) and gradients (e.g. environmental temperatures) has yet to be articulated. The TDT presented here is a formalization of essential steps for mechanistically linking and scaling functional traits for individuals with the dynamics of ecological communities and ecosystem functioning. This TDT approach builds upon and complements existing trait-based approaches in ecology (e.g. Grime, 1998; Kraft et al., 2008; Lavorel and Garnier, 2002; Suding et al., 2008a). It is appealing because it can connect individual physiology and traits with ecosystem dynamics and how both respond to climate change (Suding et al., 2008b), geographic gradients, and differing ecological processes (e.g. niche vs. neutral; see Weiher et al., 2011). Given the increasing ability to remotely sense numerous traits of terrestrial vegetation (Asner et al., 2014; Doughty et al., 2011) and the increasing access to both plant and animal trait data (Kattge et al., 2011), TDT and its elaborations are ripe for providing empirically grounded, mechanistic models of ecosystem dynamics from local to large scales.

ACKNOWLEDGEMENTS

We thank Mark Westoby who provided enthusiasm and input during the writing of initial drafts of this chapter. We also thank the ARC-NZ Vegetation Network, working group 36, led by V.M.S., for their generosity and support in bringing us all together to meet and to initiate this collaboration and chapter. We thank other members of working group 36, especially Christine Lamanna, Tony Dell, Mick McCarthy, and Graham D. Farquhar, who helped us solidify our central arguments. C.V. was supported by a Marie Curie International Outgoing Fellowship within the 7th European Community Framework Program (DiversiTraits project, no. 221060). B.J.E. and V.M.S. were supported by an NSF ATB award and B.J.E. by an NSF Macrosystems award. C.T.W. was supported by NSF Grant DEB-0618097. J.N. was supported by the Swedish Research Council and FORMAS Grant EKOKLIM. Lastly, we thank the Park Grass staff, in particular Andy Macdonald and Margaret Glendining, for assistance with and making available the Park Grass data.

APPENDIX

A.1 Using TDT to Recast Ecological Hypotheses

A unique attribute of TDT is that several differing ecological theories can now be recast in terms of how they influence the shape of trait frequency distributions ($C(z)$). Each of these theories makes differing hypotheses that influence the shape of trait distributions and the functioning of ecosystems:

(i) *H1*: *Phenotype–environment matching*: This hypothesis states that species are more successful in different parts of the landscape because individuals have different trait values across space that are better adapted to local features of that space, such that the mean 'phenotype' matches variation in the local environment (Westoby and Wright, 2006). This distinction builds upon observations that stem back to Schimper (1898, 1903) that form the foundation for understanding changes in fitness and functional traits and species composition across gradients (Levins, 1968; Westoby and Wright, 2006). A prediction of this hypothesis is that, either due to convergent evolution or abiotic filtering of relevant traits, similar environments should be dominated by species with similar trait values (see discussion in Karr and James, 1975; Mooney, 1977; Orians and Paine, 1983).

(ii) *H2*: *The competitive-ability hierarchy hypothesis*: This hypothesis states that the strength of competition between individuals is driven by the distance between individuals as measured according to their functional traits (Freckleton and Watkinson, 2001). The competitive-ability hierarchy hypothesis leads to opposite predictions than the niche-based competition-trait similarity and competition-relatedness hypotheses (see below; Mayfield and Levine, 2010). Here, the resulting competitive hierarchy will cause a reduction in the trait variance over time and increased functional clustering because individuals that share a trait value will outcompete individuals with different trait values (Freckleton and Watkinson, 2001; Kunstler et al., 2012; Mayfield and Levine, 2010).

(iii) *H3*: *Abiotic filtering*: The importance of local abiotic forces is reflected in the community trait range and variance. Abiotic filtering hypothesis states that increasingly more stressful environments will limit the range and variance influence. This hypothesis was formalized by Keddy and colleagues (Keddy, 1992; Kraft and Ackerly, 2010; Weiher and Keddy, 1999). Similarly, on ecological time scales, due to phenotype–

environment matching, increasingly more stressful environments, E, will increasingly restrict the range of trait values that could co-occur within a given environment. This 'trait filtering hypothesis' states that the abiotic environment filters trait values so as to limit the variance and range of the trait distribution. This hypothesis, which can be seen as an ecological scale version of the 'favourability' hypothesis of Terborgh (1973), predicts that more physiologically stressful environments (frost, high salinity, drought, etc.) should place especially rigid filters on the types of phenotypes (i.e. traits) that can survive and potentially co-occur (e.g. Kraft et al., 2008). Note, as discussed in (ii) and (iii), the variance and range of a trait distribution $C(z)$ are also influenced by biotic forces. Similarly, repeated disturbance or environmental variability may minimize local interactions and could also increase community trait variance (Grime, 2006).

(iv) *H4: Strength of local biotic forces is revealed via trait variance and kurtosis*: Differing biotic community assembly hypotheses can differentially influence the spacing of trait values within the range of filtered phenotypes. As stated in (ii), competition for a common limiting resource would ultimately lead to competitive exclusion (e.g. see Tilman, 1982) resulting in a convergence of 'superior competitor' phenotypes (Abrams and Chen, 2002; Mayfield and Levine, 2010; Savage et al., 2007). This convergence would be reflected by decreasing variance and an increase in 'peakedness' of the trait distribution or an increase in positive kurtosis (Navas and Violle, 2009). In contrast, according to Chesson (2000), if traits map onto niche differences, increased niche (trait) differentiation will lead to increasing coexistence of individuals with differing traits. These classical niche partitioning models predict that competition will limit functional (trait) similarity (MacArthur and Levins, 1967) and thus increase in the spacing between co-occurring phenotypes (see Diamond, 1975; MacArthur, 1958). Similarly, biological enemies (Kraft and Ackerly, 2010), facilitation (Brooker et al., 2008), and frequent disturbance (Grime, 1998) can maintain trait diversity (e.g. an over-dispersion of phenotypes). Niche packing models result in either a broader or evenly dispersed trait distribution (high variance) or even a multimodal trait (negative kurtosis) distribution.

(v) *H5: Assessing neutral forces via the shape of local and regional trait distributions*: An alternative hypothesis to (ii) and (iii) is that local communities are primarily structured by stochastic dispersal, drift, and dispersal limitation (Hubbell, 2001). Such a neutral scenario would predict on average, for traits not associated with dispersal, little to no difference in the shape of the community trait distribution when sampled across differing spatial scales. Further, for traits not associated with dispersal ability, there should be no relationship between trait distribution and changes in the environment.

A.2 Integrating Metabolic Scaling Theory into TDT

Within TDT, the total biomass associated with a trait, z, is denoted by $C(z)$. This notation avoids ever needing to account for individual mass, M, or number of individuals with mass. In contrast, the growth equations Metabolic Scaling Theory are phrased in terms of individual mass, M, where organismal growth rate, dM/dt, is given by

$$\frac{dM}{dt} = b_0(z)M^{3/4} \tag{A1}$$

where $b_0(z)$ is a coefficient that depends on a single or set (meaning z is a vector) of traits.

To integrate MST and TDT, we first recognize that, $C(z)$, the biomass associated with trait z can be expressed as $C(z) = \int dM C(z, M) = \int dM N(z, M)M$, where $C(z,M)$ is the mass density of individuals with both trait value z and individual mass M, while $N(z,M)$ is the number density of individuals that have both trait value z and individual mass M. In this expression, we have integrated over all possible values of mass, M, so we have the total biomass of *all* individuals with trait z. Furthermore, note that integrating this over all traits, z, gives the total biomass, $C_{\text{Tot}} = \int dz C(z) = \int dz \int dM N(z, M)M$. Consequently, we can multiply both sides of Eq. (A1) by $N(z, M)$, integrate both sides over $\int dz \int dM$, and multiply and divide the right side by C_{Tot} to obtain

$$\int dz \int dM N(z, M)\frac{dM}{dt} = \left[\frac{\int dz \int dM N(z, M)b_0(z)M^{3/4}}{\int dz \int dM N(z, M)M}\right] C_{\text{Tot}} \tag{A2}$$

At steady state, $N(z,M)$ is not changing in time, so we can move it inside of the derivative with respect to time, and we can also move the integrals inside of the derivative because the integration over all possible traits and masses is not a time-dependent object

$$\frac{d\left[\int dz \int dM N(z, M)M\right]}{dt} = \frac{dC_{\text{Tot}}}{dt} \tag{A3}$$

Furthermore, we can express the bracketed term on the right side of Eq. (A1) as a mass average of a growth function as follows

$$
\left[\frac{\int dz \int dM N(z,M) M\left[b_0(z)M^{-1/4}\right]}{\int dz \int dM N(z,M) M}\right] = \left[\frac{\int dz \int dM C(z,M)\left[b_0(z)M^{-1/4}\right]}{\int dz \int dM C(z,M) M}\right]
$$

$$
= \left\langle b_0(z)M^{-1/4}\right\rangle_C
$$

$$
\text{(A4)}
$$

where the C subscript denotes that the average is taken with respect to the biomass. Combining all of this, we obtain the equation for the scaling of Net Primary Productivity or NPP equation

$$
\frac{dC_{\text{Tot}}}{dt} = \left\langle b_0(z)M^{-1/4}\right\rangle_C C_{\text{Tot}} \qquad \text{(A5)}
$$

This equation is in the most generic form of a TDT equation, and the growth function $f(z) = b_0(z)M^{-1/4}$ can be expanded such that the biomass growth equation can be expressed in terms of the biomass-weighted central moments of the trait z, such as the variance, skewness, and kurtosis.

We now consider a few special cases of Eq. (A5) to relate to the TDT and scaling equations already in the literature. In the case that there is only a single mass value, M^*, or a very small range of mass values, the number density becomes $N(z,M) = N(z)\partial(M - M^*)$ in terms of a Dirac-delta function for the mass dependence. Therefore, $\int dz \int dM N(z)\partial(M - M^*)$ $M\left[b_0(z)M^{-1/4}\right] = \int dz N(z)M^*\left[b_0(z)(M^*)^{-1/4}\right] = (M^*)^{-1/4}\int dz C(z)b_0(z)$, and Eq. (A4) becomes

$$
(M^*)^{-1/4}\frac{\int dz C(z)b_0(z)}{\int dz C(z)} = (M^*)^{-1/4}\langle b_0(z)\rangle_C \qquad \text{(A6)}
$$

so the scaling of NPP equation becomes

$$
\frac{dC_{\text{Tot}}}{dt} = (M^*)^{-1/4}\langle b_0(z)\rangle_C C_{\text{Tot}} \qquad \text{(A7)}
$$

The term $(M^*)^{-1/4}$ can be thought of as an overall normalization to the growth function $f(z)$ from TDT. As such, this result reveals that TDT, as originally formulated, essentially ignores individual mass. Thus, based on Eq. (A1), growth functions within TDT should have a roughly $(M^*)^{-1/4}$ hidden with the normalization constant for their growth function.

Conversely, as a special case, we consider the function $b_0(z)$ to be a constant b_0 that occurs when $z = z^*$. In this case, $N(z, M) = N(M)\partial(z - z^*)$ and $\int dz \int dM N(M)\partial(z - z^*)M[b_0(z)M^{-1/4}] = b_0 \int dM N(M)M^{1/4}$, and Eq. (A4) becomes $b_0 \int dM N(M)M^{3/4}/C_{\text{Tot}}$. Combining these terms gives the scaling of NPP equation

$$\frac{dC_{\text{Tot}}}{dt} = \frac{b_0 \int dM N(M)M^{3/4}}{C_{\text{Tot}}} C_{\text{Tot}} = b_0 \int dM N(M)M^{3/4} \qquad (A8)$$

Following the arguments in Enquist et al. (2009), we can substitute $N(M) \propto M^{-11/8}$ to obtain $M_b \propto C_{\text{Tot}}^{8/5}$ where the subscript denotes that largest mass in the group. Using these relationships

$$\frac{dC_{\text{Tot}}}{dt} \propto b_0 \int dM M^{-11/8} M^{3/4} \propto b_0 \int dM M^{-5/8} \propto b_0 M_b^{3/8} \propto b_0 C_{\text{Tot}}^{3/5} \qquad (A9)$$

Defining a new constant b_0' to denote the product of b_0 with all of the proportionality constants, the overall scaling of NPP equation becomes

$$\frac{dC_{\text{Tot}}}{dt} = b_0' C_{\text{Tot}}^{3/5} \qquad (A10)$$

in accord with the Net Primary Productivity scaling equation derived by Enquist et al. (2009).

As a final special case, we consider the trait z and the mass M to be uncorrelated and the number density to be a separable function such that $N(z, M) = N(M)N(z)$. Therefore, using results from Eq. (A9) and the definition of an average, Eq. (A4) can be expressed as

$$\frac{\int dz N(z)b_0(z)}{\int dz N(z)} \frac{\int dM N(M)M^{3/4}}{\int dM N(M)M} = k\langle b_0(z)\rangle C_{\text{Tot}}^{-2/5} \qquad (A11)$$

where k captures the proportionality constants in deriving Eq. (A5). Substituting this into our overall growth equation (Eq. A5) yields

$$\frac{dC_{\text{Tot}}}{dt} = k \langle b_0(z) \rangle C_{\text{Tot}}^{3/5} \tag{A12}$$

where the average is now the standard abundance average and is *not* the biomass-weighted average. This is the growth equation in the scaling form for this special case. Alternatively, for this special case, we could express Eq. (A4) as

$$\frac{\int dz N(z) b_0(z)}{\int dz N(z)} \frac{\int dM N(M) M \left[M^{-1/4} \right]}{\int dM N(M) M} = k \langle b_0(z) \rangle \left\langle M^{-1/4} \right\rangle_C \tag{A13}$$

and the NPP scaling equation becomes

$$\frac{dC_{\text{Tot}}}{dt} = k \langle b_0(z) \rangle \left\langle M^{-1/4} \right\rangle_C C_{\text{Tot}} \tag{A14}$$

This equation is completely equivalent to Eq. (A13) but expresses the growth function more in terms of the TDT framework such that the right side appears to have an overall linear dependence in C_{Tot}, and as a result, we have a mixture of types of averages, with the function $b_0(z)$ being abundance averaged and the $M^{-1/4}$ being biomass averaged.

The major results of this section are Eq. (A5), which is the most general formulation of the growth equation because it does not rely on traits or mass being constant or uncorrelated. In this form, the growth equation is like the TDT formulation, but as the special cases below it reveal, $\langle b_0(z) M^{-1/4} \rangle_C$ may hide extra dependencies on C_{Tot}. Equation (A7) is the result when the mass is constant and is expressed in the form of TDT equations such that it is linear in C_{Tot} and reveals an overall $M^{-1/4}$ for adjusting the growth function across groups. Equation (A10) is the special case where the traits are constant and reduces to the exact scaling equation given in Enquist et al. (2009). Finally, when traits and mass are uncorrelated, Eqs. (A13) and (A14) are two different but completely equivalent ways to express the growth function. Equation (A13) is in the form of scaling equations by consolidating the mass average with the C_{Tot} dependence, while Eq. (A14) is in the form of TDT equations by keeping two averages around, including one that is an abundance average and one that is a biomass-weighted average. For all of these equations and cases, the functions inside the averages can be expended in terms of moments as done for TDT for biomass-weighted averages or as done in 2004 for abundance-weighted averages.

A.3 Growth Functions Across Environmental Gradients: Incorporating Trade-Offs into TDT

Importantly, as discussed in the main text, Eq. (9) predicts an unbounded growth response such that increasingly larger values of b_0 always leads to increased growth, which realistically cannot continue indefinitely. A key assumption of TDT is that there is fundamental trade-off between a given trait value and the performance of an organism across an environmental gradient, E. The final step to integrate a general TDT that can link traits, organismal performance, and environmental gradients is to specify trade-offs between underlying traits, growth, and metabolic scaling.

Within a given environment, E, an important question is what would prevent the average metabolic normalization, $<b_0>$, from becoming infinitely big or small? In the case of growth rate, possible trade-offs likely include the types of limiting resources individuals use or the environmental conditions for optimal growth. So, individuals that allocate internal resources to specific traits defined by b_0 may reduce the impact of one limiting environmental factor, but this would necessarily incur a disadvantage with respect to another environmental limiting factor.

A trade-off or cost function can be formulated within the growth function, f. Multiplying this cost function by f shows that, for a given E, the growth function has a maximum at z_{opt} or here $b_{0_{opt}}$, and as a result, the second derivative of f (the second term in Eqs. A5 and A14) will be negative, as long as $<b_0>$ is close to $b_{0_{opt}}$.

We can specify a generic form of a trade-off by following Norberg et al. (2001). We can approximate a trade-off by first invoking a general quadratic or Gaussian cost function on the community value ($<b_0> - b_{0_{opt}}$). We add a cost function to Eqs. (A5) and (A14). This provides a general form of a trade-off. That new cost function that is multiplied by f could be a general quadratic $\left[1 - \left(\frac{<b_0> - b_{0_{opt}}}{\sigma^2}\right)^2\right]$ or Gaussian function, $\exp\left[-\left(\frac{<b_0> - b_{0_{opt}}}{\sigma^2}\right)^2\right]$. Here, σ^2 is the observed standard deviation in the trait or b_0 observed within the assemblage. Both cost functions reduce to 1 when $<b_0> - b_{0_{opt}}$ (e.g. for a given environment E, the observed mean community trait, z_{opt}, and average metabolic normalization, $<b_0>$, are at the local optimum). Note both decrease in value as you go away from $b_0 - b_{0_{opt}}$. Dividing through by σ^2 defines the penalty for individual growth rate for being away from the optimum. Thus, for a given environment, E, characterized by a unique $b_{0_{opt}}$, the growth function can be made more explicit in terms of a generic trade-off where

$$f(b_0) = c_0 b_0 \langle M^{\theta-1}(b_0) \rangle \left(1 - \frac{(b_0 - b_{0,\text{opt}})^2}{\sigma^2 b_0} \right) \qquad \text{(A15)}$$

This is a modified growth function and is characterized in Fig. 2. It can be made more specific by incorporating the traits that then define b_0. Here, the first term is the general growth equation from the relative growth rate literature (Poorter, 1989) that has been more formally derived in metabolic scaling theory. The second term of Eq. (A1) is the associated trade-off. As a result, the second term in the TDT Eq. (5) in the main text gives how much of whole-community biomass production is *reduced* due to the amount of trait variance, V, in the community because of the explicit trade-off function, the second derivative of f, $\frac{d^2 f_{b_0 = <b_0>}}{db_0^2}$, would then be negative near $b_{0_{\text{opt}}}$. As a result, increasing variance in b_0 would then *decrease* total community production.

A.4 Methods: Approximating the Shape of the Community Trait Distribution via Community-Weighted Measures

In order to assess predictions of TDT, it is necessary to quantify the full distribution of traits in a community, $C(z_i)$. This involves measuring the trait values of *all individuals* and thus incorporates both inter- and intraspecific trait variability. While measuring traits of *all* individuals in a community is ideal and several studies have done so (Albert et al., 2010; Gaucherand and Lavorel, 2007; Lavorel et al., 2008), it is a time-consuming work (Baraloto et al., 2010). While there are limitations, the trait biomass distribution, $C(z_i)$, can be approximated, and predictions of TDT can be tested without explicitly measuring the traits of all individuals.

Trait distributions can be approximated in two ways. The first method is straightforward and calculates the weighted trait distribution by taking the mean species trait value and multiplying by a measure of dominance (cover, biomass, abundance; Grime, 1998). This method can be implemented by calculating the central moments of joint distribution.

This community-weighted variance or CWM is increasingly a standard metric in trait-based ecology (Garnier and Navas, 2012; Lavorel, 2013; Violle et al., 2007) and represents the trait mean calculated for all species in a community weighted by species abundances as follows:

$$\text{CWM}_{j,y} = \sum_{k=1}^{n_j} A_{k,j} \cdot z_k \qquad \text{(A16)}$$

where n_j is the number of species sampled in plot j, $A_{k,j}$ is the relative abundance of species k in plot j, and z_k is the mean value of species k. Several studies have also assessed the community-weighted variance of the trait distribution (see Lavorel et al., 2011; Ricotta and Moretti, 2011).

The assemblage variance, V, calculated via the biomass-weighted values for the community-weighted variance ($\text{CWV}_{j,y}$) is given by

$$\text{CWV}_{j,y} = \sum_{k=1}^{n_j} A_{k,j} \cdot \left(z_k - \text{CWM}_{j,y}\right)^2 \tag{A17}$$

Further, the central moments skewness and kurtosis ($\text{CWS}_{j,y}$ and $\text{CWK}_{j,y}$, respectively) are given by

$$\text{CWS}_{j,y} = \frac{\sum_{k=1}^{n_j} A_{k,j} \cdot \left(z_k - \text{CWM}_{j,y}\right)^3}{\text{CWV}_{j,y}^{3/2}};$$

$$\text{CWK}_{j,y} = \frac{\sum_{k=1}^{n_j} A_{k,j} \cdot \left(z_k - \text{CWM}_{j,y}\right)^4}{\text{CWV}_{j,y}^2} - 3 \tag{A18}$$

A limitation of this approach, however, is that it ignores the contribution of intraspecific trait variability. Community trait moments may also be sensitive to the distribution of abundances across species. For example, a highly positive community kurtosis value may just reflect the hyper-dominance of one species and not the true dispersion of traits again due to intraspecific variation.

A second method utilizes subsampling individuals to obtain a better approximation of how intraspecific variation influences the community distribution. By subsampling individuals for each species, one can begin to incorporate intraspecific variation around mean trait values for each species.

In Fig. A5, we highlight a typical example that we believe can be used to generate two approximations of the community trait distribution. We use data from Konza Prairie LTER, Kansas, USA (McAllister et al., 1998). First, data were collected for the abundance of each species. These data are illustrated in Fig. A4 to estimate the community trait distribution from mean and variance measure of species traits. We find that, consequently and counter-intuitively, the inclusion of intraspecific variation will likely simplify modelling efforts because these types of distributions are much easier to manipulate and understand analytically. For each species, the standard deviation of trait variation is equal to the reported standard error multiplied by the square root (where $n = 3$). In sum, the community trait distribution can be approximated in two ways (methods B and C). While B emphasizes inter-specific variation, C also begins to include intraspecific variation. Method B is a reasonable approximation and can easily be implemented by most ecological studies as it only requires interspecific trait information and local abundance values. Method C requires an additional standardized subsampling method to estimate the standard error for each species but will result in a more accurate moment approximation.

A.5 Methods: Rocky Mountain Biological Lab: Shifts in Trait Distributions and Ecosystem Measures Across an Elevational Gradient

A.5.1 Measuring Whole-Community Trait Distributions

We measured community trait distributions and whole-ecosystem carbon flux data along an elevational gradient near Gothic, Colorado. The elevation gradient ranged between 2460 and 3380 m and spans 39 km in geographic distance. The elevational gradient contains five long-term study sites that run from dry, shrub-dominated high desert in Almont Colorado (2475 m) through the subalpine zone, to just below tree line (3380 m). The gradient consists of five long-term study sites that were established by Enquist in 2003 and has been sampled every year since. The gradient is located within Washington Gulch and East River valleys near the Rocky Mountain Biological Lab.

Each study site along the elevation gradient has similar local slope, aspect, and vegetation physiognomy. The sample area is approximately 50 m^2 and consists of a mixture of shrubs, grasses, and forbs. As discussed in Bryant et al. (2008), there is substantial turnover of plant species between sites with very few of the 120 species sampled, occurring in more than two of the sites. Additionally, shrub cover across the gradient decreases from a high of 33% at the lowest elevation site to 0% at the highest.

We utilized carbon flux data collected during the summer months of 2010 measured across the gradient. A species list and phylogeny for species at each site are given by Bryant et al. (2008). All sites contain weather stations on-site or nearby. Each study site has a similar local slope and south–southwest aspect and contains a mixture of herbaceous perennials, grasses, and shrubs. Since 2003, each year, five 1.3×1.3 m plots have been established haphazardly along the local slope of each study site, with at least 5 m distance between plots.

In 2010, Henderson measured the SLA of each plot and collected one fully expanded leaf from every individual. Fresh leaf samples were scanned (with petiole) in the laboratory, then dried to a constant mass, and weighed. The trait values measured from every individual in each community were compiled to create individual-level trait distributions for SLA. In total, leaves from 2253 individuals across 54 species were collected and measured at the five sites. Species turnover was high, with only 11 species being found at more than one site and only one species found at more than two sites.

A.5.2 Gas Exchange and Productivity Measures

In 25 plots (5 plots per elevational site), we measured total ecosystem carbon flux. Carbon flux was measured as instantaneous daytime peak uptake (ca. 10 am) and night-time peak respiration (ca. 10 pm) (Saleska et al., 1999). Ambient CO_2 was measured by a Li-Cor 7500 infra red gas analyzer for 30 s and then the tent was put in place over the plot and the CO_2 concentration within the tent was measured for 90 s (Jasoni et al., 2005). Daytime measurements were only taken under cloudless conditions. The tent was designed to let in 75% of photosynthetically active radiation (tent fabric by Shelter Systems). Air inside the tent is well mixed by fans, and the tent chamber was sealed using a long skirt along the base of the tent that was covered with a heavy chain. The volume of the tent used along the gradient was 2.197 m^3.

Soil efflux was measured at the same time as NEP using a Li-Cor 6400 portable photosynthesis machine with the soil chamber. The soil chamber fits inside a PVC soil collar, which was placed in the plot at least 2 weeks prior to the first measurement. Soil efflux was measured in two places in each plot along the gradient and one place per plot for the manipulation.

Carbon flux measurements along the elevational gradient were taken 4 weeks after snowmelt and then again at peak season (approximately 4 weeks after the first measurement, or when the majority of plants reached maximum height). Each NEP measurement consisted of daytime peak uptake (at $\sim$10 am) and night-time respiration (at $\sim$10 pm). Following the method of Jasoni et al. (2005), ambient CO_2 was measured for 30 s and then the tent was lowered and the CO_2 concentration within the tent was measured for 90 s, under clear sky and low wind conditions. Air inside the tent was well mixed by fans, and the chamber was sealed to the ground using a heavy chain.

For data analysis, we fit the predicted TDT model, using multiple regression in R using the 'car' library we fit the following linear model

```
lm(log10(NEP) ~ CWM.SLA + CWV.SLA + log10(Bio) + as.factor(Site))
```

where CWM.SLA is the community-weighted mean SLA and CWV.SLA is the community-weighted variance SLA calculated using the above equations for CWM and CWV as presented in the vegan package in R. Here, all values of SLA were log transformed before analyses, and log10(Bio) is the $\log_{10}$ total above ground dry biomass at the time of carbon flux measurement. We use the site elevation of the sample as a factor in the model. The fit of this model explained a large fraction, ~78%, of the variation in NEP ($R^2 = 0.778$, df = 22, $F = 11.04$, $p < 0.0001$, AIC = -24.38). To evaluate potential collinearity problems that may arise from linear relationships between model covariates (Ryan, 1997) we calculated variance-inflation factors (VIFs) for each covariate in each model using vif() from the package car in R. All VIFs were generally less than 5 except for one site where the vif = 5.86 which is nonetheless still low. In this model, the effect of CWV.SLA is significant ($p = 0.023$, $t = 2.45$, SE = 0.072, parameter = 0.337), but CWV.SLA and total biomass are marginally significant ($p = 0.068$, $t = -1.921$, SE = 0.067, parameter = -0.138; $p = 0.059$, $t = 1.994$, SE = 0.067, parameter = 0.135).

Variation in NEP across the gradient appears to be primarily due to the CWM and CWV of community SLA. Removing the parameter biomass and fitting a more simplified model with only mean and variance in SLA.

```
lm(log10(NEP) CWM.SLA + CWV.SLA + as.factor(Site))
```

predicts a similar amount of variation in NEP to the full model above ($p < 0.0001$, Adjusted $R^2 = 0.67$, AIC = -21.40 and both CWM.SLA and CWV.SLA are now significant within the model ($p = 0.035$, $t = 2.247$, SE = 0.358, parameter = 0.803; $p = 0.0386$, $t = -2.194$, SE = 0.075, parameter = -0.165)).

Fitting a more simple model just using either plot biomass or plot CWM.SLA with site as a factor results in a poorer fit model when compared with the TDT-predicted model with lower R^2 and higher AIC values ((log10(posNEP) ~ log10Bio + as.factor(Site_name), $R^2 = 0.687$, AIC = 17.994; lm(log10(NEP) ~ CWM.SLA + as.factor(Site)); $R^2 = 0.684$, AIC = -17.704). Further, in both models, the effect of biomass and CWM.SLA was marginally significant ($p = 0.064$ and $p = 0.056$, respectively). These results indicate that together the CWM and CWV of community SLA are primary drivers of variation in community carbon flux.

A.6 Methods: PGE: Background, Methods, and Discussion
A.6.1 Background and Methods
The original purpose of PGE, started in 1856, was to investigate the effects of high levels of inorganic fertilizers and organic manure on hay production

relative to control treatments (see references within Crawley et al., 2005 for additional details on methodology). Our analyses mainly focused on the PGE trait dynamics of Plots 2 and 16. These plots were selected because of their contrasting botanical composition and species richness (Crawley et al., 2005; Harpole and Tilman, 2007). Plot 2 (became plot 2/2 in 1996) received farm yard manure between 1856 and 1863, but since then has received no further manure or fertilizer inputs, and is now considered to be a control plot. Plot 16, started in 1858, is a fertilized, unlimed plot that receives annual N, P, K, Na, and Mg applications (48 kg N ha^{-1} as sodium nitrate in spring; mineral applied in winter: 35 kg P ha^{-1} P as triple super-phosphate, 225 kg K ha^{-1} as potassium sulphate, 15 kg Na ha^{-1} as sodium sulphate, and 10 kg Mg ha^{-1} as magnesium sulphate). For the Park Grass dataset, we approximated the central moments of the community trait distribution, $C(z)$, for trait z within plot j and year y using Eqs. (A9)–(A11). We analyzed the time series of these plots in terms of changes in botanical composition, traits, and species richness. To focus on how experimentally paired *local* communities have responded over time, we highlighted plots 2 and 16 (the other plots also showed similar responses).

A.6.2 Assignment of Trait Values and Biomass-Weighted Trait Distributions

We assessed changes in SLA, seed size, and height. These traits have also been proposed to capture most functional and life history variation across species (Westoby, 1998; Westoby et al., 2002). Seed size is thought to characterize regenerative traits not associated with our trait-based growth model developed in Eq. (A7). Including a regeneration trait provides a basis to assess if other niche- or dispersal-based processes acting on other traits may be more important in structuring the community than traits associated with growth, dC/dt (see also discussion on effect and response traits; Suding et al., 2008). Further, variation in seed size should not directly influence our ecosystem-level predictions for dC_{Tot}/dt as this trait is not explicit in Eq. (A7). According to Eq. (A7), plant height (or size, C) can influence ecosystem NPP. So any large shifts in mean plant height would be important to note as well. Trait values are for populations sampled in United Kingdom. We used the first four central moments of $C(z)$ for plot j and year y to calculate the biomass-weighted mean, variance, skewness, and kurtosis.

Within this experiment, species abundance was measured by cutting aboveground biomass to ground level from six randomly located quadrats (50 × 25 cm) within each experimental and control plot. The plant material was then sorted into species, oven dried at 80 °C for 24 h, and the dry matter determined (Crawley et al., 2005; Williams, 1978). For each plot, yields were estimated by weighing standing biomass (t/ha at 100% dry matter) from the whole plot, harvested in mid-June. The plots were originally cut by scythe, then by horse-drawn, and then tractor-drawn mowers (Williams, 1978; see these references for additional methodological detail of the PGE; Crawley et al., 2005).

Our analysis of additional fertilization and control plots at Park Grass (plots 3 and 14) also reveal similar differences in trait distributions (mean, variance, skew, and kurtosis) between the control and fertilized plots. In sum, for all of the Park Grass plots, our central conclusions do not change. We observed coordinated shifts in the functional trait distribution. In Table A1, we show the correlations between the central moments of the trait distribution and species richness. These correlations include plots 2 (control) and 16 (fertilized) together. *None* of the central moments have significant correlation with species richness, indicating that the mechanisms and responses to environmental change captured by the shape of trait distributions are not captured by species richness. Figure A1 shows the change in the central moments of the community trait distribution for SLA. Figure A4 shows the associated changes in seed size in the 140-year long-term PGE. Figure A5 shows the change in the central moments of the community trait distribution for adult height in the 140-year long-term PGE.

A.6.3 PGE—Additional Discussion

Fertilization also changed the shape of the SLA trait distribution indicating that the underlying forces that structure these communities under differing environments changed. For example, fertilization led to a reduction in the variance (the community mean SLA was negatively related to the

Table A1 Correlations Between the Central Moments of the Community Trait Distribution of Specific Leaf Area or SLA and Species Richness (Plots 2 (Control) and 16 (Fertilized) Together)

		Species Richness
SLA	Mean	$0.00^{ns.}$
	Variance	$0.05^{ns.}$
	Skewness	$0.08^{ns.}$
	Kurtosis	$0.03^{ns.}$

None of the moments have significant correlation with species richness, indicating that the mechanisms and responses to environmental change captured by the moments are not captured by species richness. This represents one of the great advantages for Trait Driver Theory (TDT) over theories based on species richness.

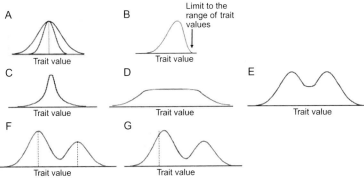

Figure A1 Trait Driver Theory and examples of the first four central moments of trait distributions: (A) *mean* (the first moment) and *variance* (the second moment). In this example, we show two communities with the same mean trait value (dashed line) but with different variances; (B) *skew* (a combination of the second and third moments). Skewness in a trait distribution can be caused by (i) time lags in community response to a new optimum trait value where a long tail of individuals expressing suboptimal trait values is present in the community (e.g. see Fig. 2), (ii) lopsided trait immigration into the community, or (iii) physical or physiological limits on trait expression (e.g. hydrological constraints on plant height); and (iv) may reflect rare species advantage. As shown in this example; and *Kurtosis* (a measure of the fourth moment relative to the second moment). Competitive exclusion and/or strong stabilizing selection will give a highly peaked (fourth moment kurtosis) distribution (C) while niche packing reflecting biotic interactions could give a more uniform distribution (D). *Note*: a normal trait distribution is defined by a skewness and kurtosis = 0. The more peaked the distribution, the more positive the kurtosis value (including the logistic, hyperbolic secant, and Laplace distributions). In contrast, processes that result in the 'spreading out' of traits will be characterized by increasingly more negative kurtosis values. In the case of a uniform distribution, kurtosis = − 1.2. An increasingly bimodal distribution (Bernoulli distribution) will have kurtosis values = − 2. (E) Bimodal distributions could arise where there are multiple optimal (dashed lines) trait values (F), or where the community is responding to a recent environmental change where the two peaks represent both an increased representation of nearly optimal individuals (the high peak) and the continued presence of individuals with optimal trait values for the historic environment (the low peak) (G). Dashed lines correspond to the optimal trait value(s).

community SLA variance, $r^2 = 0.48$, $p < 0.001$), suggesting either that fertilization was an environmental 'filter' on traits and/or competitive exclusion increased (see Table 2). The observed increase in the skewness and kurtosis of the SLA distribution with fertilization is in accord with predictions and expectations of TDT where more quick directional shifts in z_{opt} will lead to a skewed distribution. In contrast, the control plot trait distribution did not show dramatic changes in the variance or skewness of the distribution. However, the mean of SLA in the control plot did significantly

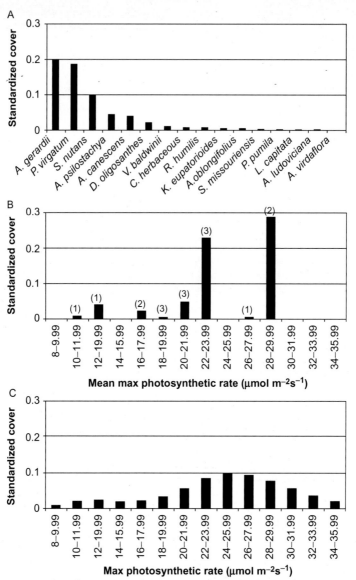

Figure A2 Examples of estimation of a community trait distribution from utilizing either mean trait values and/or intraspecific variation. In this example, we use abundance information (A) for the percent cover for 16 species from Konza Prairie LTER, Kansas, USA. Next, a trait measure, the leaf photosynthetic rate, was measured on a minimum of three leaves on three separate plants. Species means and standard errors were then calculated for this trait for each species. In (B) using the first method, the trait abundance distribution was calculated using only the mean trait data for the species in (A). Numbers in the parentheses indicate the number of species in each trait bin, and the peaks correspond to some of the dominant species in plot (A) that were rank ordered by abundance and not by photosynthetic rate. Lastly, using the second method, the community trait distribution of all individuals can be further approximated by integrating intraspecific subsampling. In (C), for each species, we incorporated intraspecific variation by using the standard error for this trait as measured in each species and then assumed a normal distribution around each species mean. We then generated the community-wide trait distribution by sampling from each species intraspecific trait distribution (defined by its mean and SE). The resulting distribution (C) is much more continuous and unimodal than in plot (B), which does not include intraspecific variation.

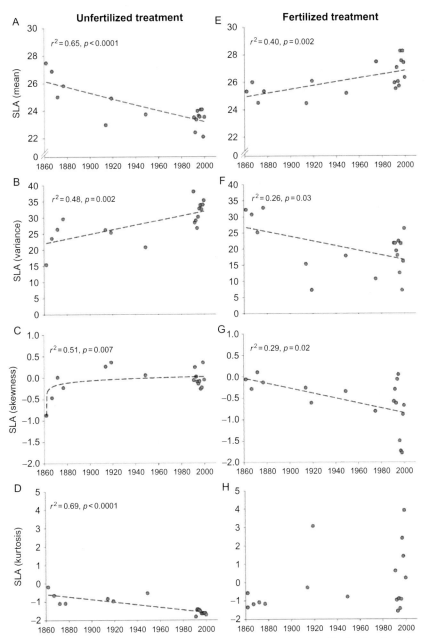

Figure A3 Change in the central moments of the community trait distribution for a key trait—specific leaf area or SLA—in the 140-year long-term Park Grass experiment. Regression lines are indicated for significant relationships. Fertilization has caused a decrease in the variance and an increase in the skewness. Fertilization increases the mean assemblage specific leaf area (SLA) but reduces the variance. This result indicates a directional shift in trait optimum, z_{opt}, and a functional shift in the composition of the community. Further, fertilized plots have become increasingly more skewed and have increasingly more positive kurtosis values indicating that communities have become increasingly dominated by a few trait values. Distributions with kurtosis values of -1.2 are characteristic of an 'overdispersed' uniform distribution, while plots with kurtosis values greater than 0 are more clumped/peaked than expected from a normal or Gaussian distribution.

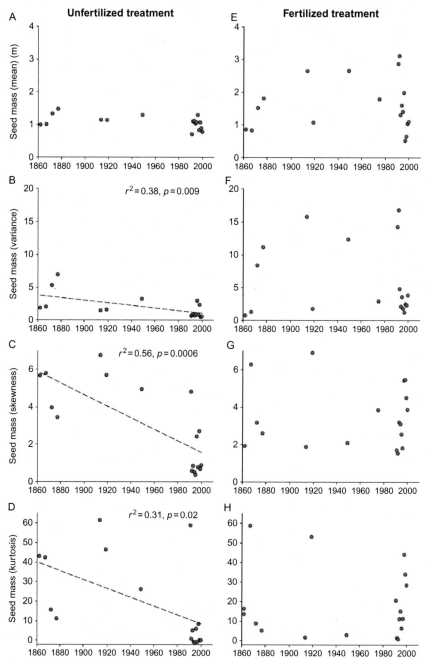

Figure A4 Change in the central moments of the community trait distribution for seed size in the 140-year long-term Park Grass experiment. Significant correlations are indicated with presence of fitted (dashed) regression lines.

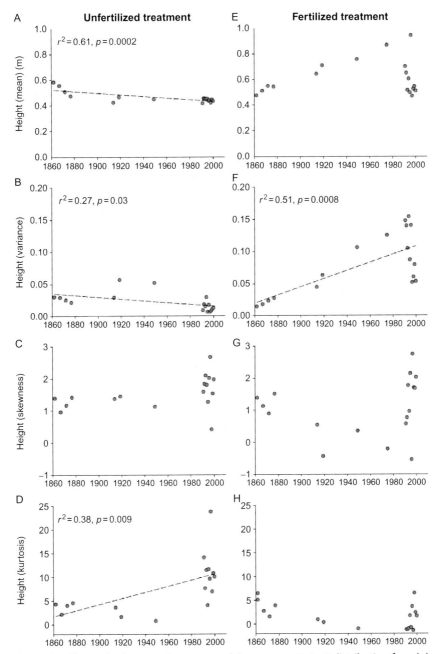

Figure A5 Change in the central moments of the community trait distribution for adult height in the 140-year long-term Park Grass experiment. Significant correlations are indicated with presence of fitted (dashed) regression lines.

decrease suggesting that natural and/or more gradual changes in the local environment (possibly due to an increase in nitrogen limitation over time and/or climate change) and/or recovery from past disturbance have influenced the control community. The kurtosis of the trait distribution in the control plot remained negative and close to -1.2 (a uniform distribution) consistent with an increased role of divergent ecological forces (niche packing and the role of biotic interactions). In contrast, in the fertilized plot, the variance of the distribution has decreased and the kurtosis tends to exhibit high positive values.

In the PGE, fertilization can be seen as a specific environmental driver, E. Fertilization changes soil resource availability and, according to TDT, differentially shifts the optimum growth rate. Indeed, in support we see a shift in z_{opt} (here being SLA) associated with fertilization. Analysis of the moments of distributions for two other community traits, seed size and adult reproductive height, shows that these trait means did not appreciably change with fertilization (Figs. A4 and A5). Importantly, the community mean of plant height did not change with fertilization supporting our assumption that observed change in community NPP was primarily due to changes in SLA, and also that the mean plant size or biomass, C, did not appreciably change. The one change with fertilization that we do observe is that the community variance of plant height increased. No other traits show any changes in the fertilized plots. Future work elaborating TDT should include the role of multiple trait drivers and their associated predictions.

Appendix references

Abrams, P.A. & Chen, X. (2002) The effect of competition between prey species on the evolution of their vulnerabilities to a shared predator. *Evolutionary Ecology Research*, 897–909.

Albert, C.H., Thuiller, W., Yoccoz, N., Soudan, A., Boucher, F., Saccone, P. & Lavorel, S. (2010) Intraspecific functional variability: extent, structure and sources of variation. *Journal of Ecology*, **98**, 604-613.

Baraloto, C., Timothy Paine, C.E., Patiño, S., Bonal, D., Hérault, B. & Chave, J. Functional trait variation and sampling strategies in species-rich plant communities. *Functional Ecology*, **24**, 208-216.

Brooker, R.W., Maestre, F.T., Callaway, R.M., Lortie, C.L., Cavieres, L.A., Kunstler, G., Liancourt, P., Tielbörger, K., Travis, J.M.J., Anthelme, F., Armas, C., Coll, L., Corcket, E., Delzon, S., Forey, E., Kikvidze, Z., Olofsson, J., Pugnaire, F., Quiroz, C.L., Saccone, P., Schiffers, K., Seifan, M., Touzard, B. & Michalet, R. (2008) Facilitation in plant communities: the past, the present, and the future. *Journal of Ecology*, **96**, 18-34.

Chesson, P. (2000) Mechanisms of maintenance of species diversity. *Annual Review of Ecology and Systematics*, **31**, 343-366.

Crawley, M.J., Johnston, A.E., Silvertown, J., Dodd, M., De Mazancourt, C., Heard, M.S., Henman, D.F. & Edwards, G.R. (2005) Determinants of species richness in the Park Grass Experiment. *American Naturalist*, **165**, 179-192.

Diamond, J.M. (1975) Assembly of species communities. *Ecology and Evolution of Communities* (ed. by M.L. Cody and J.M. Diamond), pp. 342-444. Harvard Univ. Press, Cambridge, MA.

Enquist, B.J., West, G.B. & Brown, J.H. (2009) Extensions and evaluations of a general quantitative theory of forest structure and dynamics. *Proceedings of the National Academy of Sciences of the United States of America*, **106**, 7046-7051.

Freckleton, R.P. & Watkinson, A.R. (2001) Predicting competition coefficients for plant mixtures: reciprocity; transitivity and correlations with life-history traits. *Ecology Letters*, **4**, 348-357.

Garnier, E. & Navas, M.-L. (2012) A trait-based approach to comparative functional plant ecology: concepts, methods and applications for agroecology. A review. *Agronomy for Sustainable Development*, **32**, 365-399.

Gaucherand, S. & Lavorel, S. (2007) New method for rapid assessment of the functional composition of herbaceous plant communities. *Austral Ecology*, **32**, 927-936.

Grime, J.P. (1998) Benefits of plant diversity to ecosystems: Immediate, filter and founder effects. *Journal of Ecology*, **86**, 902-910.

Grime, J.P. (2006) Trait convergence and trait divergence in herbaceous plant communities: Mechanisms and consequences. *Journal of Vegetation Science*, **17**, 255-260.

Harpole, W.S. & Tilman, D. (2007) Grassland species loss due to reduced niche dimension. *Nature*, **446**, 791-793.

Hubbell, S.P. (2001) *The Unified Neutral Theory of Biodiversity and Biogeography*. Princeton University Press, Princeton, NJ.

Karr, J.R. & James, F.C. (1975) Ecomorphological configurations and convergent evolution in species and communities. *Ecology and Evolution of Communities* (ed. by M.L. Cody and J.M. Diamond), pp. 258-291. Harvard University Press, Cambridge.

Keddy, P.A. (1992) Assembly and response rules: two goals for predictive community ecology. *Journal of Vegetation Science*, **3**, 157-164.

Kraft, N.J.B. & Ackerly, D.D. (2010) Functional trait and phylogenetic tests of community assembly across spatial scales in an Amazonian forest. *Ecological Monographs*, **80**, 401-422.

Kraft, N.J.B., Valencia, R. & Ackerly, D.D. (2008) Functional traits and niche-based tree community assembly in an Amazonian forest. *Science*, **322**, 580-582.

Kunstler, G., Albert, C.H., Courbaud, B., Lavergne, S., Thuiller, W., Vieilledent, G., Zimmermann, N.E. & Coomes, D.A. (2012) Effects of competition on tree radial-growth vary in importance but not in intensity along climatic gradients. *Journal of Ecology*, **99**, 300-312.

Lavorel, S. (2013) Plant functional effects on ecosystem services. *Journal of Ecology*, **101** 4-8.

Lavorel, S., Grigulis, K., Lamarque, P., Colace, M.-P., Garden, D., Girel, J., Pellet, G. & Douzet, R. (2011) Using plant functional traits to understand the landscape distribution of multiple ecosystem services. *Journal of Ecology*, **99**, 135-147.

Lavorel, S., Grigulis, K., McIntyr, S., Williams, N., Garden, D., Dorrough, J., Berman, S., Quétier, F., Thebault, A. & Bonis, A. (2008) Assessing functional diversity in the field: methodology matters! *Functional Ecology*, **22**, 134-147.

Levins, R. (1968) *Evolution in Changing Environments*. Princeton University Press, Princeton, N.J.

MacArthur, R.H. (1958) Population ecology of some warblers of northeastern coniferous forests. *Ecology*, **39**, 599-619.

MacArthur, R.H. & Levins, R. (1967) The limiting similarity, convergence, and divergence of coexisting species. *The American Naturalist*, **101**, 377-385.

Mayfield, M.M. & Levine, J.M. (2010) Opposing effects of competitive exclusion on the phylogenetic structure of communities. *Ecology Letters*, **13**, 1085-1093.

McAllister, C.A., Knapp, A.K. & Maragni, L.A. (1998) Is leaf-level photosynthesis related to plant success in a highly productive grassland? *Oecologia*, **117**, 40-46.

Mooney, H. (ed.) (1977) *Convergent evolution in Chile and California: Mediterranean climate ecosystems*. Dowden, Hutchinson and Ross, Stroudsburg, PA.

Navas, M.-L. & Violle, C. (2009) Plant traits related to competition: how do they shape the functional diversity of communities? *Community Ecology*, **10**, 131-137.

Orians, G.H. & Paine, R.T. (1983) Convergent evolution at the community level. *Coevolution* (ed. by D.J. Futuyma and M. Slatkin). Sinauer, Sunderland, MA.

Poorter, H. (1989) Interspecific variation in relative growth rate: on ecological causes and physiological consequences. *Causes and Consequences of Variation in Growth Rate and Productivity in Higher Plants* (ed. by H. Lambers, M.L. Cambridge, H. Konings and T.L. Pons). SPB Academic Publishing, The Hague.

Ricotta, C. & Moretti, M. (2011) CWM and Rao's quadratic diversity: a unified framework for functional ecology. *Oecologia*, **167**, 181-188.

Savage, V.M. (2004) Improved approximations to scaling relationships for species, populations, and ecosystems across latitudinal and elevational gradients. *Journal of Theoretical Biology*, **227**, 525-534.

Savage, V.M., Webb, C.T. & Norberg, J. (2007) A general multi-trait-based framework for studying the effects of biodiversity on ecosystem functioning. *Journal of Theoretical Biology*, **247**, 213-229.

Schimper, A.F.W. (1898) *Pflanzengeographie auf Physiologischer*. Grundlage, G. Fischer

Schimper, A.F.W. (1903) *Plant-geography Upon a Physiological Basis*. Clarendon, Oxford, UK.

Suding, K.N., Lavorel, S., Chapin, F.S., Cornelissen, J.H.C., DÍAz, S., Garnier, E., Goldberg, D., Hooper, D.U., Jackson, S.T. & Navas, M.-L. (2008) Scaling environmental change through the community-level: a trait-based response-and-effect framework for plants. *Global Change Biology*, **14**, 1125-1140.

Terborgh, J. (1973) On the notion of favorableness in plant ecology. *The American Naturalist*, **107**, 481-501.

Tilman, D. (1982) *Resource Competition and Community Structure*. Princeton University Press, Princeton, NJ.

Violle, C., Navas, M.-L., Vile, D., Kazakou, E., Fortunel, C., Hummel, I. & Garnier, E. (2007) Let the concept of trait be functional! *Oikos*, **116**, 882-892.

Weiher, E. & Keddy, P.A. (1999) Assembly rules as general constraints on community composition. *Ecological Assembly Rules: Perspectives, Advances, Retreats* (ed. by E. Weiher, And Keddy, P.A.). Cambridge University Press, Cambridge, MASS.

Westoby, M. (1998) A leaf-height-seed (LHS) plant ecology strategy scheme. *Plant and Soil*, **199**, 213-227.

Westoby, M. & Wright, I.J. (2006) Land-plant ecology on the basis of functional traits. *Trends in Ecology & Evolution*, **21**, 261-268.

Westoby, M., Falster, D.S., Moles, A.T., Vesk, P.A. & Wright, I.J. (2002) Plant ecological strategies: Some leading dimensions of variation between species. *Annual Review of Ecology and Systematics*, **33**, 125-159.

Williams, E.D. (1978) *Botanical Composition of the Park Grass plots at Rothamsted 1856-1976.* Rothamsted Experimental Station, Harpenden, United Kingdom.

REFERENCES

Ackerly, D.D., 2003. Community assembly, niche conservatism, and adaptive evolution in changing environments. Int. J. Plant Sci. 164, S165–S184.

Ackerly, D.D., Monson, R.K., 2003. Waking the sleeping giant: the evolutionary foundations of plant function. Int. J. Plant Sci. 164, S1–S6.

Adler, P.B., Salguero-Gómez, R., Compagnoni, A., Hsu, J.S., Ray-Mukherjee, J., Mbeau-Ache, C., Franco, M., 2014. Functional traits explain variation in plant life history strategies. Proc. Natl. Acad. Sci. U.S.A. 111, 740–745.

Albert, C.H., Thuiller, W., Yoccoz, N., Soudan, A., Boucher, F., Saccone, P., Lavorel, S., 2010. Intraspecific functional variability: extent, structure and sources of variation. J. Ecol. 98, 604–613.

Antonovics, J., 1976. The nature of limits to natural selection. Ann. Mo. Bot. Gard. 63, 224–227.

Arnold, S.J., 1983. Morphology, performance, and fitness. Am. Zool. 23, 347–361.

Asner, G.P., Martin, R.E., Tupayachi, R., Anderson, C.B., Sinca, F., Carranza-Jiménez, L., Martinez, P., 2014. Amazonian functional diversity from forest canopy chemical assembly. Proc. Natl. Acad. Sci. U.S.A. 111, 5604–5609.

Baraloto, C., Timothy Paine, C.E., Patiño, S., Bonal, D., Hérault, B., Chave, J., 2010. Functional trait variation and sampling strategies in species-rich plant communities. Funct. Ecol. 24, 208–216.

Belmaker, J., Jetz, W., 2013. Spatial scaling of functional structure in bird and mammal assemblages. Am. Nat. 181, 464–478.

Boulangeat, I., Philippe, P., Abdulhak, S., Douzet, R., Garraud, L., Lavergne, S., Lavorel, S., Van Es, J., Vittoz, P., Thuiller, W., 2012. Improving plant functional groups for dynamic models of biodiversity: at the crossroads between functional and community ecology. Glob. Chang. Biol. 18, 3464–3475.

Bryant, J.A., Lamanna, C., Morlon, H., Kerkhoff, A.J., Enquist, B.J., Green, J.L., 2008. Microbes on mountainsides: contrasting elevational patterns of bacterial and plant diversity. Proc. Natl. Acad. Sci. U.S.A. 105, 11505–11511.

Cardinale, B.J., Wright, J.P., Cadotte, M.W., Carroll, I.T., Hector, A., Srivastava, D.S., Loreau, M., Weis, J.J., 2007. Impacts of plant diversity on biomass production increase through time because of species complementarity. Proc. Natl. Acad. Sci. U.S.A. 104, 18123–18128.

Cavender-Bares, J., Kozak, K., Fine, P., Kembel, S., 2009. The merging of community ecology and phylogenetic biology. Ecol. Lett. 12, 693–715.

Chapin, F.S.I., 1980. The mineral nutrition of wild plants. Annu. Rev. Ecol. Syst. 11, 233–260.

Chapin, F.S., Shaver, G.R., 1985. Individualistic growth response of tundra plant species to environmental manipulations in the field. Ecology 66, 564–576.

Chapin, F.S.I., Zavaleta, E.S., Eviner, V.T., Naylor, R.L., Vitousek, P.M., Reynolds, H.L., Hooper, D.U., Lavorel, S., Sala, O.E., Hobbie, S.E., Mack, M.C., Diaz, S., 2000. Consequences of changing biodiversity. Nature 405, 234–242.

Choler, P., 2005. Consistent shifts in alpine plant traits along a mesotopographical gradient. Arct. Antarct. Alp. Res. 37, 444–453.

Coomes, D.A., 2006. Challenges to the generality of WBE theory. Trends Ecol. Evol. 21, 593–596.

Cornwell, W., Ackerly, D.D., 2009. Community assembly and shifts in plant trait distributions across an environmental gradient in coastal California. Ecol. Monogr. 79, 109–126.

Craine, J.M., 2009. Resource Strategies of Wild Plants. Princeton University Press, Princeton.

Craine, J.M., Froehle, J., Tilman, D.G., Wedin, D.A., Chapin, I.F.S., 2001. The relationships among root and leaf traits of 76 grassland species and relative abundance along fertility and disturbance gradients. Oikos 93, 274–285.

Damuth, J., 1981. Population-density and body size in mammals. Nature 290, 699–700.

Davis, M.B., Shaw, R.G., 2001. Range shifts and adaptive responses to quaternary climate change. Science 292, 673–679.

Dell, A.I., Pawar, S., Savage, V.M., 2013. The thermal dependence of biological traits. Ecology 94, 1205–1206.

Dell, A.I., Zhao, L., Brose, U., Pearson, R.G., Alford, R.A., 2015. Population and community body size structure across a complex environmental gradient. Adv. Ecol. Res. 52, 115–167.

DeLucia, E., Drake, J.E., Thomas, R.B., Gonzalez-Meler, M., 2007. Forest carbon use efficiency: is respiration a constant fraction of gross primary production? Glob. Chang. Biol. 13, 1157–1167.

Díaz, S., Cabido, M., 2001. Vive la difference: plant functional diversity matters to ecosystem processes. Trends Ecol. Evol. 16, 646–655.

Díaz, S., Hodgson, J.G., Thompson, K., Cabido, M., Cornelissen, J.H.C., Jalili, A., Montserrat-Marti, G., Grime, J.P., Zarrinkamar, F., Asri, S., Band, R., Basconcelo, S., Castro-Diez, P., Funes, G., Hamzehee, B., Khoshnevi, M., Perez-Harguindeguy, N., Perez-Rontome, M.C., Shirvany, F.A., Vendramini, F., Yazdani, S., Abbas-Azimi, R., Bogaard, A., Boustani, S., Charles, M., Dehghan, M., de Torres-Espuny, L., Falczuk, V., Guerrero-Campo, J., Hynd, A., Jones, G., Kowsary, E., Kazemi-Saeed, F., Maestro-Martinez, M., Romo-Diez, A., Shaw, S., Siavash, B., Villar-Salvador, P., Zak, M.R., 2004. The plant traits that drive ecosystems: evidence from three continents. J. Veg. Sci. 15, 295–304.

Diaz, S., Lavorel, S., De Bello, F., Quétier, F., Grigulis, K., Robson, M., 2007. Incorporating plant functional diversity effects in ecosystem service assessments. Proc. Natl. Acad. Sci. U.S.A. 36, 20684–20689.

Dieckmann, U., Ferrière, R., 2004. Adaptive dynamics and evolving biodiversity. In: Ferrière, R., Dieckmann, U., Couvet, D. (Eds.), Evolutionary Conservation Biology. Cambridge University Press, Cambridge, UK, pp. 188–224.

Doughty, C.E., Asner, G.P., Martin, R.E., 2011. Predicting tropical plant physiology from leaf and canopy spectroscopy. Oecologia 165, 289–299.

Elser, J.J., Fagan, W.F., Kerkhoff, A.J., Swenson, N.G., Enquist, B.J., 2010. Biological stoichiometry of plant production: metabolism, scaling and ecological response to global change. New Phytol. 186, 593–608.

Enquist, B.J., 2010. Wanted: a general and predictive theory for trait-based plant ecology. Bioscience 60, 854–855.

Enquist, B.J., Brown, J.H., West, G.B., 1998. Allometric scaling of plant energetics and population density. Nature 395, 163.

Enquist, B.J., Economo, E.P., Huxman, T.E., Allen, A.P., Ignace, D.D., Gillooly, J.F., 2003. Scaling metabolism from organisms to ecosystems. Nature 423, 639–642.

Enquist, B.J., Kerkhoff, A.J., Huxman, T.E., Economo, E.P., 2007a. Adaptive differences in plant physiology and ecosystem invariants: insights from a metabolic scaling model. Glob. Chang. Biol. 13, 591–609.

Enquist, B.J., Kerkhoff, A.J., Stark, S.C., Swenson, N.G., McCarthy, M.C., Price, C.A., 2007b. A general integrative model for scaling plant growth, carbon flux, and functional trait spectra. Nature 449, 218–222.

Enquist, B.J., Allen, A.P., Brown, J.H., Gillooly, J.F., Kerkhoff, A.J., Niklas, K.J., Price, C.A., West, G.B., 2007c. Biological scaling: does the exception prove the rule? Nature 445, E9–E10.

Enquist, B.J., West, G.B., Brown, J.H., 2009. Extensions and evaluations of a general quantitative theory of forest structure and dynamics. Proc. Natl. Acad. Sci. U.S.A. 106, 7046–7051.

Evans, G.C., 1972. The Quantitative Analysis of Plant Growth. University of California Press, Berkeley.

Ferriere, R., Legendre, S., 2013. Eco-evolutionary feedbacks, adaptive dynamics and evolutionary rescue theory. Philos. Trans. R. Soc. B Biol. Sci. 368, 20120081.

Fonseca, C.R., Overton, J.M., Collins, B., Westoby, M., 2000. Shifts in trait-combinations along rainfall and phosphorus gradients. J. Ecol. 88, 964–977.

Freckleton, R.P., Watkinson, A.R., 2001. Predicting competition coefficients for plant mixtures: reciprocity; transitivity and correlations with life-history traits. Ecol. Lett. 4, 348–357.

Frenne, P., Graae, B.J., Rodríguez-Sánchez, F., Kolb, A., Chabrerie, O., Decocq, G., Kort, H., Schrijver, A., Diekmann, M., Eriksson, O., 2013. Latitudinal gradients as natural laboratories to infer species' responses to temperature. J. Ecol. 101 (3), 784–795.

Funk, J.L., Cleland, E.E., Suding, K.N., Zavaleta, E.S., 2008. Restoration through reassembly: plant traits and invasion resistance. Trends Ecol. Evol. 23, 695–703.

Garnier, E., Navas, M.-L., 2012. A trait-based approach to comparative functional plant ecology: concepts, methods and applications for agroecology. A review. Agron. Sustain. Dev. 32, 365–399.

Garnier, E., Cortez, J., Billes, G., Navas, M.-L., Roumet, C., Debussche, M., Laurent, G., Blanchard, A., Aubry, D., Bellmann, A., Neill, C., Toussaint, J.-P., 2004. Plant functional markers capture ecosystem properties during secondary succession. Ecology 85, 2630–2637.

Gaucherand, S., Lavorel, S., 2007. New method for rapid assessment of the functional composition of herbaceous plant communities. Austral Ecol. 32, 927–936.

Ghalambor, C.K., McKay, J.K., Carroll, S.P., Reznick, D.N., 2007. Adaptive versus nonadaptive phenotypic plasticity and the potential for contemporary adaptation in new environments. Funct. Ecol. 21, 394–407.

Gilbert, J.P., DeLong, J.P., 2015. Individual variation decreases interference competition but increases species persistence. Adv. Ecol. Res. 52, 45–64.

Gillooly, J.F., Brown, J.H., West, G.B., Savage, V.M., Charnov, E.L., 2001. Effects of size and temperature on metabolic rate. Science 293, 2248–2251.

Gillooly, J., Allen, A.P., West, G.B., Brown, J.H., 2005. The rate of DNA evolution: effects of body size and temperature on the molecular clock. Proc. Natl. Acad. Sci. U.S.A. 102, 140–145.

Goldberg, D.E., Landa, K., 1991. Competitive effect and response: hierarchies and correlated traits in the early stages of competition. J. Ecol. 79, 1013–1030.

Grace, J.B., Michael Anderson, T., Smith, M.D., Seabloom, E., Andelman, S.J., Meche, G., Weiher, E., Allain, L.K., Jutila, H., Sankaran, M., 2007. Does species diversity limit productivity in natural grassland communities? Ecol. Lett. 10, 680–689.

Grime, J.P., 1977. Evidence for the existence of three primary strategies in plants and its relevance to ecological and evolutionary theory. Am. Nat. 111, 1169–1194.

Grime, J.P., 1998. Benefits of plant diversity to ecosystems: immediate, filter and founder effects. J. Ecol. 86, 902–910.

Grime, J.P., 2006. Trait convergence and trait divergence in herbaceous plant communities: mechanisms and consequences. J. Veg. Sci. 17, 255–260.

Grime, J.P., Hunt, R., 1975. Relative growth-rate: its range and adaptive significance in a local flora. J. Ecol. 63, 393–422.

Han, W., Fang, J., Guo, D., 2005. Leaf N and P stoichiometry across 753 terrestrial plant species in China. New Phytol. 168, 377–385.

Hillebrand, H., Matthiessen, B., 2009. Biodiversity in a complex world: consolidation and progress in functional biodiversity research. Ecol. Lett. 12, 1405–1419.

Hooper, D.U., Chapin, F.S.I., Ewel, J.J., Hector, A., Inchausti, P., Lavorel, S., Lawton, J.H., Lodge, D.M., Loreau, M., Naeem, S., Schmid, B., Setälä, H., Symstad, A.J., Vandermeer, J., Wardle, D.A., 2004. Effects of biodiversity on ecosystem functioning: a consensus of current knowledge. Ecol. Monogr. 75, 3–35.

Hubbell, S.P., 2001. The Unified Neutral Theory of Biodiversity and Biogeography. Princeton University Press, Princeton, NJ.

Hulshof, C.M., Violle, C., Spasojevic, M.J., McGill, B., Damschen, E., Harrison, S., Enquist, B.J., 2013. Intra-specific and inter-specific variation in specific leaf area reveal the importance of abiotic and biotic drivers of species diversity across elevation and latitude. J. Veg. Sci. 24, 921–931.

Jasoni, R.L., Smith, S.D., Arnone, J.A., 2005. Net ecosystem CO_2 exchange in Mojave Desert shrublands during the eighth year of exposure to elevated CO_2. Glob. Chang. Biol. 11, 749–756.

Jung, V., Violle, C., Mondy, C., Hoffmann, L., Muller, S., 2010. Intraspecific variability and trait-based community assembly. J. Ecol. 98, 1134–1140.

Kattge, J., Díaz, S., Lavorel, S., Prentice, I.C., Leadley, P., Bönisch, G., Garnier, E., Westoby, M., Reich, P.B., Wright, I.J., Cornelissen, J.H.C., Violle, C., Harrison, S.P., van Bodegom, P.M., Reichstein, M., Enquist, B.J., Soudzilovskaia, N.A., Ackerly, D.-D., Anand, M., Atkin, O., Bahn, M., Baker, T.R., Baldocchi, D., Bekker, R., Blanco, C.C., Blonder, B., Bond, W.J., Bradstock, R., Bunker, D.E., Casanoves, F., Cavender-Bares, J., Chambers, J.Q., Chapin III, F.S., Chave, J., Coomes, D., Cornwell, W.K., Craine, J.M., Dobrin, B.H., Duarte, L., Durka, W., Elser, J., Esser, G., Estiarte, M., Fagan, W.F., Fang, J., Fernández-Méndez, F., Fidelis, A., Finegan, B., Flores, O., Ford, H., Frank, D., Freschet, G.T., Fyllas, N.M., Gallagher, R.V., Green, W.A., Gutierrez, A.G., Hickler, T., Higgins, S.I., Hodgson, J.-G., Jalili, A., Jansen, S., Joly, C.A., Kerkhoff, A.J., Kirkup, D., Kitajima, K., Kleyer, M., Klotz, S., Knops, J.M.H., Kramer, K., Kühn, I., Kurokawa, H., Laughlin, D., Lee, T.D., Leishman, M., Lens, F., Lenz, T., Lewis, S.L., Lloyd, J., Llusià, J., Louault, F., Ma, S., Mahecha, M.D., Manning, P., Massad, T., Medlyn, B.E., Messier, J., Moles, A.T., Müller, S.C., Nadrowski, K., Naeem, S., Niinemets, Ü., Nöllert, S., Nüske, A., Ogaya, R., Oleksyn, J., Onipchenko, V.G., Onoda, Y., Ordoñez, J., Overbeck, G., Ozinga, W.A., Patiño, S., Paula, S., Pausas, J.G., Peñuelas, J., Phillips, O.L., Pillar, V., Poorter, H., Poorter, L., Poschlod, P., Prinzing, A., Proulx, R., Rammig, A., Reinsch, S., Reu, B., Sack, L., Salgado-Negret, B., Sardans, J., Shiodera, S., Shipley, B., Siefert, A., Sosinski, E., Soussana, J.F., Swaine, E., Swenson, N., Thompson, K., Thornton, P., Waldram, M., Weiher, E.,

White, M., White, S., Wright, S.J., Yguel, B., Zaehle, S., Zanne, A.E., Wirth, C., 2011. TRY—a global database of plant traits. Glob. Chang. Biol. 17, 2905–2935.

Kingsolver, J.G., Huey, R.B., 2003. Introduction: The evolution of morphology, performance, and fitness. Integr. Comp. Biol. 43, 361–366.

Kleyer, M., Bekker, R.M., Knevel, I.C., Bakker, J.P., Thompson, K., Sonnenschein, M., Poschlod, P., van Groenendael, J.M., Klimes, L., Klimesova, J., Klotz, S., Rusch, G.M., Hermy, M., Adriaens, D., Boedeltje, G., Bossuyt, B., Dannemann, A., Endels, P., Gotzenberger, L., Hodgson, J.G., Jackel, A.K., Kuhn, I., Kunzmann, D., Ozinga, W.A., Romermann, C., Stadler, M., Schlegelmilch, J., Steendam, H.J., Tackenberg, O., Wilmann, B., Cornelissen, J.H.C., Eriksson, O., Garnier, E., Peco, B., 2008. The LEDA Traitbase: a database of life-history traits of Northwest European flora. J. Ecol. 96, 1266–1274.

Knops, J.M., Reinhart, K., 2000. Specific leaf area along a nitrogen fertilization gradient. Am. Midl. Nat. 144, 265–272.

Kraft, N.J.B., Valencia, R., Ackerly, D.D., 2008. Functional traits and niche-based tree community assembly in an Amazonian forest. Science 322, 580–582.

Kunstler, G., Albert, C.H., Courbaud, B., Lavergne, S., Thuiller, W., Vieilledent, G., Zimmermann, N.E., Coomes, D.A., 2012. Effects of competition on tree radial-growth vary in importance but not in intensity along climatic gradients. J. Ecol. 99, 300–312.

Lamanna, C.A., Blonder, B., Violle, C., Kraft, N.J.B., Sandel, B., Simova, I., Donoghue II, J.C., Svenning, J.-C., McGill, B.J., Boyle, B., Dolins, S., Jørgensen, P.M., Marcuse-Kubitza, A., Morueta-Holme, N., Peet, R.K., Piel, W., Regetz, J., Schildhauer, M., Spencer, N., Theirs, B.M., Wiser, S.K., Enquist, B.J., 2014. Functional trait space and the latitudinal diversity gradient. Proc. Natl. Acad. Sci. U.S.A. 111 (38), 13745–13750.

Lambers, H., Freijsen, N., Poorter, H., Hirose, T., van der Werff, H., 1989. Analyses of growth based on net assimilation rate and nitrogen productivity: their physiological background. In: Lambers, H., Cambridge, M.L., Konings, H., Pons, T.L. (Eds.), Variation in Growth Rate and Productivity of Higher Plants. SPB Academic Publishing, The Hague, The Netherlands, pp. 1–17.

Laskowski, K.L., Pearish, S., Bensky, M., Bell, A.M., 2015. Predictors of individual variation in movement in a natural population of threespine stickleback (Gasterosteus aculeatus). Adv. Ecol. Res. 52, 65–90.

Lavorel, S., Garnier, E., 2002. Predicting changes in community composition and ecosystem functioning from plant traits: revisiting the Holy Grail. Funct. Ecol. 16, 545–556.

Lavorel, S., Díaz, S., Cornelissen, J.H.C., Garnier, E., Harrison, S.P., McIntyre, S., Pausas, J.G., Pérez-Harguindeguy, N., Roumet, C., Urcelay, C., 2007. Plant functional types: are we getting any closer to the Holy Grail? In: Terrestrial Ecosystems in a Changing World. Springer, Berlin Heidelber, pp. 149–164.

Lavorel, S., Grigulis, K., McIntyr, S., Williams, N., Garden, D., Dorrough, J., Berman, S., Quétier, F., Thebault, A., Bonis, A., 2008. Assessing functional diversity in the field: methodology matters!. Funct. Ecol. 22, 134–147.

Lavorel, S., Grigulis, K., Lamarque, P., Colace, M.-P., Garden, D., Girel, J., Pellet, G., Douzet, R., 2011. Using plant functional traits to understand the landscape distribution of multiple ecosystem services. J. Ecol. 99, 135–147.

Levins, R., 1968. Evolution in Changing Environments. Princeton University Press, Princeton, NJ.

Litchman, E., Klausmeier, C.A., 2008. Trait-based community ecology of phytoplankton. Annu. Rev. Ecol. Evol. Syst. 39, 615–639.

Loreau, M., Naeem, S., Inchausti, P., Bengtsson, J., Grime, J.P., Hector, A., Hooper, D.U., Huston, M.A., Raffaelli, D., Schmid, B., Tilman, D., Wardle, D.A., 2001. Biodiversity and ecosystem functioning: current knowledge and future challenges. Science 294, 804–808.

MacArthur, R.H., 1972. Geographical Ecology: Patterns in the Distribution of Species. Princeton University Press, Princeton, NJ.

Marks, C., Lechowicz, M., 2006. Alternative designs and the evolution of functional diversity. Am. Nat. 167, 55–67.

Mason, N.W.H., Mouillot, D., Lee, W.G., Wilson, J.B., 2005. Functional richness, functional evenness and functional divergence: the primary components of functional diversity. Oikos 111, 112–118.

Mayfield, M.M., Levine, J.M., 2010. Opposing effects of competitive exclusion on the phylogenetic structure of communities. Ecol. Lett. 13, 1085–1093.

Maynard Smith, J., 1976. What determines the rate of evolution? Am. Nat. 110, 331–338.

McGill, B.J., Enquist, B.J., Weiher, E., Westoby, M., 2006. Rebuilding community ecology from functional traits. Trends Ecol. Evol. 21, 178–185.

Messier, J., McGill, B.J., Lechowicz, M.J., 2010. How do traits vary across ecological scales? A case for trait-based ecology. Ecol. Lett. 13, 838–848.

Michaletz, S.T., Cheng, D., Kerkhoff, A.J., Enquist, B.J., 2014. Convergence of terrestrial plant production across global climate gradients. Nature 512, 39–43.

Moses, M.E., Hou, C., Woodruff, W.H., West, G.B., Nekola, J.C., Zuo, W., Brown, J.H., 2008. Revisiting a model of ontogenetic growth: estimating model parameters from theory and data. Am. Nat. 171, 632–645.

Muller, E.B., Nisbet, R.M., Kooijman, S., Elser, J.J., McCauley, E., 2001. Stoichiometric food quality and herbivore dynamics. Ecol. Lett. 4, 519–529.

Naeem, S., Wright, J.P., 2003. Disentangling biodiversity effects on ecosystem functioning: deriving solutions to a seemingly insurmountable problem. Ecol. Lett. 6, 567–579.

Naeem, S., Thompson, L.J., Lawler, S.P., Lawton, J.H., Woodfin, R.M., 1994. Declining biodiversity can alter the performance of ecosystems. Nature 368, 734–737.

Naeem, S., Bunker, D.E., Hector, A., Loreau, M., Perrings, C., 2009. Can we predict the effects of global change on biodiversity loss and ecosystem functioning? In: Naeem, S., Bunker, D.E., Hector, A., Loreau, M., Perrings, C. (Eds.), Biodiversity, Ecosystem Functioning, and Human Wellbeing: An Ecological and Economic Perspective. Oxford University Press, Oxford, pp. 290–298.

Norberg, J., 2004. Biodiversity and ecosystem functioning: a complex adaptive systems approach. Limnol. Oceanogr. 49, 1269–1277.

Norberg, J., Swaney, D.P., Dushoff, J., Lin, J., Casagrandi, R., Levin, S.A., 2001. Phenotypic diversity and ecosystem functioning in changing environments: a theoretical framework. Proc. Natl. Acad. Sci. U.S.A. 98, 11376–11381.

Paine, C., Baraloto, C., Chave, J., Hérault, B., 2011. Functional traits of individual trees reveal ecological constraints on community assembly in tropical rain forests. Oikos 120, 720–727.

Pawar, S. 2015. The role of body size variation in community assembly. Adv. Ecol. Res. 52, 201–248.

Petchey, O.L., Gaston, K.J., 2002. Functional diversity (FD), species richness and community composition. Ecol. Lett. 5, 402–411.

Peters, R.H., 1983. The Ecological Implications of Body Size. Cambridge University Press, Cambridge.

Pettorelli, N., Hilborn, A., Duncan, C., Durant, S.M., 2015. Individual variability: the missing component to our understanding of predator-prey interactions. Adv. Ecol. Res. 52, 19–44.

Poorter, H., 1989. Interspecific variation in relative growth rate: on ecological causes and physiological consequences. Causes and consequences of variation in growth rate and productivity of higher plants, 45–68.

Poorter, H., Niinemets, Ü., Poorter, L., Wright, I.J., Villar, R., 2009. Causes and consequences of variation in leaf mass per area (LMA): a meta-analysis. New Phytol. 182 (3), 565–588.

Pulliam, R.H., 1988. Sources, sinks, and population regulation. Am. Nat. 132, 652–661.

Reich, P.B., 2005. Global biogeography of plant chemistry: filling in the blanks. New Phytol. 168, 263–266.

Reich, P.B., 2014. The world-wide 'fast–slow' plant economics spectrum: a traits manifesto. J. Ecol. 102, 275–301.

Reich, P.B., Oleksyn, J., 2004. Global patterns of plant leaf N and P in relation to temperature and latitude. Proc. Natl. Acad. Sci. U.S.A. 101, 11001–11006.

Reich, P.B., Walters, M.B., Ellsworth, D.S., 1997. From tropics to tundra: global convergence in plant functioning. Proc. Natl. Acad. Sci. U.S.A. 94, 13730–13734.

Ricker, W.E., 1979. Growth rates and models. In: Fish Physiology. Academic Press, New York, pp. 677–743.

Roscher, C., Schumacher, J., Gubsch, M., Lipowsky, A., Weigelt, A., Buchmann, N., Schmid, B., Schulze, E.-D., 2012. Using plant functional traits to explain diversity-productivity relationships. PLoS One 7, e36760.

Rosenzweig, M.L., 1995. Species Diversity in Space and Time. Cambridge University Press, Cambridge.

Ryan, T.P., 1997. Modern Regression Methods. Wiley.

Safi, K., Cianciaruso, M.V., Loyola, R.D., Brito, D., Armour-Marshall, K., Diniz-Filho, J.A. F., 2011. Understanding global patterns of mammalian functional and phylogenetic diversity. Philos. Trans. R. Soc. B Biol. Sci. 366, 2536–2544.

Saleska, S.R., Harte, J., Torn, M.S., 1999. The effect of experimental ecosystem warming on CO_2 fluxes in a montane meadow. Glob. Chang. Biol. 5, 125–141.

Savage, V.M., 2004. Improved approximations to scaling relationships for species, populations, and ecosystems across latitudinal and elevational gradients. J. Theor. Biol. 227, 525–534.

Savage, V.M., Gillooly, J.F., Brown, J.H., West, G.B., Charnov, E.L., 2004. Effects of body size and temperature on population growth. Am. Nat. 163, E429–E441.

Savage, V.M., Webb, C.T., Norberg, J., 2007. A general multi-trait-based framework for studying the effects of biodiversity on ecosystem functioning. J. Theor. Biol. 247, 213–229.

Savage, V.M., Deeds, E.J., Fontana, W., 2008. Sizing up allometric scaling theory. PLoS Comput. Biol. 4, e1000171.

Schmitz, O.J., Buchkowski, R.W., Burghardt, K.T., Donihue, C.M., 2015. Functional traits and trait-mediated interactions: connecting community-level interactions with ecosystem functioning. Adv. Ecol. Res. 52, 319–343.

Shipley, B., 2000. Plasticity in relative growth rate and its components following a change in irradiance. Plant Cell Environ. 23, 1207–1216.

Shipley, B., 2010. From Plant Traits to Vegetation Structure. Cambridge University Press, Cambridge, UK.

Sides, C.B., Enquist, B.J., Ebersole, J.J., Smith, M.N., Henderson, A.N., Sloat, L.L., 2014. Revisiting Darwin's hypothesis: does greater intraspecific variability increase species' ecological breadth? Am. J. Bot. 101 (1), 56–62.

Silvertown, J., Poulton, P., Johnston, A.E., Edwards, G., Heard, M., Biss, P.M., 2006. The Park Grass Experiment 1856–2006: its contribution to ecology. J. Ecol. 94, 801–814.

Šímová, I., Violle, C., Kraft, N.J.B., Storch, D., Svenning, J.-C., Boyle, B., Donoghue, J., Jorgensen, P., McGill, B., Moruueta-Holme, N., Peet, R.K., Wiser, S.K., Piel, W., Regetz, J., Shildhauer, M., Thiers, B., Enquist, B.J., 2014. Shifts in trait means and variances in North American tree assemblages: species richness patterns are loosely related to the functional space. Ecography, in press.

Sloat, L.L., Henderson, A., Lamanna, C.A., Enquist, B.J., 2014. The effect of the foresummer drought on carbon exchange in subalpine meadows. Ecosystems 18 (3), 533–545.

Stegen, J.C., Enquist, B.J., Ferrière, R., 2009. Advancing the metabolic theory of biodiversity. Ecol. Lett. 12, 1001–1015.

Stephenson, N.L., Das, A.J., Condit, R., Russo, S.E., Baker, P.J., Beckman, N.G., Coomes, D.A., Lines, E.R., Morris, W.K., Rüger, N., 2014. Rate of tree carbon accumulation increases continuously with tree size. Nature 507, 90–93.

Stevens, R.D., Cox, S.B., Strauss, R.E., Willig, M.R., 2003. Patterns of functional diversity across an extensive environmental gradient: vertebrate consumers, hidden treatments and latitudinal trends. Ecol. Lett. 6, 1099–1108.

Suding, K.N., Goldstein, L.J., 2008. Testing the Holy Grail framework: using functional traits to predict ecosystem change. New Phytol. 180, 559–562.

Suding, K.N., Collins, S.L., Gough, L., Clark, C., Cleland, E.E., Gross, K.L., Milchunas, D.G., Pennings, S., 2005. Functional- and abundance-based mechanisms explain diversity loss due to N fertilization. Proc. Natl. Acad. Sci. U.S.A. 102, 4387–4392.

Suding, K., Lavorel, S., Chapin, F., Cornelissen, J., Diaz, S., Garnier, E., Goldberg, D., Hooper, D., Jackson, S., Navas, M.-L., 2008a. Scaling environmental change through the community-level: a trait-based response-and-effect framework for plants. Glob. Chang. Biol. 14, 1125–1140.

Suding, K.N., Lavorel, S., Chapin, F.S., Cornelissen, J.H.C., Díaz, S., Garnier, E., Goldberg, D., Hooper, D.U., Jackson, S.T., Navas, M.-L., 2008b. Scaling environmental change through the community-level: a trait-based response-and-effect framework for plants. Glob. Chang. Biol. 14, 1125–1140.

Swenson, N.G., 2013. The assembly of tropical tree communities—the advances and shortcomings of phylogenetic and functional trait analyses. Ecography 36, 264–276.

Swenson, N.G., Enquist, B.J., 2007. Ecological and evolutionary determinants of a key plant functional trait: wood density and its community-wide variation across latitude and elevation. Am. J. Bot. 94, 451–459.

Swenson, N.G., Enquist, B.J., 2009. Opposing assembly mechanisms in a Neotropical dry forest: implications for phylogenetic and functional community ecology. Ecology 90, 2161–2170.

Swenson, N.G., Enquist, B.J., Pither, J., Kerkhoff, A.J., Boyle, B., Weiser, M.D., Elser, J.J., Fagan, W.F., Forero-Montaña, J., Fyllas, N., Kraft, N.J.B., Lake, J.K., Moles, A.T., Patiño, S., Phillips, O.L., Price, C.A., Reich, P.B., Quesada, C.A., Stegen, J.C., Valencia, R., Wright, I.J., Wright, S.J., Andelman, S., Jørgensen, P.M., Lacher Jr., T.E., Monteagudo, A., Núñez-Vargas, M.P., Vasquez-Martínez, R., Nolting, K.M., 2012. The biogeography and filtering of woody plant functional diversity in North and South America. Glob. Ecol. Biogeogr. 21, 798–808.

Tilman, D., 1982. Resource Competition and Community Structure. (Mpb–17). Princeton University Press.

Tilman, D., 1999. The ecological consequences of changes in biodiversity: a search for general principles. Ecology 80, 1455–1474.

Tilman, D., 2001. An evolutionary approach to ecosystem functioning. Proc. Natl. Acad. Sci. U.S.A. 98, 10979–10980.

Tilman, D., Lehman, C.L., Thomson, K.T., 1997. Plant diversity and ecosystem productivity: theoretical considerations. Proc. Natl. Acad. Sci. U.S.A. 94, 1857–1861.

Tilman, D., Hillerislambers, J., Harpole, S., et al., 2004. Does metabolic theory apply to community ecology? It's a matter of scale. Ecology 85, 1797–1799.

Vile, D., Shipley, B., Garnier, E., 2006. Ecosystem productivity can be predicted from potential relative growth rate and species abundance. Ecol. Lett. 9 (9), 1061–1067.

Violle, C., Jiang, L., 2009. Towards a trait-based quantification of species niche. J. Plant Ecol. 2, 87–93.

Violle, C., Navas, M.-L., Vile, D., Kazakou, E., Fortunel, C., Hummel, I., Garnier, E., 2007. Let the concept of trait be functional!. Oikos 116, 882–892.

Violle, C., Enquist, B.J., McGill, B.J., Jiang, L., Albert, C.c.H., Hulshof, C., Jung, V., Messier, J., 2012. The return of the variance: intraspecific variability in community ecology. Trends Ecol. Evol. 27, 244–252.

Violle, C., Reich, P., Pacala, S., ENquist, B.J., Kattge, J., 2014. The emergence and promise of functional biogeography. Proc. Natl. Acad. Sci. U.S.A. 111 (38), 13690–13696.

von Allmen, E., Sperry, J., Smith, D., Savage, V., Enquist, B., Reich, P., 2012. A species' specific model of the hydraulic and metabolic allometry of trees II: testing predictions of water use and growth scaling in ring-and diffuse-porous species. Funct. Ecol. 26, 1066–1076.

Wardle, D.A., 2002. Ecology by the numbers. Trends Ecol. Evol. 11, 533–534.

Webb, C.T., Hoeting, J.A., Ames, G.M., Pyne, M.I., Poff, N.L., 2010. A structured and dynamic framework to advance traits-based theory and prediction in ecology. Ecol. Lett. 13, 267–283.

Weiher, E., Keddy, P.A., 1995. Assembly rules, null models, and trait dispersion: new questions from old patterns. Oikos 74, 159–164.

Weiher, E., Freund, D., Bunton, T., Stefanski, A., Lee, T., Bentivenga, S., 2011. Advances, challenges and a developing synthesis of ecological community assembly theory. Philos. Trans. R. Soc. B Biol. Sci. 366, 2403–2413.

West, G.B., Brown, J.H., Enquist, B.J., 1997. A general model for the origin of allometric scaling laws in biology. Science 276, 122–126.

West, G.B., Brown, J.H., Enquist, B.J., 2001. A general model for ontogenetic growth. Nature 413, 628–631.

West, G.B., Woodruff, W.H., Brown, J.H., 2002. Allometric scaling of metabolic rate from molecules and mitochondria to cells and mammals. Proc. Natl. Acad. Sci. U.S.A. 99, 2473–2478.

West, G.B., Enquist, B.J., Brown, J.H., 2009. A general quantitative theory of forest structure and dynamics. Proc. Natl. Acad. Sci. U.S.A. 106, 7040–7045.

Westoby, M., Wright, I.J., 2006. Land-plant ecology on the basis of functional traits. Trends Ecol. Evol. 21, 261–268.

Whittaker, R.H., Levin, S.A., Root, R.B., 1973. Niche, habitat, and ecotope. Am. Nat. 107, 321–338.

Wright, I.J., Reich, P.B., Cornelissen, J.H.C., Falster, D.S., Groom, P.K., Hikosaka, K., Lee, W., Lusk, C.H., Niinemets, U., Oleksyn, J., Osada, N., Poorter, H., Warton, D.I., Westoby, M., 2005. Modulation of leaf economic traits and trait relationships by climate. Glob. Ecol. Biogeogr. 14, 411–421.

Yvon-Durocher, G., Caffrey, J.M., Cescatti, A., Dossena, M., del Giorgio, P., Gasol, J.M., Montoya, J.M., et al., 2012. Reconciling the temperature dependence of respiration across timescales and ecosystem types. Nature 487, 472–476.

Zhang, L., Thygesen, U.H., Knudsen, K., Andersen, K.H., 2013. Trait diversity promotes stability of community dynamics. Theor. Ecol. 6, 57–69.

Zhao, M., Running, S.W., 2010. Drought-induced reduction in global terrestrial net primary production from 2000 through 2009. Science 329, 940–943.

Zhao, M., Heinsch, F.A., Nemani, R.R., Running, S.W., 2005. Improvements of the MODIS terrestrial gross and net primary production global data set. Remote Sens. Environ. 95, 164–176.

CHAPTER TEN

Functional Traits and Trait-Mediated Interactions: Connecting Community-Level Interactions with Ecosystem Functioning

Oswald J. Schmitz*,†,1, Robert W. Buchkowski*, Karin T. Burghardt†, Colin M. Donihue*

*School of Forestry and Environmental Studies, Yale University, New Haven, Connecticut, USA
†Department of Ecology and Evolutionary Biology, Yale University, New Haven, Connecticut, USA
1Corresponding author: e-mail address: oswald.schmitz@yale.edu

Contents

Abstract

Concerted effort in ecology is focused on developing synthetic frameworks that quantify general trends between diversity of organismal functional traits and ecosystem functioning. Yet much variation about the general trend routinely remains unexplained by trait diversity alone. We argue that this arises because these approaches fail to consider flexibility in trait expression as organisms adaptively respond to different environmental contexts (e.g., changes in resource quality or consumer pressure). We present here a framework for resolving how flexibility in functional trait expression is related to ecosystem functioning. We propose an approach that considers focal species, their resources and their consumers as a modular trophic unit. The approach then examines functional trait expressions of focal species when juxtaposed between different resource species (and associated traits) and consumer species (and associated traits). In such cases, focal species not only directly respond to different resource qualities and consumer pressure, but also mediate the indirect effects of consumer pressure on resource quality, causing feedbacks that ramify through ecosystems. Using case studies, we illustrate the utility of our modular approach for understanding how functional traits determine ecosystem

Advances in Ecological Research, Volume 52
ISSN 0065-2504
http://dx.doi.org/10.1016/bs.aecr.2015.01.003
319

functioning in a variety of aboveground and belowground trophic modules within eco-systems. We offer some general principles for explaining how variation in interactions among species determines variation in ecosystem functioning through a lens of flexible functional traits expression.

1. INTRODUCTION

Ecologists' original conception of ecological systems considered the biotic and biophysical components of the environment and their functional interplay as an integrated whole (Leopold, 1939; Lindeman, 1942; Tansley, 1935). It follows then that biotic characteristics of ecological systems, rep-resented broadly by species diversity, should instrumentally influence eco-system processes, such as, for example, nutrient cycling (Cardinale et al., 2012; Chapin et al., 1997, 2000; Hooper et al., 2005). However, work to date that focuses on species diversity exclusively only gets us so far towards elucidating the mechanisms by which species determine ecosystem function (Díaz et al., 2006; McGill et al., 2006; Naeem and Wright, 2003; Schmitz, 2010). This is because much residual variation often remains unexplained when quantifying statistical trends between species diversity and ecosystem processes (Fig. 1A) (McGill et al., 2006; Schmitz, 2010).

A proposed alternative is to characterize species in terms of their 'func-tional traits'—defined as any organismal character or phenotype associated with a biotic interaction or ecosystem function of interest (Naeem and Wright, 2003)—and then determine how variation in functional traits between systems is associated with variation in levels of a function (de Bello et al., 2010; Eviner and Chapin, 2003; Lavorel and Garnier, 2002; McGill et al., 2006; Mlambo, 2014; Naeem and Wright, 2003; Petchey and Gaston, 2006). The rationale is that species' roles are not deter-mined by their taxonomic identity but rather by their morphological, behavioural and physiological traits (McGill et al., 2006; Naeem and Wright, 2003), which can differ between species even within the same tro-phic level. Most approaches to linking functional traits and ecosystem func-tioning typically identify and catalogue suites of candidate functional traits, quantify their diversity within ecosystems and finally relate that diversity to ecosystem functioning (de Bello et al., 2010; Díaz et al., 2006; Eviner and Chapin, 2003; Lavorel et al., 2013; Naeem and Wright, 2003). These syn-thetic frameworks improve the resolution of species diversity approaches (Fig. 1B), by characterizing broad predictive relationships between

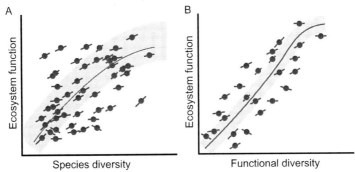

Figure 1 Hypothetical relationship between species diversity (A) or functional trait diversity (B) and level of ecosystem functioning based on syntheses of multiple studies. Points on each graph relate to the average values from individual studies used as part of the metadata in the synthesis. Lines through the points correspond to the directional trend (slope) between diversity and function within an individual study. The figures illustrate that much residual variation can arise (and often does arise) because the individual mean values deviate considerably from the overall average trend and relationships (slope) within a site may not correspond with the overall relationship between diversity and function. In this chapter, we argue that a modular functional traits approach can increase our power to explain the residual variation by explaining how flexibility in trait expression varies with biophysical context. *Adapted from Schmitz (2010).*

functional diversity and ecosystem processes (Eviner and Chapin, 2003; Lavorel et al., 2013). Nevertheless, much residual variation still often remains unexplained by the broad relationship (Fig. 1B).

We argue that much of this residual variation results from variation in the direction of relationships within sites or studies (Fig. 1B) due to flexibility or variation in functional trait expression in response to different biotic and abiotic contexts (e.g., thermal environment, resource quality and abundance, or predators). Moreover, ecological communities can be fundamentally depicted as species interconnected through trophic relationships. So, flexibility in trait expression often causes species to directly mediate trophic interactions between other species causing a host of indirect effects to propagate through the ecosystem (Burghardt and Schmitz, 2015; Hawlena and Schmitz, 2010b; Schmitz, 2008; Schmitz et al., 2008, 2010; Trussell and Schmitz, 2012). Thus, a fundamental challenge in relating functional traits to ecosystem functioning is to resolve how context-dependency in trait expression accounts for variation about any broadly divined trend (Schmitz, 2010). We elaborate here on an approach that shows how to explain context-dependency in ecosystem functioning based on flexibility in expression of functional traits.

Our approach requires a methodological shift from conventional approaches that develop *a priori* synthetic classification schemes that map functional trait diversity onto levels of ecosystem function. We argue for a more modular experimental approach that identifies a focal ecosystem function first and then observes key functional traits in action among different ecosystem contexts to quantify their effect on ecosystem functioning. For example, conventional classification schemes that identify important functional traits focus on the plant level and typically ignore the fact that plant traits may have evolved in response to pressures from consumers in higher trophic levels (e.g., herbivore-induced changes in growth form and chemical elemental content, anti-herbivore defence composition, etc.; Burghardt and Schmitz, 2015; Schmitz, 2010). Such traits that significantly affect a plant species' functional role may remain latent until the plant confronts its consumers.

Consider as a case in point the goldenrod *Solidago rugosa*, which is a competitively dominant plant in New England old-field ecosystems. Analyses of biodiversity–ecosystem function relationships that exclusively focus on the old-field plant community would quickly conclude that this species pre-empts resources and light and thus dictates ecosystem function—particularly production and elemental cycling—because it dominates the community. This interpretation is consistent with the mass-ratio hypothesis (Grime, 1998; Mulder et al., 2013). But, in this case, the mass-ratio hypothesis alone fails to predict context-dependence in ecosystem functioning, as top-down trophic control by grasshopper herbivores determine whether or not this plant species comes to dominate in the first place (Schmitz, 2010). This prediction is impossible before making the experimental observation that the traits that make it strongly competitive—its erect, leafy structure that allow it to attenuate light to the lower canopy and secure space (Schmitz, 2003)—also makes it highly desirable as a predation refuge for grasshopper herbivores. It is the fact that *S. rugosa* serves as a refuge for grasshoppers from predation that *ultimately* determines the ecosystem functioning within this context; its competitive ability due to its mass effect is a secondary determinant (Schmitz, 2003).

Resolving how functional traits vary according to ecological context begins with the starting premise that organisms ultimately try to maximize individual fitness measured in terms of either or all of survival, growth and reproduction (Holt, 1995). The drive to maximize fitness then determines the strategies used by organisms when they interact with other organisms (either predators or resources) or in different environmental contexts (Grime and Pierce, 2012; Schmitz, 2010). Clearly, not all strategies have an adaptive basis and so exceptions do occur (Miner et al., 2005).

Nevertheless, potentially explaining variation in the nature and strength of species interactions using principles from evolutionary ecology is a useful starting point that can help to identify generalities (Agrawal, 2001; Berg and Ellers, 2010; Miner et al., 2005; Schmitz et al., 2008). Through examination of how and why the same organisms express their traits in different ways in different environmental contexts, we hope to show that the nature of organismal responses to their biotic and abiotic environment through adaptive changes in traits can lead to a complementary predictive understanding of the structure and functioning of ecosystems called for in a functional traits research programme. It also responds to the larger call to link evolutionary ecology to ecosystem functioning (Grime and Pierce, 2012; Holt, 1995; Loehle and Pechmann, 1988; Schmitz, 2010; Schmitz et al., 2008).

2. TOWARD A PROCEDURAL FRAMEWORK

We propose that in order to move towards a predictive functional traits framework it is necessary to step back and consider not just the direct effects of these traits, but also how genetic variation within and between species and adaptive plasticity in functional traits can result in indirect feedbacks between community-level trophic interactions and ecosystem processes. This approach hinges on the fact that functional traits can play important roles in communities and ecosystems through two different pathways: adaptive and functional processes. These different pathways have received unequal consideration between community and ecosystem ecology (Fig. 2). Historically, community ecologists, particularly animal ecologists, have considered functional traits to be fitness–related traits and have spent most of their effort in elucidating adaptive trait dynamics as they scale from the individual to the community (Ohgushi et al., 2012).

In contrast, ecosystem ecologists have focused on connecting easily measured, aka 'soft', species traits (e.g., leaf area; *sensu* Hodgson et al., 1999) to ecosystem processes in an attempt to predict how environmental change will alter ecosystem functioning (Lavorel and Garnier, 2002). Both approaches have separately provided useful insight. We argue that these approaches can be combined into a common conceptual framework that can then be applied to any focal organism to highlight potential feedbacks between community and ecosystem processes. This reconciled conceptual framework reveals that traits that are acted on by adaptive processes may also indirectly alter ecosystem processes via their changes to community interactions (Fig. 2). Similarly, changes in ecosystem processes due to 'effect traits' (*sensu* Lavorel and Garnier, 2002; Lavorel et al., 1997) may indirectly alter

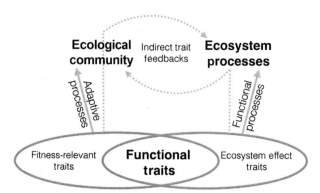

Figure 2 Illustration of the multiple pathways through which functional traits can affect species interactions and ecosystem properties. A suite of behavioural, morphological and physiological traits defines the life history characteristics of an organism. Adaptive processes affect a subset of those traits (fitness-relevant traits), while another subset have direct effects on ecosystem processes (ecosystem effect traits; *sensu* Lavorel and Garnier, 2002). However, effect traits may also have indirect effects on community interactions by altering ecosystem processes. In addition, an organism's fitness-relevant traits may indirectly affect ecosystem processes through the alteration of community composition and interactions. Thus, both categories of traits are potentially connected through indirect feedbacks (dashed lines). The functional traits expressed here may be altered through phenotypic plasticity (or genetic variation) (illustrated in Fig. 3), which makes the expression of traits dependent on the context the organism experiences. Therefore, the same organism placed within a different environmental context may have differing effects on community and ecosystem processes.

community dynamics by changing the environmental context of an organism and thus community interactions (Fig. 2).

To illustrate the point, an adaptive trait of an herbivore such as a shift in foraging behaviour in response to predators can impact plant community dominance structure, which in turn results in changes in ecosystem productivity (Schmitz, 2003). Alternatively, it is also possible for a non-adaptive plant trait such as leaf tissue nitrogen content, to change ecosystem properties altering community dynamics and thus the adaptive landscape (Wardle et al., 2004). This in turn changes selection on other traits indirectly. Within the case studies below we emphasize phenotypic plasticity as the mode of adaptation that creates environmental context-dependence in the role of functional traits and associated feedbacks. This is simply due to our particular focus on plasticity in our research over the years. Nevertheless, this framework can also apply to adaptation to environmental context through genetic trait variation between populations within a species or between species characterized by different functional traits.

Feedbacks can arise because indirect effects propagate through top-down effects within communities to influence ecosystem processes, which in turn exert bottom-up effects on communities. These indirect effects occur because organisms must continually reconcile the competing demands of consuming resources to meet requirements (bottom-up forcing of resource abundance and quality) and avoiding becoming resources for other consumers (top-down forcing from consumers). The ecosystem impacts of these interactions may drastically change as organisms reconcile the trade-offs between bottom-up and top-down factors in different environmental contexts. It follows that these changes can be best predicted by understanding the trade-off made by the middle trophic level. This leads to a universal, and disarmingly simple, modular rule. Namely, individuals are continually balancing trade-offs between resource consumption and avoiding being consumed in order to maximize individual fitness (Schmitz, 2010). As environmental context changes, species should flexibly adjust their trait expressions to re-balance fitness gains from foraging with fitness losses from their consumers. Such adaptive responses to community-level interactions can and do precipitate changes to whole-ecosystem functioning (Fig. 2; Schmitz, 2010).

2.1 Conceptualizing an ecological system

As stated above, our conception of a modular approach does not start with a specific set of traits. Instead, it develops a simple abstraction of ecosystem structure and processes in terms of their most fundamentally important components (Schmitz, 2010). This can be done by organizing the biotic and abiotic components of ecosystems into linear chains in which consumers of resources are themselves resources for other consumers and energy and materials flow along the consumption chains (Fig. 3). Greater biological realism can be infused by recognizing that two basic kinds of chains exist in ecosystems: those whose basal resources are live plant biomass (plant-based, 'green' chains) and those whose basal resources are non-living organic matter (detritus-based, 'brown' chains). Plant-based chains are composed of plants, herbivores and carnivores (Fig. 3). Detritus-based chains are slightly branched being composed of detritus, detritivores and carnivores or microbes, microbivores and carnivores (Fig. 3). These chains are not independent; they are coupled at their base because most plant production enters the non-living detrital pool (Cebrian, 1999; Hairston and Hairston, 1993; Moore et al., 2004). They could also be coupled at the apex should carnivores that feed on herbivores also feed

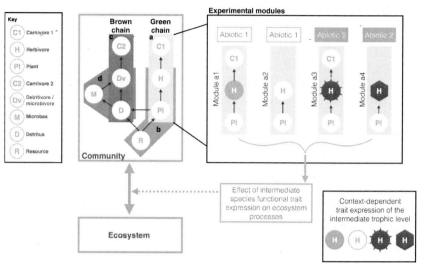

Figure 3 The central goal of functional traits is to clarify the feedbacks between communities and ecosystems (double arrow). Typically, research is focused on a particular ecosystem process such as nutrient cycling. These feedbacks can change due to differences in community or environment. Our approach picks dominant, focal species within trophic modules in this complex community and considers the main forces that control their ecology and evolution: abiotic factors, resources and predators. Functional traits are those traits that control a species response to these forces in different contexts. They determine how the group of species as a whole influences the ecosystem processes (solid arrow from experimental modules). As a result, the impact of these particular species on the ecosystem process is an approximation of the impact of the entire community on the ecosystem process (dotted arrow). The approximation becomes successively better as we consider more species and their interactions. However, this approximation differs from traditional synthetic approaches that divine a single broad-scale trend. Traditional approaches produce a coarse approximation because they assume species or functional traits act the same everywhere. We are proposing a means of getting a finer approximation by being sensitive to context.

on detritivores or microbivores (Schmitz, 2010). The chains are further linked belowground because fungal and bacterial decomposers mineralize nitrogen and carbon and contribute to the release of other materials. To properly understand these intricate interplays, we propose to focus attention on variation in trait expression of a dominant focal species to its consumers, resources and abiotic environment (Fig. 3).

2.2 Illustration of the modular approach

As a practical starting point, we will focus on species in the middle trophic levels of four illustrative chains (Fig. 3a–d, corresponding to text

Sections 2.2.1–2.2.4, respectively) because they are important mediators of indirect effects through adaptive trait expression (e.g., 'trait-mediated indirect effects') (Werner and Peacor, 2003). Our proposed modular approach then focuses on the functional traits of a species in the middle of that tritrophic chain—the mediating species—and quantifies how they are influenced by abiotic conditions and the functional traits of species in the next higher and lower trophic levels. Finally, using these case studies, we discuss how to implement this modular approach by linking functional traits, community-level interactions and ecosystem functioning through experimental methods that address context-dependency.

2.2.1 Carnivore—herbivore—plant module

A dominant herbivore in old-field ecosystem functioning (Schmitz, 2010) is the grasshopper *Melanoplus femurrubrum*, a generalist feeder (a functional trait) that consumes a mixture of grasses and herbs. In the absence of predation, the grasshopper prefers grass because of its comparative high nitrogen content needed to meet metabolic demands for maintenance and production (Fig. 4). But, grass is a risky place to feed because its simple structure does not allow the grasshopper to hide from predators. The grasshopper also faces several species of hunting spider predators that differ in their hunting mode ranging from active pursuit to sit-and-wait ambush (Miller et al., 2014).

Sit-and-wait spiders hunt from fixed locations in the upper canopy of the field providing a persistent threat. Grasshoppers respond to these predators via adaptive plasticity in two ways. They change their foraging behaviour (a functional trait) by trading-off the risk of consuming grass to hiding and foraging in structurally more complex herbs like the competitively

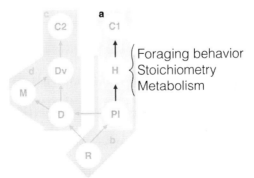

Figure 4 The carnivore—herbivore—plant module highlighting context-dependence of herbivore foraging behaviour, stoichiometry and metabolism.

dominant goldenrod (*S. rugosa*) that serves as a refuge from predation (Schmitz, 2010). Perceived predation risk also induces chronic physiological stress responses in the herbivores (a functional trait) that elevate their metabolic rate (Hawlena and Schmitz, 2010a). Such stress responses keep the grasshoppers in a heightened state of alertness to increase the chance they can escape their predators in the face of persistent risk (Hawlena and Schmitz, 2010b). But, elevated metabolism (respiration) shifts demand to energy containing soluble carbohydrates to meet heightened maintenance demands at the expense of consuming nitrogen for production. Soluble carbohydrate is readily supplied by the same goldenrod species that provides refuge. This adaptive foraging shift leads to higher plant species diversity because the grasshopper suppresses this competitively dominant plant (Schmitz, 2003). It also changes the C:N content of the plant community, and hence the C:N content of plant biomass entering the detrital pool to be decomposed (Hawlena and Schmitz, 2010a). Moreover, the change in physiological demand for nutrients causes the herbivore to excrete more N, to avoid toxicity and respire more C to the atmosphere relative to conditions where the predator is absent (Hawlena and Schmitz, 2010a). The herbivore adaptive foraging shift also impacts plant physiology because it causes plants to alter photosynthesis and respiration rates and aboveground–belowground C allocation in plant tissue (Strickland et al., 2013). Finally, the difference in grasshopper body C:N content between predation and predation-free conditions causes differences in priming of soil microbes such that there is up to a sixfold difference (e.g., a multiplier effect; Schmitz et al., 2014) in microbial carbon mineralization rate of subsequent plant detrital inputs (Hawlena et al., 2012). Collectively, the large multiplier effect of the sit-and-wait spider on elemental cycling, mediated by trait changes in the herbivore prey, came about because significant changes occurred to the C:N balance of organic matter of the largest trophic compartments (plants and detritus) and the ability of another large trophic compartment (microbes) to decompose that organic matter for recycling.

These effects are not observed when this species is in a different context, such as when it co-occurs with widely roaming active hunting spiders. Actively hunting spiders impose low predation risk because they have an ephemeral presence in any one location. Consequently, grasshopper prey would waste considerable energy and nutrient intake, and thereby unnecessarily compromise fitness, if they remained chronically stressed (Schmitz, 2010). Instead, grasshoppers only respond to the imminent threat that arises during the rare encounters with the predator. This results in altogether different plant species diversity, plant elemental content and levels of eco-system functioning than when grasshoppers face sit-and-wait predators (Schmitz, 2008). Certainly, spider predator body size is important because it determines whether or not they can capture and consume the grasshopper prey to begin with (Schmitz and Suttle, 2001). But predator hunting mode, not body size, is what explains context-dependency in the nature and strength of the community- and ecosystem-level effects. Prey responses to predator hunting mode enables prediction of how changes in herbivore trait expression (Fig. 2) mediate predator effects to impact both community and ecosystem dynamics (reviewed in Miller and Rudolf, 2011).

The way indirect effects of carnivores on communities and ecosystems are mediated by herbivores also depends on an important herbivore func-tional trait, feeding mode, which may also determine the nature of herbivore species responses to different environmental contexts (Schmitz, 2010; Singer et al., 2014). Herbivores can be highly specialized on one or a few plants or be broad generalists; and they can engage in leaf chewing (grazing and browsing), sap-feeding or leaf mining feeding behaviour (Schmitz, 2010; Singer et al., 2014). Dietary specialization, especially, appears to be an evolved response to generalist predators because specialist herbivores can more effectively enlist characteristics of their host plants for defence or ref-uge than generalist herbivores (Singer et al., 2014). For example, specialist herbivorous insects often are able to sequester plant toxins that make them unpalatable or toxic to their predators, or they can mimic structural traits or coloration of their host plants to become cryptic, something that would be difficult for a generalist species to do given the variety of different plant types they rely on. Moreover, many specialist species are leaf miners and sap feeders that tend to be sedentary. Thus, the nature and strength of predator indirect effects become predictably contingent on the hunting mode of the predator in relation to the feeding mode and movement behaviour of the prey (Schmitz, 2010). Predator indirect effects on plants are likely to be much weaker when specialist species dominate the herbivore community than when generalist species predominate (Schmitz, 2010; Singer et al., 2014). Furthermore, the sign of the indirect effect on plants may differ between specialist and generalist herbivores even if they have similar func-tional body mass (Schmitz and Price, 2011).

Specialist herbivores also can incur elevated metabolic rates in response to stress from predation risk (Thaler et al., 2012). But, specialists have no recourse to shift their plant selection and thus compensate not by decreasing foraging effort but by altering the passage rate of food, resulting in altered

N assimilation efficiency (Thaler et al., 2012). This in turn alters the chemical elemental composition of their body tissues in opposite ways to generalist herbivores.

2.2.2 Herbivore—plant—soil elements module

Much of the work linking plant traits to ecosystem functioning is on the direct pathway between plant biomass and detritus-based food chains through litter inputs. Herbivores are not often considered within terrestrial plant functional trait research because they consume a relatively small fraction of plant biomass (Cebrian and Lartigue, 2004). However, the modular approach emphasizes that herbivores may also indirectly influence ecosystem processes through the lingering presence of induced plant defensive traits within uneaten leaf tissue (Choudhury, 1988). Here, we use plant-induced defensive strategies as a case study for a focal functional trait that may provide predictive ability for understanding the effect of herbivory on nutrient cycling (Fig. 5).

Plants engage in a wide variety of anti-herbivore defence strategies. The way plants express those traits depends on the interplay between soil nutrient (element) availability and herbivory (Power, 1992). Across nutrient environments the cost-benefit trade-off of defending tissue with N- and C-rich defensive compounds changes, creating situations where the best-performing allocation strategy in one environment may be maladaptive in another (Coley et al., 1985). Phenotypic plasticity in plant allocation to defence across nutrient environments alleviates this problem by allowing flexibility in allocation patterns to the presence of another trophic level (e.g., herbivory; Agrawal, 2001). These defensive traits are often organized into plant defensive syndromes, suites of co-occurring structural,

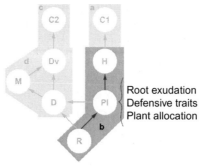

Figure 5 The herbivore—plant—soil elements module highlighting context-dependence in plant root exudation, defensive traits and allocation.

physiological and allocational patterns within the plant (Agrawal and Fishbein, 2006). Similar to Grime's C-S-R strategies (Grime and Pierce, 2012), selection and allocation trade-offs within the plant limit the combinations of traits that are likely to co-occur.

One syndrome is an induced resistance strategy which is associated with plant traits that are known to decrease decomposition rates in ecosystems (e.g., increased leaf toughness, chemical defences, C:N ratio), while the alternative syndrome is induced tolerance associated with changes in plant traits that are known to increase nutrient cycling in the presence of herbivores (Belovsky and Slade, 2000; Ritchie et al., 1998; Schweitzer et al., 2005). Herbivore feeding may also alter ecosystem processes by changing plant inputs belowground (e.g., litter, root exudates) or through alteration in dominance and interactions across the landscape (Fig. 3). Most current plant functional trait frameworks cannot account for this contingency because they focus exclusively on direct effects of plant litter traits on ecosystem impacts rather than considering how environmental context through alterations in community dynamics or species interactions determine the level of trait expression in the first place.

Thus, contingency as to whether herbivores increase or decrease nutrient cycling in ecosystems may be explained by differential defence induction across nutrient gradients (Burghardt and Schmitz, 2015; O'Donnell et al., 2013). A shift in nutrient availability can change the opportunity costs of induced defences and potentially the outcome of plant–herbivore or plant competitive interactions (Cipollini et al., 2003). Thus, the efficacy of and selection for the plant defensive traits outlined above are influenced by the environmental context in which they are expressed (Belovsky and Schmitz, 1994). This means that negative or positive feedbacks are possible between nutrient environments and selection for the expression of particular plant defensive strategies (Burghardt and Schmitz, 2015). A functional plant traits framework can only become predictive once the flexible interplay between nutrient availability, plant defence syndrome and the nature of herbivory are explored through systematic experimentation.

2.2.3 Carnivore—detritivore/microbivore—detritus module

Detritivores and microbivores are a highly diverse group of species that influence elemental cycling through their effects on (1) the physical structure of detritus (Bastow, 2011; Salmon, 2004; Seastedt, 1984), (2) microbial community structure and biomass (Crowther et al., 2011a, 2012) and (3) soil C and N availability (Bouché et al., 1997; Carrillo et al., 2011; De

Ruiter et al., 1993; Holtkamp et al., 2008; Seastedt, 1984; Teuben and Verhoef, 1992) (Fig. 6).

Research on species interactions within detrital chains is not widely given to consideration of trait-mediated effects (Moore et al., 2004). Yet, it is altogether conceivable that animal species within detrital chains are subject to the same foraging-predation avoidance trade-offs faced by species within plant-based chains, even though, unlike plants, detritus is non-living and therefore, will not have countervailing adaptive responses to consumption (Moore et al., 2004). Still, synthesis that applies a trait-mediated perspective to understand mechanisms of interactions along detritus-based chains (Schmitz, 2010) reveals considerable parallels. Like herbivores, detritivore species selectively feed on resources in order to maximize fitness (Hättenschwiler et al., 2005; Scheu and Folger, 2004). They also alter their foraging behaviour in response to predation risk through changes in the time spent foraging or through shifts in habitat use (Grear and Schmitz, 2005; Schmitz, 2010; Sitvarin and Rypstra, 2014).

Some trade-offs can involve detritivores moving from surface to subsurface soil layers to avoid predation. This can change the distribution of organic matter in the soil with cascading effects on the degree to which microbes can access organic matter and decompose it into elemental form (Coleman et al., 1983; Moore et al., 2004; Seastedt, 1984). Moreover, the nature of the detritivore-mediated indirect effect of predators on microbes may depend on the elemental content of soil. For example, Lenoir et al. (2007) found that the nature of cascading effects of predators

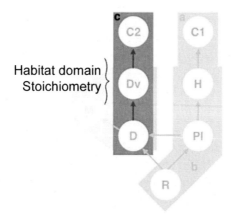

Figure 6 The carnivore—detritivore/microbivore—detritus module highlighting changes in detritivores, habitation domain and stoichiometry according to context.

on fungal biomass, mediated by fungivores (collembolans and orabatid mites) depended on the nitrogen content of the humus. In high nitrogen humus, predators had no effect on fungal biomass, while in low nitrogen they decreased it. Lenoir et al. (2007) suggest that the differences may have arisen because fungi can exhibit compensatory growth under high nutrient conditions (a potentially adaptive trait response) but not in low nutrient conditions.

The potential for adaptive trait responses is revealed in a second example, involving effects of predaceous and microbivorous nematodes on nutrient cycling (Mikola and Setälä, 1998). Microbivorous nematodes can be strongly limited by the presence of their predator, which can counteract their enhancing effects on nitrogen cycling. These effects were not mediated by a strong impact of the microbivore on microbial abundance in this particular case (Mikola and Setälä, 1998), while other experiments have found that microbial abundance is strongly dependent on microbivores (Allen-Morley and Coleman, 1989). Clearly, the importance of interactions between microbivores and their prey is context-dependent.

There are other traits of detritivores and microbivores that could mediate interactions along trophic chains. Temperature and moisture tolerance is particularly important because detritivores subsist in environments where both variables change across small spatial extents (Brady and Weil, 2009). The level of tolerance to changing temperature and moisture regimes determines detritivore movement behaviour and hence the ability to evade predators or redistribute detritus within the soil layers with cascading effects on litter heterogeneity and the distribution of microbes (e.g., Fujii and Takeda, 2012). Also, detritivore body morphology can influence how they move through soil and litter environments, what litter types they degrade (fine root litter or leaf litter) and whether they are specialist or generalist consumers (Fujii and Takeda, 2012). Hence, body morphology can determine how detritivores mediate indirect effects of predators.

Like herbivores, detritivores and microbivores must meet homeostatic elemental balances determined by physiological processes that govern their element uptake and excretion (Martinson et al., 2008; Reiners, 1986; Sterner and Elser, 2002). This suggests that detritivores and microbivores could exhibit the same predation-induced physiological stress responses as herbivores, given that the physiological and hormonal machinery that drives such adaptive responses is highly evolutionarily conserved among animal taxa (Hawlena et al., 2012).

Despite the potential for widespread trade-offs between foraging and predation-risk avoidance (Schmitz, 2010; Sitvarin and Rypstra, 2014), it remains uncertain whether this trade-off can be a dominant driver of cascading effects within the detrital chain. Only further experimentation (perhaps using a modular approach) will reveal whether or not these changes will have strong effects detritus-based systems.

2.2.4 Detrivore/microbivore—soil microbe—detritus module

Soil microbial communities are diverse and interconnected in a complex web of trophic interactions (Clarholm, 1994; Neutel et al., 2002), so that identifying all of the individual species in these food webs is currently impractical. Such diversity begs for beginning with a more modular approach that aggregates species into trophic compartments (Moore et al., 2004; Schmitz, 2010) in which microbes are middle trophic levels that interact with detrital resources and predators. A useful starting point is either to focus on dominant microbial species or taxa (e.g., Clarholm, 1985; Crowther et al., 2011a) or examine whole microbial community function using changes in biomass or exoenzyme activity (Ritz et al., 1994). Both of these methods have been used to explore how microbes participate in trophic interactions that include detritus and detritivores (Crowther et al., 2011a,c; Lenoir et al., 2007; Mikola and Setälä, 1998) (Fig. 7).

Soil microbial communities are influenced by bottom–up factors like the quality and structural properties of their detrital resources (Allison, 2006). They are particularly limited by the quality and the, often heterogeneous, spatial distribution of their detrital resources (Allison, 2006). The distribution of microbial species is also spatially heterogeneous, because individuals

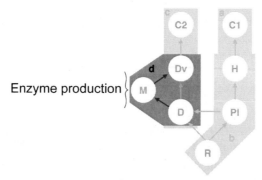

Figure 7 Detritivore/microbivore—soil microbe—detritus module with a focus on context-dependent enzyme production among microbes.

are filtered at local sites based on relatively low nutrient availability, harsh environmental conditions or competition (Goberna et al., 2014; Moore et al., 2004). The diversity of microbial communities, which results from these local selective pressures, creates very different functional capacities across soil conditions. For example, Keiser et al. (2014) argued that some communities have a strong home field advantage, wherein they degrade litter from their environment better than foreign litter. Thus, microbial communities likely have a large capacity to exhibit context-dependent changes in their functional traits based on the quality of their resources.

Microbe functioning also may be determined by the capacity to shift foraging strategies and take up organic nutrients in the rhizosphere when plants release root exudates (Drake et al., 2013; Hamilton and Frank, 2001). In fact, the biomass and exoenzyme production of the microbial community generally changes when nutrients are added to the soil (Bardgett and McAlister, 1999; Carreiro et al., 2000; Sinsbaugh et al., 2002, 2005). There is clearly strong context-dependency in responses of microbial communities to their resource base that may interact with top-down effects to determine how soil microbial communities function in different contexts.

Grazers (detritivores/microbivores) also exert top-down effects on microbes. High grazing pressure by large or abundant soil fauna can reduce microbial biomass (Crowther et al., 2011b; Lenoir et al., 2007), with microbes compensating by increasing their growth rates to maintain the same or higher biomass when nutrients are not limiting (Coûteaux and Bottner, 1994; Mikola and Setälä, 1998; Vedder et al., 1996). Hence, the magnitude of the compensatory growth response depends on interplay between the strength of the grazing impact and nutrient availability. Microbial biomass (Lenoir et al., 2007) and function (Coûteaux and Bottner, 1994; Coûteaux et al., 1991) may remain high under grazing pressure in nutrient-rich environments but are more likely to be depressed in nutrient poor environments. Given that microbial biomass is related to exoenzyme production, microbes thus can mediate the cascading effects of predators on organic matter decomposition rate (Crowther et al., 2011a,c).

Despite the variability in microbial communities within soils and their response to environmental contexts, some generalities are beginning to emerge when examining processes through the lens of a modular approach. First, the response of the microbial community to grazing pressure is highly dependent on the resource environment, with high resource environments leading to compensatory growth and low resource environments leading to net biomass loss. Second, the effect of grazing pressure likely has a greater influence on microbial community composition and function than on bio-mass per se. However, microbial communities are rarely studied using this modular perspective (Schmitz, 2010). More empirical examples of how the resource environment and grazers impact microbial community interactions are required to build the predictive framework we are proposing.

3. MOVING FORWARD

Historically, the study of organisms within different trophic groups has been concentrated in disparate, isolated areas of our framework (Fig. 2). Detritivores and microbes are generally viewed as primarily impacting func-tional processes (often to the exclusion of plasticity and adaptive processes; but see Crowther et al., 2011a), while herbivores and predators are exam-ined primarily through the lens of community rather than ecosystem impacts. Plant ecologists have focused most equitably between these path-ways, but rarely account for the indirect connections between them, and generally ignore the interacting effects of higher trophic levels. Some pioneering work on plants is beginning to fill this gap (Baxendale et al., 2014; Bezemer et al., 2013; Schweitzer et al., 2008). We believe particularly ripe places for future progress lie within those areas of the framework not previously explored that link community and ecosystem feedbacks. We have elucidated some of these mechanisms in our old-field ecosystem. For exam-ple, goldenrod dominance exerts strong bottom-up control and slows ele-mental cycling, yet, the presence of sit-and-wait predators cause grasshopper herbivores to consume more goldenrod. This consumption changes the plant community and elemental cycling, which has a new degree of bottom-up control on these communities. However, we argue that further empirical examples that include other ecosystems or other components of the ecosystem (e.g., detritivores) are necessary to develop a predictive framework.

We focus here on elemental cycling because of its fundamental impor-tance to ecosystem function (DeAngelis, 1992; Loreau, 2010). The princi-ples discussed nonetheless can assist in understanding other kinds of ecosystem processes and properties (e.g., trophic transfer efficiency, number of trophic levels; Trussell and Schmitz, 2012). Once a basic set of interacting species are identified through the proposed modular approach, additional detail and complexity can be added to our conceptual model (Fig. 8). Draw-ing from the examples used in this text, one can begin to predict essential aspects of elemental cycling (DeAngelis, 1992; Loreau, 2010; Moore et al., 2004). Here, we depict elemental uptake by plants from the abiotic environment (i.e., carbon uptake from the atmosphere and nitrogen uptake from soils) and elemental transfer and loss to and from all compartments through trophic interactions, respiration, excretion, egestion and leaching out of the ecosystem (Fig. 8). The flow of elements up plant-based and detritus-based chains results from trophic (consumptive) interactions between a consumer and a resource trophic level. The strength of this

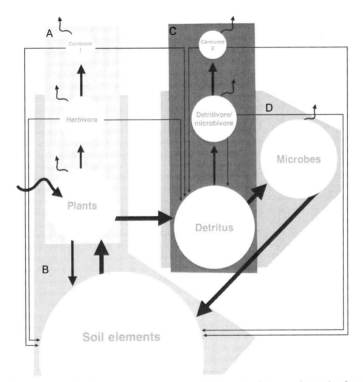

Figure 8 Conceptualization of ecosystem structure needed to combine the functional traits approach to understand whole-ecosystem functioning. This conceptualization reveals key trophic compartments that can control functioning. It depicts different processes related to elemental cycling, including atmospheric CO_2 uptake into the plant compartment, respiratory CO_2 release from all living trophic compartments to the atmosphere, nutrient (elemental) uptake by one compartment from an adjacent lower compartment, inorganic inputs from trophic compartments to the soil elemental pool, organic inputs from living trophic compartments to the detrital pool and conversion of organic matter into elements via foundational cycling of matter through plant, detrital, microbial and soil elemental compartments and pools. The modular approach examines trait responses of species within a trophic compartment when explicitly juxtaposed between at least two other trophic compartments.

consumptive interaction determines the flux rate of elements between trophic compartments. In essence, we propose that using experimentation within trophic modules facilitates incorporating biological detail about organismal functional traits that can help to determine the magnitude of the flux and thus connect consumptive interactions to ecosystem functioning.

The approach we describe here acknowledges that ecosystems are dominated by plant, detrital and microbial biomass and that biomass of animals in higher trophic levels are typically orders of magnitude less (Fig. 8). It also recognizes that the majority of plant biomass in ecosystems does not flow up plant-based chains but enters the soil as detritus where it is decomposed by microbes into constituent minerals that are released to the soil and then recycled back to plants (Cebrian, 1999; Hairston and Hairston, 1993; Moore et al., 2004). Nevertheless, it also shows that biomass representation may not always adequately quantify the contribution of a species to ecosystem functioning when considering mediation of species effects by flexible expression of functional traits.

For example, smaller biomass pools, including animals, may have disproportionately strong effects on ecosystems because, as consumers, they induce flexible, adaptive trait responses in their resource species (Schmitz et al., 2008; Trussell and Schmitz, 2012). Adaptive trait responses can in turn precipitate a sequence of responses between species in adjacent trophic compartments that can cause a host of indirect effects that propagate through ecosystems.

By quantifying these trait-mediated indirect effects, the framework we propose advances the predictive ability of traditional community and ecosystem ecology, which both struggle to explain context-dependency. By focusing on how differences in trait expression of a middle trophic level changes in response to predators, resources or abiotic context, this approach provides the needed mechanistic basis to predict context-dependent outcomes in ecosystem processes. Therefore, when one understands the important players within an ecosystem and their respective fitness trade-offs across environments, then the patterns of context-dependency in the link between functional traits, community-level interactions and ecosystem functioning become clear. Given that the evolutionary ecological principles we apply here are foundational to all taxa (Agrawal, 2001; Grime and Pierce, 2012; Miner et al., 2005; Schmitz, 2010), our modular approach is broadly applicable to a wide variety of species and food-web modules.

ACKNOWLEDGEMENTS

This research was supported by NSF DEB-1354762 to O.J.S, GRFP under NSF Grant DGE-1122492 to K.T.B and an NSERC PGS-D to R.W.B.

REFERENCES

Agrawal, A.A., Fishbein, M., 2006. Plant defense syndromes. Ecology 87, S132–S149.

Agrawal, A.A., 2001. Phenotypic plasticity in the interactions and evolution of species. Science 294, 321–326.

Allen-Morley, C.R., Coleman, D.C., 1989. Resilience of soil biota in various food webs to freezing perturbations. Ecology 70, 1127–1141.

Allison, S.D., 2006. Brown ground: a soil carbon analogue for the green world hypothesis? Am. Nat. 167, 619–627.

Bardgett, R.D., Mcalister, E., 1999. The measurement of soil fungal: bacterial biomass ratios as an indicator of ecosystem self-regulation in temperate meadow grasslands. Biol. Fertil. Soils 29, 282–290.

Bastow, J.L., 2011. Facilitation and predation structure a grassland detrital food web: the responses of soil nematodes to isopod processing of litter. J. Anim. Ecol. 80, 947–957.

Baxendale, C., Orwin, K.H., Poly, F., Pommier, T., Barkgett, R.D., 2014. Are plant-soil feedback responses explained by plant traits? New Phytol. 204, 408–423.

Belovsky, G., Slade, J., 2000. Insect herbivory accelerates nutrient cycling and increases plant production. Proc. Natl. Acad. Sci. U.S.A. 97, 14412–14417.

Belovsky, G.E., Schmitz, O.J., 1994. Plant defenses and optimal foraging by mammalian herbivores. J. Mammal. 816–832.

Berg, M., Ellers, J., 2010. Trait plasticity in species interactions: a driving force of community dynamics. Evol. Ecol. 24, 617–629.

Bezemer, T.M., van der Putten, W.H., Martens, H., van de Voorde, T.F.J., Mulder, P.P.J., Kostenko, O., 2013. Above- and below-ground herbivory effects on below-ground plant-fungus interactions and plant-soil feedback responses. J. Ecol. 101, 325–333.

Bouché, M.B., Al-Addan, F., Cortez, J., Hammed, R., Heidet, J.-C., Ferrière, G., Mazaud, D., Samih, M., 1997. Role of earthworms in the N cycle: a falsifiable assessment. Soil Biol. Biochem. 29, 375–380.

Burghardt, K.T., Schmitz, O.J., 2015. Influence of plant defenses and nutrients on trophic control of ecosystems. In: Hanley, T., La Pierre, K.J. (Eds.), Trophic Ecology: Bottom-Up and Top-Down Interactions Across Aquatic and Terrestrial Systems. Cambridge University Press, Cambridge, MA.

Brady, N.C., Weil, R.R., 2009. Elements of the Nature and Properties of Soils, third ed. Prentice and Hall, New Jersey.

Cardinale, B.J., Duffy, J.E., Gonzalez, A., Hooper, D.U., Perrings, C., Venail, P., Narwani, A., Mace, G.M., Tilman, D., Wardle, D.A., Kinzig, A.P., Daily, G.C., Loreau, M., Grace, J.B., Larigauderie, A., Srivastava, D.S., Naeem, S., 2012. Biodiversity loss and its impact on humanity. Nature 486, 59–67.

Carreiro, M.M., Sinsabaugh, R.L., Repert, D.A., Parkhurst, D.F., 2000. Microbial enzyme shifts explain litter decay responses to simulated nitrogen deposition. Ecology 81, 2359–2365.

Carrillo, Y., Ball, B.A., Bradford, M.A., Jordan, C.F., Molina, M., 2011. Soil fauna alter the effects of litter composition on nitrogen cycling in a mineral soil. Soil Biol. Biochem. 43, 1440–1449.

Cebrian, J., Lartigue, J., 2004. Patterns of herbivory and decomposition in aquatic and terrestrial ecosystems. Ecol. Monogr. 74, 237–259.

Cebrian, J., 1999. Patterns in the fate of production in plant communities. Am. Nat. 154, 449–468.

Chapin, F.S., Zavaleta, E.S., Eviner, V.T., Naylor, R.L., Vitousek, P.M., Reynolds, H.L., Hooper, D.U., Lavorel, S., Sala, O.E., Hobbie, S.E., Mack, M.C., Diaz, S., 2000. Consequences of changing biodiversity. Nature 405, 234–242.

Chapin, F.S., Walker, B.H., Hobbs, R.J., Hooper, D.U., Lawton, J.H., Sala, O.E., Tilman, D., 1997. Biotic control over the functioning of ecosystems. Science 277, 500–504.

Choudhury, D., 1988. Herbivore induced changes in leaf-litter resource quality: a neglected aspect of herbivory in ecosystem nutrient dynamics. Oikos 389–393.

Cipollini, D., Purrington, C.B., Bergelson, J., 2003. Costs of induced responses in plants. Basic Appl. Ecol. 4, 79–89.

Clarholm, M., 1985. Interctions of bacteria, protozoa and plants leading to mineralization of soil-nitrogen. Soil Biol. Biochem. 17, 181–187.

Clarholm, M., 1994. The microbial loop in soil. In: Ritz, K., Dighton, J., Giller, K.E. (Eds.), Beyond Biomass. Wiley-Sayce Publication, New York, NY.

Coleman, D.C., Reid, C.P.P., Cole, C.V., 1983. Biological strategies of nutrient cycling in soil systems. In: MacFyaden, A., Ford, E.D. (Eds.), Advances in Ecological Research, vol. 13. Academic Press, London, UK, pp. 1–55.

Coley, P.D., Bryant, J.P., Chapin, F.S., 1985. Resource availability and plant antiherbivore defense. Science (Washington) 230, 895–899.

Coûteaux, M.-M., Bottner, P., 1994. Biological interactions between fauna and the microbial community in soils. In: Ritz, K., Dighton, J., Giller, K.E. (Eds.), Beyond Biomass. Wiley-Sayce, New York, NY.

Coûteaux, M.-M., Mousseau, M., Célérier, M.-L., Bottner, P., 1991. Increased atmospheric CO_2 and litter quality: decomposition of sweet chestnut leaf litter with animal food webs of different complexities. Oikos 61, 54–64.

Crowther, T.W., Boddy, L., Jones, T.H., 2011a. Outcomes of fungal interactions are determined by soil invertebrate grazers. Ecol. Lett. 14, 1134–1142.

Crowther, T.W., Boddy, L., Jones, T.H., 2011b. Species-specific effects of soil fauna on fungal foraging and decomposition. Oecologia 167, 535–545.

Crowther, T.W., Boddy, L., Jones, T.H., 2012. Functional and ecological consequences of saprotrophic fungus-grazer interactions. ISME J. 6, 1992–2001.

Crowther, T.W., Jones, T.H., Boddy, L., Baldrian, P., 2011c. Invertebrate grazing determines enzyme production by basidiomycete fungi. Soil Biol. Biochem. 43, 2060–2068.

De Bello, F., Lavorel, S., Díaz, S., Harrington, R., Cornelissen, J.C., Bardgett, R., Berg, M., Cipriotti, P., Feld, C., Hering, D., Martins Da Silva, P., Potts, S., Sandin, L., Sousa, J., Storkey, J., Wardle, D., Harrison, P., 2010. Towards an assessment of multiple ecosystem processes and services via functional traits. Biodivers. Conserv. 19, 2873–2893.

De Ruiter, P.C., Vanveen, J.A., Moore, J.C., Brussaard, L., Hunt, H.W., 1993. Calculation of nitrogen mineralization in soil food webs. Plant Soil 157, 263–273.

DeAngelis, D.L., 1992. Dynamics of Nutrient Cycling and Food Webs. Chapman and Hall, New York.

Díaz, S., Fargione, J., Chapin, F.S., Tilman, D., 2006. Biodiversity loss threatens human well-being. PLoS Biol. 4, e277.

Drake, J.E., Darby, B.A., Giasson, M.A., Kramer, M.A., Phillips, R.P., Finzi, A.C., 2013. Stoichiometry constrains microbial response to root exudation-insights from a model and a field experiment in a temperate forest. Biogeosciences 10, 821–838.

Eviner, V.T., Chapin, F.S., 2003. Functional matrix: A conceptual framework for predicting multiple plant effects on ecosystem processes. Annu. Rev. Ecol. Evol. Syst. 34, 455–485.

Fujii, S., Takeda, H., 2012. Succession of collembolan communities during decomposition of leaf and root litter: effects of litter type and position. Soil Biol. Biochem. 54, 77–85.

Goberna, M., Navarro-Cano, J.A., Valiente-Banuet, A., García, C., Verdú, M., 2014. Abiotic stress tolerance and competition-related traits underlie phylogenetic clustering in soil bacterial communities. Ecol. Lett. 17, 1191–1201.

Grear, J.S., Schmitz, O.J., 2005. Effects of grouping behavior and predators on the spatial distribution of a forest floor arthropod. Ecology 86, 960–971.

Grime, J., Pierce, S., 2012. The Evolutionary Strategies that Shape Ecosystems. Wiley-Blackwell, Oxford.

Grime, J., 1998. Benefits of plant diversity to ecosystems: immediate, filter and founder effects. J. Ecol. 86, 902–910.

Hairston Jr., N.G., Hairston Sr., N.G., 1993. Cause-effect relationships in energy flow, trophic structure, and interspecific interactions. Am. Nat. 142, 379–411.

Hamilton, E.W., Frank, D.A., 2001. Can plants stimulate soil microbes and their own nutrient supply? Evidence from a grazing tolerant grass. Ecology 82, 2397–2402.

Hättenschwiler, S., Tiunov, A.V., Scheu, S., 2005. Biodiversity and litter decomposition in terrestrial ecosystems. Annu. Rev. Ecol. Evol. Syst. 191–218.

Hawlena, D., Schmitz, O.J., 2010a. Herbivore physiological response to predation risk and implications for ecosystem nutrient dynamics. Proc. Natl. Acad. Sci. U.S.A. 107, 15503–15507.

Hawlena, D., Schmitz, O.J., 2010b. Physiological stress as a fundamental mechanism linking predation to ecosystem functioning. Am. Nat. 176, 537–556.

Hawlena, D., Strickland, M.S., Bradford, M.A., Schmitz, O.J., 2012. Fear of predation slows plant-litter decomposition. Science 336, 1434–1438.

Hodgson, J.G., Wilson, P.J., Hunt, R., Grime, J.P., Thompson, K., 1999. Allocating C-S-R plant functional types: a soft approach to a hard problem. Oikos 85, 282–294.

Holt, R.D., 1995. Linking species and ecosystems: where is Darwin? In: Jones, C., Lawton, J.H. (Eds.), Linking Species and Ecosystems. Chapman and Hall, London.

Holtkamp, R., Kardol, P., Van Der Wal, A., Dekker, S.C., Van Der Putten, W.H., De Ruiter, P.C., 2008. Soil food web structure during ecosystem development after land abandonment. Appl. Soil Ecol. 39, 23–34.

Hooper, D., Chapin Iii, F., Ewel, J., Hector, A., Inchausti, P., Lavorel, S., Lawton, J., Lodge, D., Loreau, M., Naeem, S., 2005. Effects of biodiversity on ecosystem functioning: a consensus of current knowledge. Ecol. Monogr. 75, 3–35.

Keiser, A.D., Keiser, D.A., Strickland, M.S., Bradford, M.A., 2014. Disentangling the mechanisms underlying functional differences among decomposer communities. J. Ecol. 102, 603–609.

Lavorel, S., Garnier, E., 2002. Predicting changes in community composition and ecosystem functioning from plant traits: revisiting the Holy Grail. Funct. Ecol. 16, 545–556.

Lavorel, S., Mcintyre, S., Landsberg, J., Forbes, T.D.A., 1997. Plant functional classifications: from general groups to specific groups based on response to disturbance. Trends Ecol. Evol. 12, 474–478.

Lavorel, S., Storkey, J., Bardgett, R.D., Bello, F., Berg, M.P., Roux, X., Moretti, M., Mulder, C., Pakeman, R.J., Díaz, S., 2013. A novel framework for linking functional diversity of plants with other trophic levels for the quantification of ecosystem services. J. Veg. Sci. 24, 942–948.

Lenoir, L., Persson, T., Bengtsson, J., Wallander, H., Wiren, A., 2007. Bottom-up or top-down control in forest soil microcosms? Effects of soil fauna on fungal biomass and C/N mineralisation. Biol. Fertil. Soils 43, 281–294.

Leopold, A., 1939. A biotic view of land. J. For. 37, 727–730.

Lindeman, R.L., 1942. The trophic-dynamic aspect of ecology. Ecology 23, 399–418.

Loehle, C., Pechmann, J.H.K., 1988. Evolution—the missing ingredient in systems ecology. Am. Nat. 132, 884–899.

Loreau, M., 2010. From Population to Ecosystems: Theoretical Foundation for a New Ecological Synthesis. Princeton University Press, New Jersey.

Martinson, H.M., Schneider, K., Gilbert, J., Hines, J.E., Hambäck, P.A., Fagan, W.F., 2008. Detritivory: stoichiometry of a neglected trophic level. Ecol. Res. 23, 487–491.

McGill, B.J., Enquist, B.J., Weiher, E., Westoby, M., 2006. Rebuilding community ecology from functional traits. Trends Ecol. Evol. 21, 178–185.

Mikola, J., Setälä, H., 1998. No evidence of trophic cascades in an experimental microbial-based soil food web. Ecology 79, 153–164.

Miller, J.R.B., Ament, J.M., Schmitz, O.J., 2014. Fear on the move: predator hunting mode predicts variation in prey mortality and plasticity in prey spatial response. J. Anim. Ecol. 83, 214–222.

Miller, T.E.X., Rudolf, V.H.W., 2011. Thinking inside the box: community-level consequences of stage-structured populations. Trends Ecol. Evol. 26, 457–466.

Miner, B.G., Sultan, S.E., Morgan, S.G., Padilla, D.K., Relyea, R.A., 2005. Ecological consequences of phenotypic plasticity. Trends Ecol. Evol. 20, 685–692.

Mlambo, M.C., 2014. Not all traits are 'functional': insights from taxonomy and biodiversity-ecosystem functioning research. Biodivers. Conserv. 23, 781–790.

Moore, J.C., Berlow, E.L., Coleman, D.C., De Ruiter, P.C., Dong, Q., Hastings, A., Johnson, N.C., McCann, K.S., Melville, K., Morin, P.J., Nadelhoffer, K., Rosemond, A.D., Post, D.M., Sabo, J.L., Scow, K.M., Vanni, M.J., Wall, D.H., 2004. Detritus, trophic dynamics, and biodiversity. Ecol. Lett. 7, 584–600.

Mulder, C., Ahrestani, F.S., Lewis, O.T., Mancinelli, G., Naeem, S., Penuelas, J., Poorter, H., Reich, P.B., Rossi, L., Rusch, G.M., 2013. Connecting the green and brown worlds. Allometric and stoichiometric predictability of above-and below-ground networks. Adv. Ecol. Res. 49, 69–175.

Naeem, S., Wright, J.P., 2003. Disentangling biodiversity effects on ecosystem functioning: deriving solutions to a seemingly insurmountable problem. Ecol. Lett. 6, 567–579.

Neutel, A.M., Heesterbeek, J.A., De Ruiter, P.C., 2002. Stability in real food webs: weak links in long loops. Science 296, 1120–1123.

O'Donnell, D.R., Fey, S.B., Cottingham, K.L., 2013. Nutrient availability influences kairomone-induced defenses in Scenedesmus acutus (Chlorophyceae). J. Plankton Res. 35, 191–200.

Ohgushi, T., Schmitz, O.J., Holt, R.D., 2012. Trait-Mediated Indirect Interactions: Ecological and Evolutionary Perspectives. Cambridge University Press, Cambridge.

Petchey, O.L., Gaston, K.J., 2006. Functional diversity: back to basics and looking forward. Ecol. Lett. 9, 741–758.

Power, M.E., 1992. Top-down and bottom-up forces in food webs: do plants have primacy. Ecology 73, 733–746.

Reiners, W.A., 1986. Complementary models for ecosystems. Am. Nat. 127, 59–73.

Ritchie, M.E., Tilman, D., Knops, J.M., 1998. Herbivore effects on plant and nitrogen dynamics in oak savanna. Ecology 79, 165–177.

Ritz, K., Dighton, J., Giller, K.E., 1994. Beyond Biomass. Wiley-Sayce Publication, New York, NY.

Salmon, S., 2004. The impact of earthworms on the abundance of Collembola: improvement of food resources or of habitat? Biol. Fertil. Soils 40, 323–333.

Scheu, S., Folger, M., 2004. Single and mixed diets in Collembola: effects on reproduction and stable isotope fractionation. Funct. Ecol. 18, 94–102.

Schmitz, O.J., Price, J.R., 2011. Convergence of trophic interaction strengths in grassland food webs through metabolic scaling of herbivore biomass. J. Anim. Ecol. 80, 1330–1336.

Schmitz, O.J., Suttle, K.B., 2001. Effects of top predator species on direct and indirect interactions in a food web. Ecology 82, 2072–2081.

Schmitz, O.J., 2003. Top predator control of plant biodiversity and productivity in an old-field ecosystem. Ecol. Lett. 6, 156–163.

Schmitz, O.J., 2008. Effects of predator hunting mode on grassland ecosystem function. Science 319, 952–954.

Schmitz, O.J., 2010. Resolving Ecosystem Complexity. Princeton University Press, Princeton, NJ.

Schmitz, O.J., Grabowski, J.H., Peckarsky, B.L., Preisser, E.L., Trussell, G.C., Vonesh, J.R., 2008. From individuals to ecosystem function: toward an integration of evolutionary and ecosystem ecology. Ecology 89, 2436–2445.

Schmitz, O.J., Hawlena, D., Trussell, G.C., 2010. Predator control of ecosystem nutrient dynamics. Ecol. Lett. 13, 1199–1209.

Schmitz, O.J., Raymond, P.A., Estes, J.A., Kurz, W.A., Holtgrieve, G.W., Ritchie, M.E., Schindler, D.E., Spivak, A.C., Wilson, R.W., Bradford, M.A., 2014. Animating the carbon cycle. Ecosystems 17, 344–359.

Schweitzer, J.A., Bailey, J.K., Fischer, D.G., Leroy, C.J., Lonsdorf, E.V., Whitham, T.G., Hart, S.C., 2008. Plant-soil-microorganisms interactions: heritable relationship between plant genotype and associated soil microorganisms. Ecology 89, 773–781.

Schweitzer, J.A., Bailey, J.K., Hart, S.C., Wimp, G.M., Chapman, S.K., Whitham, T.G., 2005. The interaction of plant genotype and herbivory decelerate leaf litter decomposition and alter nutrient dynamics. Oikos 110, 133–145.

Seastedt, T.R., 1984. The role of microarthropods in decomposition and mineralization processes. Annu. Rev. Entomol. 29, 25–46.

Singer, M.S., Lichter-Marck, I.H., Farkas, T.E., Aaron, E., Whitney, K.D., Mooney, K.A., 2014. Herbivore diet breadth mediates the cascading effects of carnivores in food webs. Proc. Natl. Acad. Sci. U.S.A. 111, 9521–9526.

Sinsabaugh, R.L., Carreiro, M.M., Repert, D.A., 2002. Allocation of extracellular enzymatic activity in relation to litter composition, N deposition, and mass loss. Biogeochemistry 60, 1–24.

Sinsabaugh, R.L., Gallo, M.E., Lauber, C., Waldrop, M.P., Zak, D.R., 2005. Extracellular enzyme activities and soil organic matter dynamics for northern hardwood forests receiving simulated nitrogen deposition. Biogeochemistry 75, 201–215.

Sitvarin, M.I., Rypstra, A.L., 2014. Fear of predation alters soil carbon dioxide flux and nitrogen content. Biol. Lett. 10.

Sterner, R., Elser, J., 2002. Ecological Stoichiometry: The Biology of Elements From Molecules to The Biosphere. Princeton University Press, Princeton, NJ.

Strickland, M.S., Hawlena, D., Reese, A., Bradford, M.A., Schmitz, O.J., 2013. Trophic cascade alters ecosystem carbon exchange. Proc. Natl. Acad. Sci. U.S.A. 110 (27), 11035–11038.

Tansley, A.G., 1935. The use and abuse of vegetational concepts and terms. Ecology 16, 284–307.

Teuben, A., Verhoef, H.A., 1992. Direct contribution by soil arthropods to nutrient availability through body and fecal nutrient content. Biol. Fertil. Soils 14, 71–75.

Thaler, J.S., Mcart, S.H., Kaplan, I., 2012. Compensatory mechanisms for ameliorating the fundamental trade-off between predator avoidance and foraging. Proc. Natl. Acad. Sci. U.S.A. 109, 12075–12080.

Trussell, G.C., Schmitz, O.J., 2012. Species functional traits, trophic control, and the ecosystem consequences of adaptive foraging in the middle of food chains. In: Ohgushi, T., Schmitz, O.J., Holt, R.D. (Eds.), Trait-Mediated Indirect Interactions: Ecological and Evolutionary Perspectives. Cambridge University Press, New York, NY.

Vedder, B., Kampichler, C., Bachmann, G., Bruckner, A., Kandeler, E., 1996. Impact of faunal complexity on microbial biomass and N turnover in field mesocosms from a spruce forest soil. Biol. Fertil. Soils 22, 22–30.

Wardle, D.A., Bardgett, R.D., Klironomos, J.N., Setala, H., Van Der Putten, W.H., Wall, D.H., 2004. Ecological linkages between aboveground and belowground biota. Science 304, 1629–1633.

Werner, E.E., Peacor, S.D., 2003. A review of trait-mediated indirect interactions in ecological communities. Ecology 84, 1083–1100.

INDEX

Note: Page numbers followed by "*f*" indicate figures and "*t*" indicate tables.

ADVANCES IN ECOLOGICAL RESEARCH VOLUME 1–52

CUMULATIVE LIST OF TITLES

Aerial heavy metal pollution and terrestrial ecosystems, **11**, 218

Age determination and growth of Baikal seals (*Phoca sibirica*), **31**, 449

Age-related decline in forest productivity: pattern and process, **27**, 213

Allometry of body size and abundance in 166 food webs, **41**, 1

Analysis and interpretation of long-term studies investigating responses to climate change, **35**, 111

Analysis of processes involved in the natural control of insects, **2**, 1

Ancient Lake Pennon and its endemic molluscan faun (Central Europe; Mio-Pliocene), **31**, 463

Ant-plant-homopteran interactions, **16**, 53

Anthropogenic impacts on litter decomposition and soil organic matter, **38**, 263

Arctic climate and climate change with a focus on Greenland, **40**, 13

Arrival and departure dates, **35**, 1

Assessing the contribution of micro-organisms and macrofauna to biodiversity-ecosystem functioning relationships in freshwater microcosms, **43**, 151

A belowground perspective on Dutch agroecosystems: how soil organisms interact to support ecosystem services, **44**, 277

The benthic invertebrates of Lake Khubsugul, Mongolia, **31**, 97

Big data and ecosystem research programmes, **51**, 41

Biodiversity, species interactions and ecological networks in a fragmented world **46**, 89

Biogeography and species diversity of diatoms in the northern basin of Lake Tanganyika, **31**, 115

Biological strategies of nutrient cycling in soil systems, **13**, 1

Biomanipulation as a restoration tool to combat eutrophication: recent advances and future challenges, **47**, 411

Biomonitoring of human impacts in freshwater ecosystems: the good, the bad and the ugly, **44**, 1

Bray-Curtis ordination: an effective strategy for analysis of multivariate ecological data, **14**, 1

.

Edwards Brothers Malloy
Thorofare, NJ USA
May 12, 2015